# ULTIMATE COMPUTING

Biomolecular Consciousness and NanoTechnology

# ULTIMATE COMPUTING

## Biomolecular Consciousness and NanoTechnology

Stuart R. HAMEROFF
Department of Anesthesiology
College of Medicine
University of Arizona
Tucson, Arizona, U.S.A

1987

NORTH-HOLLAND
AMSTERDAM · NEW YORK · OXFORD · TOKYO

ISBN: 0 444 70283 0

*Published by:*

ELSEVIER SCIENCE PUBLISHERS B.V.
P.O. Box 1991
1000 BZ Amsterdam
The Netherlands

*Sole distributors for the U.S.A. and Canada:*

ELSEVIER SCIENCE PUBLISHING COMPANY, INC.
52 Vanderbilt Avenue
New York, N.Y. 10017
U.S.A.



PRINTED IN THE NETHERLANDS

Billionth scale activities in biomolecular assemblies
could define life itself, and provide
a frontier for the evolution of technology.

# Contents

# List of Figures

# List of Tables

# Prelude

What is this book, and why has it been written by an anesthesiologist? This book is a view of the co-evolution of consciousness and technology—past, present and future.

This book has been written by an anesthesiologist because of a confluence of two fascinations. The first is the nature of consciousness, which anesthesiologists routinely erase and restore in their patients. The second is a fifteen year trail of notions that would not go away. While a third year medical student in 1972, I spent a summer research elective in a cancer laboratory. For some reason I became fascinated and fixated by one particular question. When cells divided, the chromosomes were separated and daughter cell architecture established by wispy strands called mitotic spindles ("microtubules") and cylindrical organelles called centrioles. Somehow, the centrioles and spindles *"knew"* when to move, where to go, and what to do. The uncanny guidance and orientation mechanism of these tiny biomolecular structures seemed to require some kind of motorized intelligence. At about the same time, electron microscopy techniques were revealing the interior of all living cells to be densely filled with wispy strands, some of which were identical to mitotic spindles. Interconnected in dynamic parallel networks, these structures were thought to serve a purely supportive, or mechanical structural role and were collectively termed the "cytoskeleton."

But several factors suggested that the cytoskeleton was more than the structural "bones" of the cell: they manipulated dynamic activities, orchestrating complex and highly efficient processes such as cell growth, mitosis and transport. Another factor was a lack of any other candidate for "real time" dynamic organization within cells. Long term blueprints and genetic information clearly resided in DNA and RNA, and membranes performed dynamic functions at cell surfaces. However, a mechanism for the moment to moment execution, organization, and activities within cells remained unknown. Where was the nervous system within the cell? Was there a biological controller? This book is based on the premise that the cytoskeleton *is* the cell's nervous system, *the* biological controller/computer. In the brain this implies that the basic levels of cognition are *within* nerve cells, that cytoskeletal filaments are the roots of consciousness. The small size and rapid conformational activities of cytoskeletal proteins are just beyond the resolution of current technologies, so their potential dynamics remain unexplored and a cytoskeletal controlling capability untested. Near future technologies will be able to function in the nanoscale (nano $= 10^{-9}$; nanometer = one billionth meter, nanosecond = one billionth second) and will hopefully resolve these questions. If indeed cytoskeletal dynamics are the texture of intracellular information processing, these same "nanotechnologies" should enable direct monitoring, decoding and interfacing between biological and technological information devices. This in turn could result in important biomedical applications and perhaps a merger of mind and machine: *Ultimate Computing*.

A thorough consideration of these ideas involves a number of disciplines, all of which are at least tangentially related to anesthesiology. These include biochemistry, cognitive science, computer science, engineering, mathematics, microbiology, molecular biology, pharmacology, philosophy, physics, physiology, and psychology. As an expert in none, but a dabbler in all, I hope true experts in these fields will find my efforts never-the-less interesting.

Starting from a cytoskeletal perspective, this book flings metaphors at the truth. Perhaps one or more will land on target, or at least come close.

## Acknowledgements

This book is a collective effect of the following people: Ralph Abraham, Ross Adey, Sigma Alpha, Fred Anderson, Amy Barnes, Joanne Barnes, Michel Bornens, Burnell Brown, George Carlson, Forrest Carter, Peter Christiansen, Jim Clegg, John Condeelis, Leonor Cruzeiro, Marc DeBrabander, Yves Engelborghs, Lawrence Fried, Herbert Fröhlich, Kit Grantham, Jamie, Harrison, Lillian, Amy, and the entire Hameroff/Kaplan/ Bowman family, Max Headroom, Emery Hetrick, Dixie Holmes, Paul Jablonka, Martha Juarez, Jan Julianus, Ben Kahn, Charles Kiselyak, Djuro Koruga, Julio Kuperman, Teruo Matsumoto, Leo Martin, Kathleen McAuliffe, Claris Nelson, Michio Okuma, Karl Pribram, Mary Quimby, Steen Rasmussen, Conrad Schneiker, Alwyn Scott, Steven Smith, Branko Soucek, Arthur Villa, Rich Watt, Juli Weiss and Arthur Winfree.

Most of the artwork was thoughtfully done by scientist/systems engineer/artist Paul Jablonka. Conrad Schneiker supplied most of the material on nanotechnology and replicators for Chapter 10, and compiled the appendices. I am indebted to the Laboratory for Advanced Mathematics and Physics at the Technical University of Denmark and the Danish Camping Union who hosted me, Jamie and Harrison during our sabbatical. I sincerely appreciate the efforts of my colleagues in the Department of Anesthesiology and the College of Medicine, University of Arizona who permitted me time and mental latitude. Thanks to Personal TEX's port of Don Knuth's TEX, plus Leslie Lamport's LATEX, Textset's DVILASER/PS, Boreland's Turbo Lightning, Adobe Systems' PostScript, Apple's Macintosh and LaserWriter, IBM's PC/AT and its clones, QMS's PS800 laser printer and Xerox's machines and their clones. The ever friendly and competent technical support from Textset, Personal TEX and HDS Systems is also greatly appreciated.

Finally, Jan Julianus and Elsevier North-Holland deserve credit for their instigation and patience.

## Dedication

To H. H., who loved to gamble.

# Chapter 1

# Toward Ultimate Computing

## 1.1 Mind/Tech: Merger in the Nanoscale

Biology and technology are both evolving toward more efficient methods of information processing. With a head start of a billion years, biology has evolved human consciousness; technology appears to be catching up rapidly.

*Ultimate Computing* is the common destination for the evolution of information processing systems in both biology and technology. At this point it is an extrapolation of converging trajectories, but Ultimate Computing may soon exist in the nanoscale. Nano = $10^{-9}$, one nanometer is a billionth of a meter, and one nanosecond is a billionth of a second. Subunits within biological protein assemblies (cytoskeletal polymers, organelles, membrane proteins, virus coats) are of nanometer size scale and undergo conformational oscillations in the nanosecond time scale. Nanoscale excitations, which may be *coherent* and coupled to intraprotein dipole shifts, can generate communicative "collective modes" within protein assemblies and provide a substrate for biological information processing. Thus the "nanoscale" (Figure 1.1) may be where living intelligence has evolved. Coincidentally, nanoscale devices including molecular computers, Feynman machines and von Neumann replicators are becoming feasible through technologies such as scanning tunneling microscopy. A nanoscale marriage of biomolecules and nanotech devices, providing direct communication and information transfer, could have profound benefits for biomedicine and our culture in general.

Comingling of consciousness and computer technology is a prevalent

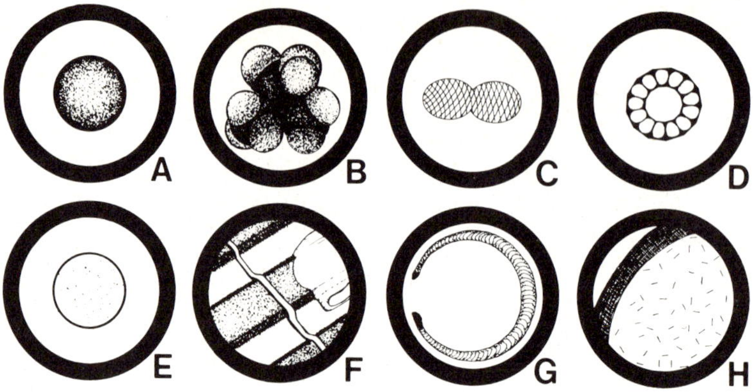


Figure 1.1: Sizing the Nanoworld. The diameter of each circle is given in nanometers (nm). A) 0.30 nm—a carbon atom, 0.15 nm in diameter. B) 0.50 nm—alanine, an amino acid with 13 atoms including 3 carbons, is about .33 nm in diameter. C) 12 nm—a tubulin dimer protein, the subunit of microtubules, is 8 nm long. It is composed of 2 similar monomers (alpha and beta tubulin), each made of about 440 amino acids. Cross hatching suggests the approximate amount of space available for each amino acid. D) 50 nm—a microtubule, 13 sided tube with an outside cross-sectional diameter of 25 nm. E) 1900 nm—a small 1000 nm diameter nerve axon might contain 100 microtubules (shown) and 1000 smaller filaments (not shown). Microtubules associate in informal clumps of 1 to 5 microtubules each, represented by dots. F) 40,000 nm—a nerve cell grown on the surface of a Motorola 68000 computer chip. The wire thickness is 15,000 nm wide. G) 170,000 nm—a nematode is a small worm of less than 1000 cells, 300 of which are neurons. Nematodes have a brain, teeth, muscles, gut, and sex lives. This one is about 450,000 nm long. H) 5,000,000 nm—one quarter of a human thumbnail with 50 nematodes represented to scale. By Paul Jablonka.

dream. Artificial intelligence based on brain/mind organization is a tentative step in this direction, as is the proposed use of self assembling protein arrays as switching circuits or "biochips." The Japanese effort towards the "Sixth Generation Computer" aims to "integrate biology and technology" by merging research in artificial intelligence and the functions of living organisms (Corcoran, 1987). By attempting to understand the conditions required to maintain biological "homeostasis", the Japanese are hoping to embark on a symbiosis between intelligent biological structures and technological de-

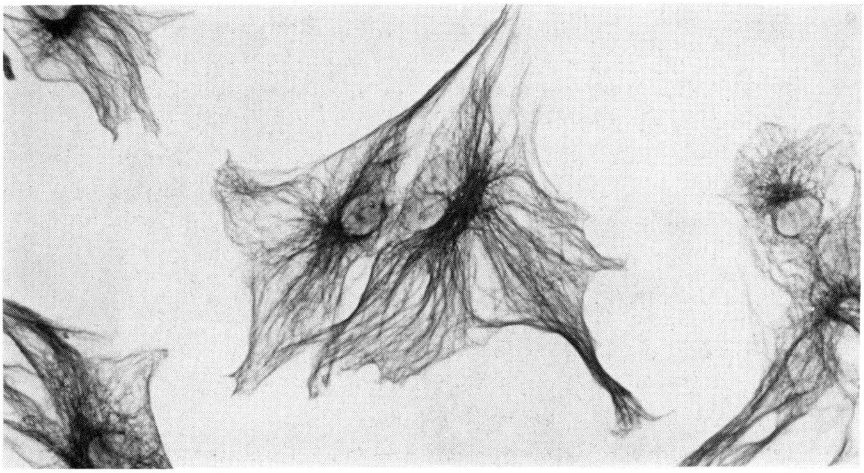

Figure 1.2: Cytoskeleton within cells who have just divided. Intracellular microtubules are visualized by immunostaining. Spherical areas are cell nuclei, adjacent to which are the dense microtubule organizing centers (MTOC). With permission from DeBrabander, Geuens, DeMey and Joniav (1986), courtesy of Marc DeBrabander.

vices, and even predict an "artificial brain"! One missing ingredient for such a Mind/Tech merger is an understanding of the mechanism of consciousness. Most models of brain organization consider nerve cells and their connections to be the brain's fundamental units of information processing. However, profoundly complex and intelligent activities occur *within* nerve cells. Further, simple organisms like single cell amoeba and paramecium perform complex tasks without benefit of brain or nervous system. In this book we view the *cytoskeleton*—networks of protein polymers which occupy and organize the interiors of all living cells (Figure 1.2)—as a highly evolved information processing system operating at nanoscale levels. Collective nanoscale activities of the cytoskeleton and related structures can explain biological organization, information processing, and consciousness, and be the target for the future evolution of technology.

## 1.2 Evolution of Technology

Technological emulation of life since the 13th century has been reviewed by author Claris Nelson (1985). Albertus Magnus is said to have created a life-like mechanical servant out of metal, wood, glass, leather and wax that could open doors and greet visitors. It was considered blasphemous work of the devil by Magnus' student Saint Thomas Aquinas who destroyed it. Science fiction writers predicted computers and robots long before they existed. In 1879, Edward Page Mitchell's *The Ablest Man in the World* featured a mechanical brain and in Edmund Hamilton's 1928 *The Metal Giants* an artificial brain turned against its creators.

Computers descended from calculating machines, the earliest of which was the abacus. In 1642 French mathematician and philosopher Pascal made a mechanical calculator that used the decimal system to add and subtract. In 1694, German mathematician/philosopher Leibniz created a "Stepped Reckoner," which was supposed to multiply, divide and take square roots. It didn't work, but utilized principles later essential to modern computers. Tasks were broken down into a great many simple mathematical steps using binary numbers and were performed sequentially. When computers later came to be operated by electricity, binary zero and one became represented by off and on. In the early 1800's George Boole developed "Boolean algebra," the mathematical logic by which computer circuits are designed. Charles Babbage and Ada Lovelace—Lord Byron's eldest daughter—designed an "analytical engine" using punched cards. Their contemporary technology could not construct the machine accurately enough, but it *was* built and functioned in the twentieth century.

The first electronic computer was apparently constructed and operated in 1939 by John Vincent Atanasoff, a theoretical physicist at Iowa State University (Mackintosh, 1987). Shortly thereafter, Alan Turing and colleagues in Bletchley, England designed a computer to perform all possible mathematical calculations. It was based on Turing's work proving the logical limits of computability and was used to decipher the German "Enigma" code during World War II. In a masterful presentation of key ideas previously developed by other pioneers, John von Neumann further advanced computer design by separating the machine from its problems. Prior to von Neumann, a computer would have to be rewired for each new task. With enough time, memory and software, computers could solve the problems that could be broken down into finite sequences of logical steps. Most current computers use "serial" processing based on von Neumann's design. In the 1940's, the University of Pennsylvania developed the first electronic computer, the Electronic Numerical Integrator and Calculator or "ENIAC." It weighed 30 tons, took up 3,000 cubic feet of space, and contained 18,000 vacuum tubes, one of which failed every seven minutes. It could calculate nuclear physics problems in two hours that would have taken 100 engineers a year to complete. Today, the same capacity is available on one chip. In 1950 Remington Rand marketed UNIVAC, which dealt with words and numbers stored by their binary equivalent. Since that time, roughly four generations of computers have evolved due to increased demand and advances in design, chip size, materials and other factors. For the same reasons further advances seem inevitable.

Von Neumann and Turing hoped that computers could duplicate our ability to think, so that our minds could be amplified just as our muscles had been by industrial machines. However further evolution of computers using serial processing seems limited. Computers and artificial intelligence are now evolving to parallel systems based on brain architecture and neural net models; a future step may be nanoscale, self organizing intelligence.

Von Neumann is one of several "fathers of the computer." In the "serial" processing which he skillfully formalized, information flows in one dimension. In the 1950's and 1960's, von Neumann (1966) and Stanislav Ulam developed the mathematics of computing in multiple dimensions. They considered two dimensional information spaces with discrete subunits ("cells") whose states could vary depending on the states of neighboring cells. Each cell and its neighbor relations were identical. Relatively simple rules among neighbors and discrete time intervals ("generations") led to evolving patterns and self-organization which were exquisitely sensitive to initial conditions. They called these systems "cellular automata." Von Neumann described a "universal computer" automaton which could solve any problem if given sufficient area and time. Today, computer technologists are considering the profound advantages of implementing molecular scale automata (Milch, 1986).

Edward Fredkin of Massachusetts Institute of Technology has considered multidimensional automata and the discreteness of time and matter. He argues that the universe is a cellular automaton whose "cells" are atomic and subatomic particles (Wright, 1985). The universe is made of information, Fredkin reasons. Cellular automata may be generalized "primordial computers" of which all other computers and complex systems are particular examples. Cellular automata in conformational states of cytoskeletal subunits could process biological information and be the substrate of consciousness.

The current trend in computer design and artificial intelligence or "AI" is parallel connectedness, emulating the brain. Many types of problems can be solved by breaking them down into serial mathematical steps. Today's electronic computers serially process very rapidly and can solve complex mathematical problems far faster than can humans alone. However qualitative functions which the brain performs naturally—recognizing patterns, or making judgments—are extremely difficult for computers. Consider the letter "a." We recognize it automatically, in any typeface, in all but the worst handwriting. To our brains it's simple, quick, obvious even if it's missing. If we see, "Sally 'red' a newspaper," we mentally insert the absent "a." Computer/AI scientist Jerome Feldman (1985) cites the example of interpreting the statement "John threw a ball for charity." The inherent ambiguities of this type of statement can be resolved in a highly parallel system in which multiple simultaneous interpretations are processed and evaluated. Hurling a sphere versus hosting a dance can be resolved by the qualifier "for charity" which is much more consistent with a dance than with a sphere. Human brains commonly resolve conflicts among differing drives or input, although failure to do so may cause psychiatric or emotional problems. At least according to science fiction, computers can suffer similar disturbances. In Arthur C. Clarke's and Stanley Kubrick's *2001: Space Odyssey* and its sequel *2010*, the computer "Hal 9000" becomes psychotic because of conflicting instructions and reacts by killing the space voyagers because their mission was too important to be entrusted to them. The brain/mind can perform "cognitive" functions including resolution of conflict by "subcognitive" processes such as recognizing patterns, making assumptions and performing imaginative leaps. The net effect is consciousness: a *collective* effect of simpler processes.

## 1.3 Collective Intelligence

A collective phenomenon is *more the product of, rather than the sum of, its parts*, and has been explained by Cal Tech biophysicist John Hopfield (1982) whose "neural net" models are collective.

> Suppose you put two molecules in a box, every once in a while they collide and that's an exciting event. ... If we'd put ten or even a thousand more molecules in the box all we'd get is more collisions. But if we put a billion billion molecules in the box, there is a new phenomenon—sound waves.

Observation of two, or ten, or thousands of those molecules would not suggest the Mozart or Madonna that can arise in a collection of more than a trillion trillion molecules. Other examples of collective phenomena may be

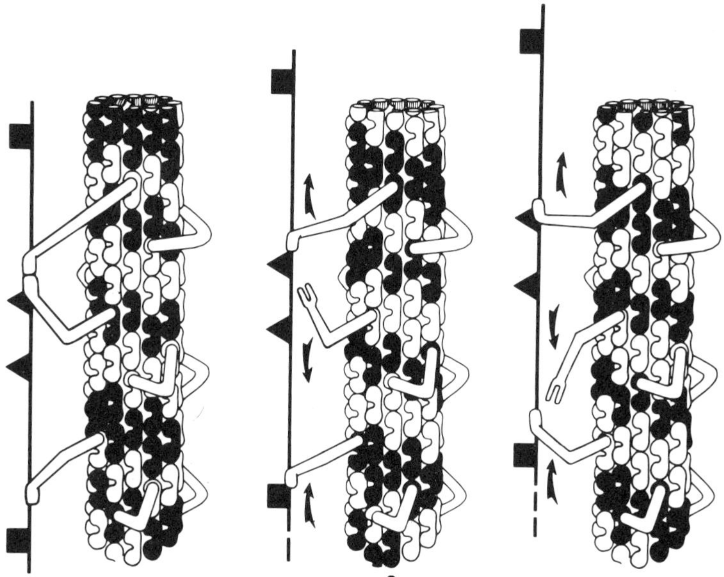


Figure 1.3: Axoplasmic transport occurs by coordinated activities of microtubule attached sidearm, contractile proteins ("dynein") which cooperatively pass material in a "bucket brigade." The orchestration mechanism is unknown, but shown here as the consequence of signaling by "soliton" waves of microtubule subunit conformational states. By Fred Anderson.

seen in beehives, ant colonies, football teams, governments and various types of material phase transitions. For example, superconductivity and magnetism are collective effects which occur in certain metals as their individual atoms come into alignment. By cooling these metals, thermal fluctuations cease, atoms become highly aligned, and below a critical temperature totally different qualitative properties of superconductivity or magnetism emerge.

How might collective phenomena be tied to consciousness? Brain neuron synaptic transmissions are relatively slow at several milliseconds per computation–they are about 100,000 times slower than a typical computer switch. Nevertheless vision and language problems can be solved in a few hundred milliseconds or what would appear to be about 100 serial steps. Artificial Intelligence (AI) researchers conclude that this computational richness is accounted for by collective effects of parallelism and rich interconnectedness. With billions of neurons, and with each neuron connected to up to hundreds of thousands of other neurons, AI "connectionists" view the brain as a collective phenomenon of individually stupid neurons. Groups of highly connected neurons are thought to attain intelligent behavior through properties of feedback and reverberation. Walter Freeman (1972, 1975, 1983) of the University of California at Berkeley contends that a "critical mass" of about 100,000 neurons yields intelligent behavior. However, intelligent behavior occurs within nematode worms of 1000 cells and 300 neurons, within cytoplasm in single cell organisms and *within* single neurons. Individual neurons with tens to hundreds of thousands of connections cannot be stupid and fulfill their multiple functions, integrate input/output and modulate synaptic connection strength. Each nerve cell is a sophisticated information processing system in and of itself! The cytoskeleton within neurons and all living cells is a parallel connected network which can utilize its own collective phenomena to organize and process subcellular information (Figure 1.3). The cytoskeleton can convey analog patterns which may be connected symbols (Chapter 8). Although overlooked by AI researchers, the cytoskeleton may take advantage of the same attributes used to describe neural level networks. Properties of networks which can lead to collective effects among both neurons and cytoskeletal subunits include parallelism, connectionism, and coherent cooperativity.

### 1.3.1 Parallelism

The previous generations of computer architecture have been based on the von Neumann concept of sequential, serial processing. In serial processing, computing steps are done consecutively which is time consuming. One false bit of information can cascade to chaotic output. The brain with its highly parallel nerve tracks shines as a possible alternative. In parallel computing, information enters a large number of computer pathways which process the data simultaneously. In parallel computers information processors may be independent of each other and proceed at individual tempos. Separate processors, or groups of processors, can address different aspects of a given problem asynchronously. As an example, Reeke and Edelman (1984) have described a computer model of a parallel pair of recognition automata which use complementary features (Chapter 4). Parallel processing requires reconciliation of multiple outputs which may differ due to individual processors being biased differently than their counterparts, performing different functions, or because of random error. Voting or reconciliation must occur by lateral connection, which may also function as associative memory. Output from a parallel array is a collective effect of the input and processing, and is generally a consensus which depends on multiple features of the original data input and how it is processed. Parallel and laterally connected tracks of nerve fibers inspired AI researchers to appreciate and embrace parallelism. Cytoskeletal networks *within* nerve cells are highly parallel and interconnected, a thousand times smaller, and contain millions to billions of cytoskeletal subunits *per nerve cell*!

Present day evolution of computers toward parallelism has engendered the "Connection Machine" (Thinking Machines, Inc.) which is a parallel assembly of 64,000 microprocessors. Early computer scientists would have been impressed with an assembly of 64,000 switches without realizing that each one was a microprocessor. Similarly, present day cognitive scientists are impressed with the billions of neurons within each human brain without considering that each neuron is itself complex.

Another stage of computer evolution appears as multidimensional network parallelism, or "hypercubes." Hypercubes are processor networks whose *interconnection topology* is seen as an "n-dimensional" cube. The "vertices" or "nodes" are the processors and the "edges" are the interconnections. Parallelism in "n-dimensions" leads to hypercubes which can maximize available computing potential and, with optimal programming, lead to collective effects. Complex interconnectedness observed among brain neurons and among cytoskeletal structures may be more accurately described as hypercube architecture rather than simple parallelism. Hypercubes are exemplified in Figures 1.4, 1.5, and 1.6.

AI/Roboticist Hans Moravec (1986) of Carnegie-Mellon University has attempted to calculate the "computing power" of a computer, and of the human brain. Considering the number of "next states" available per time in binary digits, or bits, Moravec arrives at the following conclusions. A microcomputer has a capacity of about $10^6$ bits per second. Moravec calculates the brain "computing" power by assuming 40 billion neurons which can change states hundreds of times per second, resulting in $40 \times 10^{11}$ bits per second. Including the cytoskeleton *increases the potential capacity for information processing immensely.* Microtubules are the most visible cytoskeletal structures. Making some rough assumptions about cytoskeletal density (i.e. microtubules spaced about 1000 nanometers apart) and the volume of brain which is neuronal cytoplasm leads to about $10^{14}$ microtubule subunits in a human brain (ignoring neurofilaments and other cytoskeletal elements). As described in Chapters 5 and 6, the frequency of cytoskeletal subunit state changes may be greater than billions per second! The cytoskeleton is capable not only of immense information capacity, but appears to be designed such that interacting conformational state patterns may perform computing

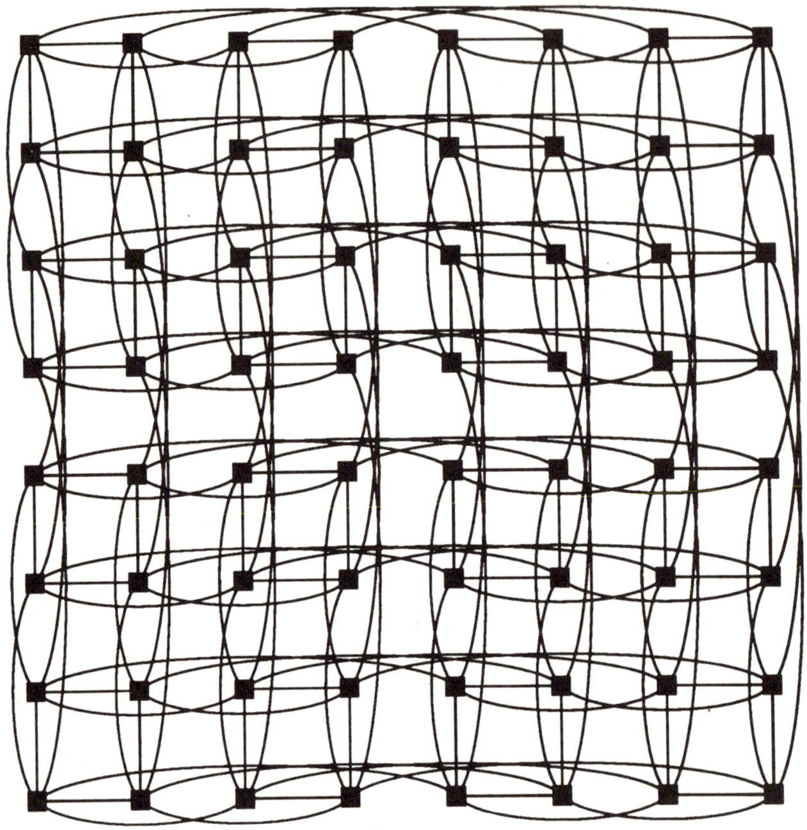

Figure 1.4: Six dimensional hypercube with 64 nodes, and 6 connections per node. Computer generation by Conrad Schneiker.

functions. Several theories which propose such mechanisms will be described in Chapter 8.

The brain is a *continuous* system. Classical computers have operated on recursive repetitive functions to process information in batches and the output is obtained as the final product. Similarly, most parallel processing designs have discrete input and output points. Carl Hewitt (1985) has described open systems within computers in which processing may never halt, which can provide output while computing is still in operation, and can accept input from sources not anticipated when the computation began. Like the human brain/mind, open continuous systems can interact with the environment and adapt to new situations. Hewitt describes an asynchronous

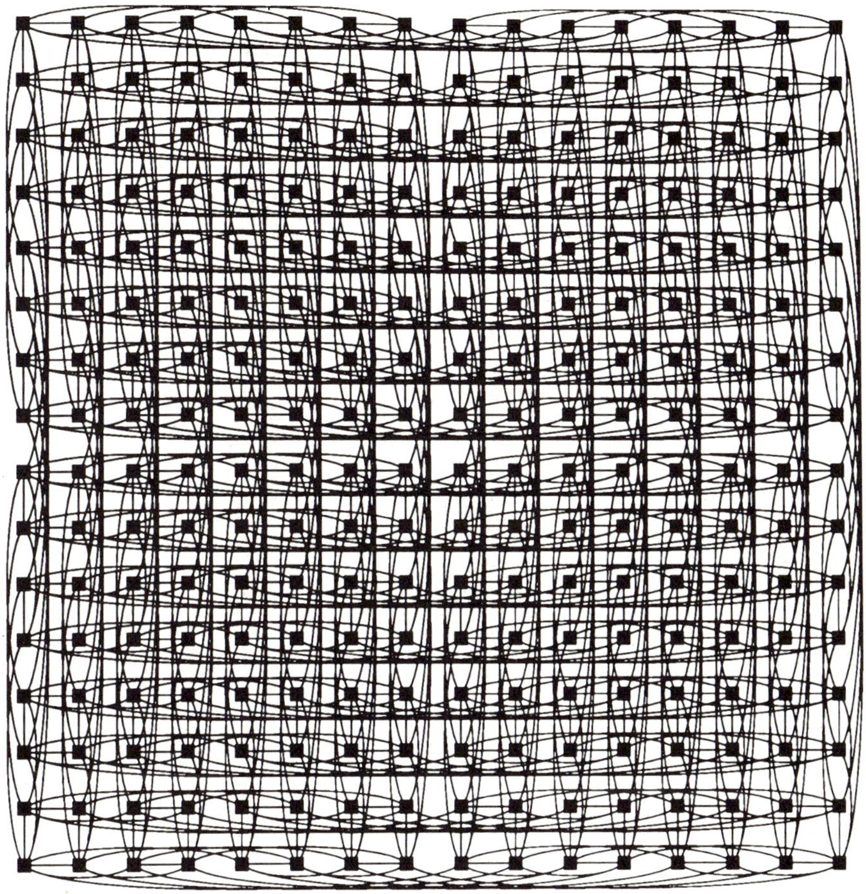

Figure 1.5: Eight dimensional hypercube with 256 nodes, and 8 connections per node. Computer generation by Conrad Schneiker.

parallel computer system which can make use of multiple inputs and outputs and whose parallel elements are connected by "arbiters" which "weigh" and reconcile differing content, and can provide continuous input and output. Among brain neurons, "arbiters" would appear to be synaptic connections among laterally connected parallel neurons. Within the cytoskeleton, laterally connecting filaments and microtubule associated proteins ("MAPs") could serve as logical arbiters.

Hewitt argues that parallel, open systems are "non-hierarchical" because input and output are continuously processed throughout the system. Early views of brain/mind organization assumed a hierarchical arrangement of processing units. Sensory input was thought to be processed and relayed to higher and higher levels of cognition until reaching a single "Grandfather neuron" or "Mind's Eye" which comprehended the input's "essence." Classical brain research by Lashley (1929, 1950) and others (Chapter 4) strongly suggest that memory and information are distributed throughout the brain and that specific anatomical hierarchical arrangements leading to "Grandfather neurons" do not exist. The "Mind's Eye" is not localized to a given site but is mobile over wide volumes of brain. Assuming that humans actually do comprehend the essence of at least some things, who or what is comprehending? The site and nature of attention, "self," consciousness or the Mind's Eye remains a philosophical issue and barrier to Mind/Tech merger. Neuroanatomical structure and the distributed storage of brain information point toward highly parallel, open brain/mind computing systems which may occur both at the neural level, and within neurons in the cytoskeleton. The perception component of consciousness, the "Mind's Eye" may be a mobile hierarchy determined by collective dynamics.

### 1.3.2 Connectionism

The Mind's Eye may be the apex of a *collective* hierarchy of parallel systems in which the cytoskeleton and related structures are the ground floor. Parallel systems in both computers and biological systems rely on lateral connections and networks to provide the richness and complexity required for sophisticated information processing. Computer simulations of parallel connected networks of relatively simple switches ("neural nets") develop "cognitive-like functions" at sufficient levels of connectedness complexity—a "collective phenomenon" (Huberman and Hogg, 1985). Philosopher John Searle (Pagels, 1984), who has an understandable bias against the notion that computer systems can attain human consciousness equivalence, points out that computers can do enormously complex tasks without appreciating the essence of their situation. Searle likens this to an individual sorting out Chinese characters into specific categories without understanding their meaning, being unable to speak Chinese. He likens the computer to the individual sorting out information without comprehending its essence.

It would be difficult to prove that human beings comprehend the essence of anything. Nevertheless, even the simulation of cognitive-like events is interesting. Neural net models and connectionist networks (described further

Figure 1.6: Ten dimensional hypercube with 1,024 nodes, and 10 connections per node. Computer generation by Conrad Schneiker.

in Chapter 4) have been characterized mathematically by Cal Tech's John Hopfield (1982) and others. His work suggests that solutions to a problem can be understood in terms of minimizing an associated energy function and that isolated errors or incomplete data can, within limits, be tolerated. Hopfield describes neural net energy functions as having contours like hills and valleys in a landscape. By minimizing energy functions, information (metaphorically) flows like rain falling on the landscape, forming streams and rivers until stable states ("lakes") occur. A new concept in connectionist neural net theory has emerged with the use of multilevel networks. Geoffrey Hinton (1985) of Carnegie-Mellon University and Terry Sejnowski of Johns Hopkins University have worked on allowing neural nets to find optimal solutions, like finding the lowest particular lake in an entire landscape. According to Sejnowski (Allman, 1986; Hinton, Sejnowski and Ackley, 1984) the trick is to avoid getting stuck in a tiny depression between two mountains:

> Imagine you have a model of a landscape in a big box and you want to find a lowest point on the terrain. If you drop a marble into the box, it will roll around for a while and come to a stop. But it may not be the lowest point, so you shake the box. After enough shaking you usually find it.

Hinton and Sejnowski have used this concept of mathematically shaking their neural net simulations to find optimal solutions. It requires a multilevel hierarchy of parallel systems so that one level can "shake" or tune a lower level. Such an arrangement can perhaps explain the relationship between hierarchical layers of parallel systems within the brain. For example, neural networks based on synaptic connection may regulate (and be regulated by) smaller, faster, more comprehensive networks in the intracellular cytoskeleton.

Extensive comparisons between information processing in the brain and artificial intelligence have been reviewed by A. M. Decallatay (1986) who feels the laws of thought described in philosophy have been rediscovered by AI: "The mental world of Plato is reproduced in the physical symbols of Newell and Simon." DeCallatay observes that AI represents data by virtual pointers which connect symbols. In computers these virtual relations are actual wires with potential gate connection; in the brain they appear to be neuronal synaptic connections. *Within* neurons they may be cross-bridge filaments connecting cytoskeletal microtubules. As a computer expert evaluating the brain, DeCallatay states that the brain learns by opening gates to build new connections between elements simultaneously activated. He sees the presence or absence of dendritic spines playing the role of an "all or none" switch at the neural level. Dendritic spines are knobby projections of membrane covered cytoplasm on neuronal dendrites which are generated and maintained by the cytoskeleton and form synapses with other neurons. The most accepted theory for learning and memory in the brain is that of strengthening of specific synapses within neural circuits, an idea generated by Donald Hebb (1949). As will be described in Chapters 4 and 5, dynamic structural activities of the cytoskeleton are responsible for all cytoplasmic rearrangements including formation and regulation of dendritic spines and synapses. The spines are branchings of dendrites which themselves are branchings of neurons. A further dimension of complexity, these cytoskeletal appendages are prime candidates for "synaptic plasticity," the cornerstone for prevalent models of brain learning and memory.

### 1.3.3 Cooperativity and Coherence

Collective effects manifest as diffuse reverberation, sustained oscillation, phase transitions, and deterministic chaos have been observed in computer simulation of parallel networks (Choi and Huberman, 1984). Collective mechanisms can exert long-range cooperativity and an executive level of organization within parallel arrays. Collective phase transitions in brain parallel arrays could be a fabric of consciousness, an "idea" emerging like the property of superconductivity from a large number of simple, "aligned" subunits. In most views the neuronal synapse is the brain's fundamental subunit, however synaptic activities are the net result of dynamic processes orchestrated by the cytoskeleton. Layers of cytoskeletal organization are evident within neurons, and their participation in cognitive functions appears unavoidable. Thus the highly branched cytoskeleton may be another dimension of brain organization, perhaps related to neuronal networks as a "fractal." Many natural processes manifest fractals, growth patterns in which local areas are scaled down images of the entire pattern. This occurs through some form of long range correlation in the pattern: components "know about each other over distances far in excess of the range of the forces between them" (Sander, 1986). Fractal relationships are one type of long range cooperativity (Figures 1.7 and 1.8). Densely parallel interconnected networks of cytoskeletal structures resemble larger scale networks of neurons, and may be viewed as fractal subdimensions of neural networks.

Long range cooperativity and collective mechanisms are favored by the property of coherence which means peak energy excitations within an area occur "in phase," or simultaneously as in a laser. How may coherence arise in distributed processes? DeCallatay (1986) proposes that coherence in the brain and AI need to be imparted from the top of a hierarchy downward, like the chief executive of a corporation setting goals and intentions. A different view is that of an underlying rhythm or beat to which all elements are tuned. Rhythmic coupling among neurons may be important, and some interpreters of brain electrical activity (EEG) believe regional brain wave entrainment leads to functional regions of mental representation. A more fundamental coherence at the level of protein assemblies may be universally important

Figure 1.7: Tree fractal in which branching patterns are the same at every scale, or dimension. Long range order is present. Computer generation by Conrad Schneiker.

for biological cooperativity and communication.

Proteins and their components oscillate among specific conformational

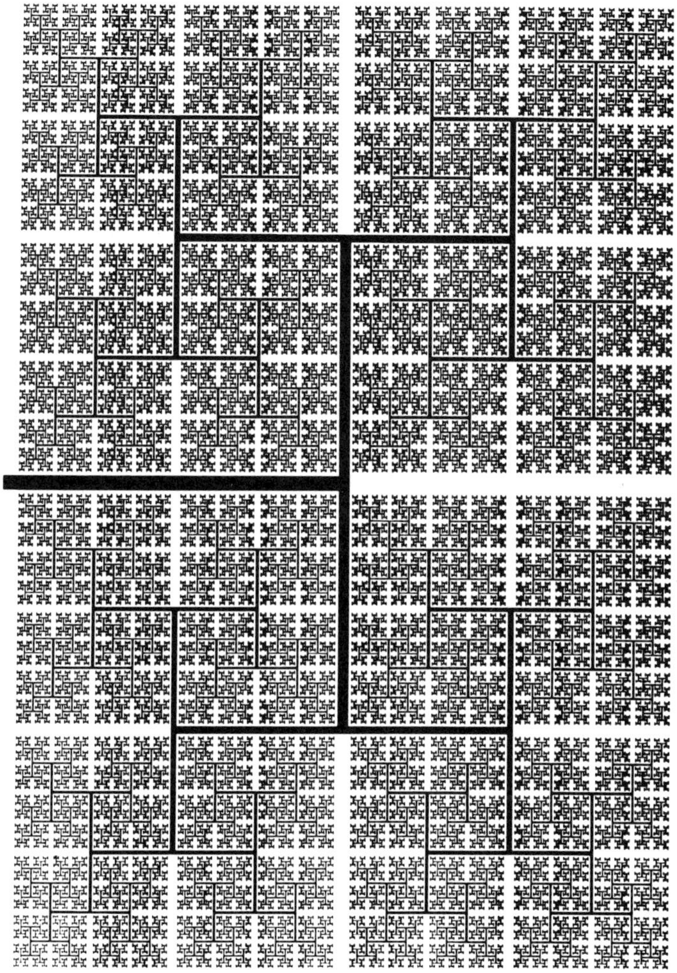

Figure 1.8: Branching box fractal in which patterns are identical at every scale, or dimension. Long range order is present. Computer generation by Conrad Schneiker.

states which exist transiently for durations ranging from femtoseconds ($10^{-15}$ sec) to minutes or longer. As will be described in Chapter 6, *functional* conformational states appear coupled to nanosecond ($10^{-9}$ sec) oscillations and more prolonged "metastable states." Herbert Fröhlich, an eminent physicist who helped develop the theory of superconductivity in the 1950's, has devoted recent efforts to the question of cooperativity in biological systems. Fröhlich (1970, 1975, 1984) argues that biochemical energy supplied to biomolecular assemblies can result in *coherent* elastic vibrations of individual subunits in the sub-nanosecond time range. The effect presupposes a voltage effect in the biomolecule (i.e. an "electret") and an organized spatial structure whose geometry favors coupling among subunits. Coherent oscillations in an appropriate medium like the cytoskeleton can lead to collective phenomena such as long range cooperativity, communication, and holography.

Another model can help explain long range cooperativity in biomolecules. Soviet biophysicist A. S. Davydov has considered almost lossless energy transfer in biomolecular chains or lattices as wave-like propagations of coupled conformational and electronic disturbances: "solitons." Davydov used the soliton concept to explain molecular level events in muscle contraction, however solitons in the cytoskeleton may do what electrons do in computers.

The Fröhlich and Davydov approaches may be seen as complementary (Tuszynski, Paul, Chatterjee, and Sreenivasan, 1984). Fröhlich's coherency model focuses on time-independent effects (stable states) leading to order whereas Davydov's model looks at time-dependent effects which propagate order through the system. These and other theories of collective effects applied to information processing in cytoskeletal lattices will be described in Chapters 6 and 8.

## 1.4 Molecular Computing

To approach the cognitive capabilities of the human brain, AI must emulate brain structure at the nanoscale. Computer hardware is indeed evolving to smaller switching components, and advantages of proteins themselves are being considered. The smallward evolution of technological computing elements embraces a number of concepts and material collectively known as "molecular computing."

The potential advantages of molecular computers have been described by D. Waltz (1982) of Thinking Machines Corporation. 1) Current "planar" computer design is limited in overall density and use of three dimensional space. 2) Further miniaturization is limited with silicon and gallium arsenide technologies. Chips and wires cannot be made much smaller without becoming vulnerable to stray cosmic radiation or semiconductor impurities. 3) Biomolecular based devices may offer possibilities for self-repair or self-regeneration. 4) Certain types of analog, patterned computation may be particularly suited to molecular computers.

Forrest L. Carter (1984) of the Naval Research Laboratory has catalyzed the molecular computing movement through his own contributions and by sponsoring a series of meetings on *Molecular Electronic Devices* (in 1981, 1983, 1986). Strategies described by Carter and others at his meetings have been aimed at implementing *nanoscale* computing through switching in material arrays of polyacetylenes, Langmuir-Blodgett films, electro-optical molecules, proteins and a number of other materials. Interfacing between nanoscale devices and macroscale technologies is an obstacle with several possible solutions: 1) engineering upward, self assembling components, 2) optical communication, 3) molecular wires, 4) don't interface; build systems that are totally nanoscale (though they'd have to be somehow developed and tested), and 5) a sensitive bridge between macroscale and nanoscale. Technologies which may fulfill this latter possibility include ion beam nanolithography, molecular spectroscopy, quantum well devices, and scanning tunneling microscopy (STM). In STM, piezoceramic positioners control an ultra sharp conductor with a monoatomic tip which can probe and image material surfaces with atomic level resolution. STM related nanotools may soon be capable of ultraminiature fabrication and interfacing: "nanotechnology" (Chapter 10).

The medium of information flow in conventional computers is electronic current flow, but electron transfer may be too energetically expensive and unnecessary at the molecular nanoscale. Many of the projected modes of molecular computing rely on propagation of nonlinear coupling waves called "solitons" similar to what Davydov proposed for linear biomolecules. Carter (1981) proposed that solitons could propagate through switching circuits made of branched polyacetylene chains. He has also considered molecular computing in periodic arrays using electron tunneling, soliton "valving" and photo-activated conformational changes in lattice materials. He envisions three dimensional molecular scale memory and switching densities of $10^{15}$ to $10^{18}$ elements per cubic centimeter, near the theoretical limit for charge separation. A number of materials *may* be suitable for soliton switching and biological propagation of solitons in proteins has been suggested. Several authors have argued for cytoskeletal solitons mediating information processing (Chapter 8).

Wayne State University's Michael Conrad has defined his vision of a molecular computer in which proteins integrate multiple input modes to perform a functional output (Conrad, 1986). In addition to smaller size scale, protein based molecular computing offers different architectures and computing dimensions. Conrad suggests that "non-von Neumann, nonserial and non-silicon" computers will be "context dependent," with input processed as dynamical physical structures, patterns, or analog symbols. Multidimensional conditions determine the conformational state of any one protein: temperature, pH, ionic concentrations, voltage, dipole moment, electroacoustical vibration, phosphorylation or hydrolysis state, conformational state of bound neighbor proteins, etc. Proteins integrate all this information to determine output. Thus *each protein is a rudimentary computer* and converts a complex analog input to an output state or conformation.

Conrad and Liberman (1982) have defined an "extremal computer" as one which uses physical resources as effectively as possible for computation. They suggest that an extremal computer should be a molecular computer, with individual switches or information representation subunits composed of molecules. The state of each information subunit should be coupled to an energy event near the quantum limit. Protein conformational states leveraged to dipole oscillations in the nanoscale may be that limit. Conrad and Liberman conclude that, within biological systems, macromolecular computing occurs by conformational changes generating "reaction diffusion patterns" of concentrations of biochemical energy molecules (cyclic AMP).

A 1984 conference (Yates, 1984) considered *Chemically Based Computer Designs* (Yates, 1984) and attempted to answer 6 relevant questions. 1) Are there fundamental, quantum mechanical limitations on computation? This question deals with energy loss due to friction or other factors in computation. The work of Benioff (1980, 1982), Landauer (1982) and Feynman (1986) lead to the conclusion that, in principle, computation can be achieved by a frictionless, energy conserving system. Thus there appear to be no quantum mechanical limitations on computation. 2) Are there fundamental, thermodynamic limitations on computation? Although there are some computing operations that are irreversible and dissipative, the work of Landauer (1982) and Bennett (1982) show that there are no fundamental thermodynamic limitations on computation per se. 3) Are there fundamental limits to serial processing on digital computers based on binary switches? This question has philosophical implications (does the universe function through continuous or discrete processes?) and so cannot be answered assuredly. The consensus of the conference was that there are *probably* limits on serial, digital computing. 4) What are the practical physical limitations on computer design? There are several practical limitations to the further miniaturization of digital switching circuits. However those limits probably won't be reached for decades. 5) What are the potential contributions of molecular electronics to digital computer design? The conference considered molecular conformational changes, solitons, charge flow and other approaches. Molecular gates, wires and switches may be worth trying to build, although redundancy and parallelism may be necessary. 6) Do biochemical systems inspire technological imitations for the purpose of computer design? Many biological systems (DNA, antibodies, receptors, enzymes) were reviewed and a major conclusion was that,

> None of these materials is as rich in chemoelectric physical phenomena as are (cytoskeletal) microscopic biological objects. *Microtubules* offer the most possibilities for inspiring chemically based computation! (Yates, 1984)

## 1.5 Dynamic Pattern Representation

Processing of patterns or symbols is conducive to optimal computing. Patterns can be dynamically represented by a number of descriptive mechanisms which would be useful in both AI and biological systems. These include reaction-diffusion systems, holograms, macrons, and cellular automata.

### 1.5.1 Reaction Diffusion Systems

Reaction diffusion systems are evolving patterns which result from various types of reactions and product diffusion within a dynamic medium. Biological reaction diffusion systems within the submembrane cytoplasm have been suggested by Conrad and Liberman (1982) as a mechanism of biological information representation. In their model, reaction diffusion patterns of the energy rich nucleotide, cyclic AMP, which are regulated by the membrane are the texture of cytoplasmic information. Propagation and interaction of chemical, nonlinear waves lead to pattern formation in a number of chemical and biological media (Winfree and Strogatz, 1984). In the well studied "Belousov-Zhabotinsky reaction," spiral chemical reaction waves propagate at uniform speed and interact with other waves to produce complex patterns. Waves radiate from spiral centers at a rate of a few millimeters per minute as the spirals turn in about one minute. Several chemical reactions with suitable diffusion rates and visible color changes of reaction products show these characteristic patterns, as do cultured amoeba cells responding

Figure 1.9: Self organizing spatial and temporal patterns described by the chemical reaction-diffusion system known as the Belousov-Zhabotinsky reaction and emulated by biological systems. With permission from Arthur Winfree.

to pulses of cyclic AMP (Figure 1.9). Similar phenomena have also been reported in retinal and cortical nerve nets and in heart muscle. Smaller scale reaction diffusion patterns are accordingly faster.

Winfree and Strogatz (1984) have studied the 3 dimensional behavior of reaction diffusion systems. They find that reaction diffusion waves commonly appear as involute spirals or scrolls radiating from tiny rotating activity patterns called "organizing centers." The scrolls emanate from their central organizing axis which typically forms a closed ring or toroidal vortex. The origin of the waves is defined as a phase singularity whose immediate

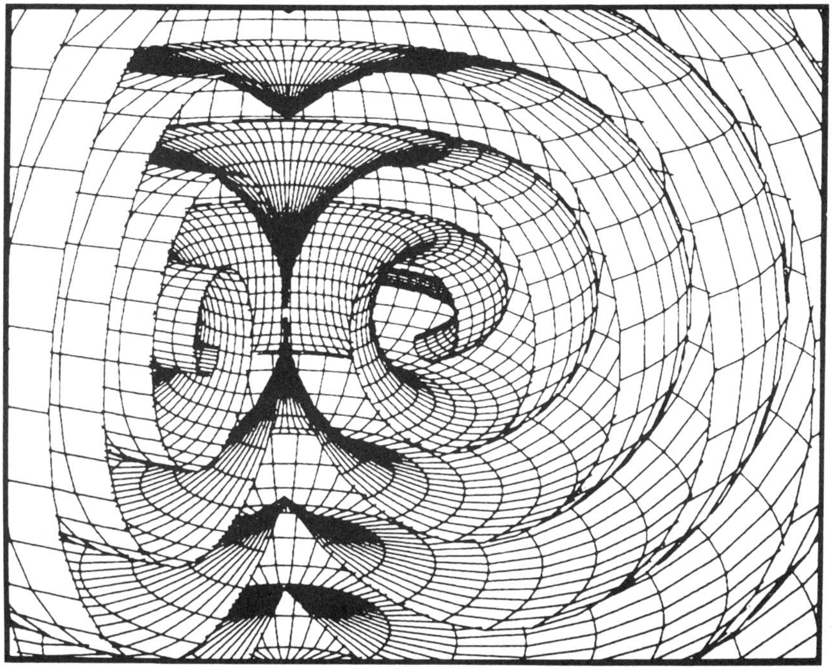

Figure 1.10: Three dimensional computer simulation of a reaction-diffusion system. A filamentous organizing center emanates "scroll ring" patterns. With permission from Arthur Winfree.

neighborhood is a rotating pattern of chemical activities, the pivot of the rotating spiral wave from which it radiates. The ostensibly flat spiral is actually a cross section of a three dimensional wave shaped like a scroll which emerges from a *filament* of singularity in 3 dimensions (Figure 1.10).

Cytoplasmic microtubules and centrioles are organizing centers which could behave like the singularities described by Winfree and Strogatz. Dynamic activities of the cytoskeleton may release diffusing waves of calcium ions which can alter the nature of surrounding cytoplasm by sol-gel transformations (Chapter 5). Coding by microtubule associated proteins (MAPs) and other factors could result in reaction-diffusion patterns specific to the dynamic state of the organizing center. Such patterns could suffice as short term memory in cells ranging from simple protozoa to human brain neurons. Another type of interactive, 3 dimensional pattern with interesting properties is the hologram.

### 1.5.2 Holograms

The brain stores image files in a "distributed" manner which is resistant to local damage and allows for correct retrieval even when variable cues or addresses are presented. One explanation is that memory, learning and real time cognitive functions are represented in the brain by interference patterns which are the convergence of two or more wave trains: signal and reference information sources (Hudspeth and Jones, 1975). Interference patterns can be dynamic, expressive, ordered or chaotic; one example is the ocean surf as an interface and monitor of the collective effects of wind, current, beach, tides, water bonds, etc. Interference patterns are used in information and imaging technologies such as interferometry, coherent processing, autocorrelation filtering, pattern recognition and many others whose capabilities are limited by their coupling medium (Dolgoff, 1975). All space in the universe is, as 17th century German philosopher Leibniz said, "the result of harmonious coexistence of forces." Consciousness as well may be described as the dynamic coexistence of forces within the brain, although the harmony may vary over time.

A method of recording and reconstructing wavefronts associated with interference patterns is call "holography," a technology whose mechanism has inspired numerous speculations of "holographic" brain function and consciousness. Holography is a method of information storage employing coherent beams of electromagnetic radiation. It was invented in the late 1940's by Denis Gabor (1948) who won the Nobel prize, and achieved technical importance with the arrival of the laser as a convenient source of coherent light in the 1950's. A hologram is a permanent record of the pattern of interference between two sources of coherent light (or any coherent waveforms) in localized regions of space, usually a photographic film plate. Subsequent reference waves unlock the patterns from storage. The record of both the original interfering waves are stored and the relevant information used as an address to retrieve patterns. Each portion of the hologram contains information about each part of both original interfering waves. Consequently reillumination of any small fragment of a hologram will recreate the entire image stored there, losing only focus or clarity. Holograms thus store image files in a "distributed" manner, much like the brain is thought to function, and are also "fractal," in that small portions are scaled down versions of the whole. By exposing a hologram to time varying sets of interfering waves, it can function as a distributed memory. These properties led to a flurry of holographic brain models (Westlake, 1970; Longuet-Higgins, 1968; Pribram,

1971). Among these, van Heerdon (1968) discussed methods of optical information storage in solids using coherent light. Van Heerdon pointed out that such systems can store large amounts of information although they require a calibrating system to maintain exact phase relations between waves. Requirements for well tuned filters or coherent resonators to maintain phase relations between patterns in the spatial domain remain a major question regarding holographic models of brain function and memory. Consequently the biological existence of holograms has been questioned, based on the assumption that the coherence and phase relation would have to be provided *at the cellular or neural level.* However, nanoscale coherence may have the required spatial and temporal periodicity to generate *cytoplasmic* holograms. Photo-refractive crystals can produce dynamic, real time holography (Gower, 1985). Conformational dynamics of the cytoskeleton could tune and generate coherent standing waves and interference patterns of calcium gradient fields, sol-gel states, and structure of the cytoskeletal microtrabecular lattice (Chapters 6 and 8). Dynamic and deterministic intracellular patterns would be useful in biological activities of all sorts. Holographic models of consciousness including a cytoskeletal approach will be described further in later chapters.

### 1.5.3 Macrons

The evolution of form and information from chaos has been termed "morphogenesis" and related to philosophical literature from many cultures. Mathematician Ralph Abraham (1976) has compared mathematical descriptions of the dynamic evolution of biological form to the Rigveda, I-Ching, Kabala, and Heraclitus. Using the catastrophe theory of Rene Thom (1973) and an observational device, the macroscope of Hans Jenny, Abraham has studied collective vibrational patterns which occur widely in nature and which he calls "macrons." Abraham describes physical, chemical, and electrical categories of macrons which may be further subdivided according to the material state of the macron medium. For example, physical macrons may occur within a solid, isotropic liquid, liquid crystal, or gas. Abraham cites one example of a solid macron: if a flat metal plate is vibrated transversely by an external force such as coupled electromechanical transducers, a vibrational pattern may be observed as a "spider-web" of motionless curves (the "Chladni" nodal lines). Originally observed by sprinkling sand on a vibrating plate, these patterns more recently have been observed by laser interferometry. The pattern is the "macron" and depends upon intrinsic dimensions and elasticity of the medium, and extrinsic frequency and amplitude of the driving force. The plate is a two dimensional example, however a simple rubber ball may be visualized with stable vibrational modes characterized by symmetric distortions of shape separated by motionless nodal surfaces. Another macron example is a round dish filled with a thin layer of isotropic liquid. If the bottom of the dish is heated, the liquid will soon begin to simmer; careful observation reveals nodal lines and packed hexagons called Benard cells within which the liquid convects toroidally. This Benard phenomenon, also seen as wind induced patterns in the sands of the Sahara and other deserts, is also considered by Abraham as a macron. These macrons or stable modes also depend on intrinsic controls such as shape, compressibility and viscosity, and external controls such as frequency and amplitude of the driving force.

Other forms of macrons described by Abraham include smoke rings, opalescences like abalone shell, and the aurora borealis or Northern Lights. Turning to the brain, Abraham conjectures: "a thought is a macron of the brain bioplasma." He proposes that spatial patterns of EEG are electrical macrons at dendritic surfaces or that macrons occur within nerve cells. He suggests that repetitive reinforcement of specific macrons "hardens" them into a structural form in a learning mechanism. Abraham's macrons may be compared to standing waves, reaction diffusion systems, and holograms which can all manifest 3 dimensional analog patterns of interactive information suitable to the cytoskeleton. Another "digital" system of interactive patterns in dynamic lattices is the "cellular automaton."

### 1.5.4 Cellular Automata

Complex behavior resulting from collective activities of simple subunits occurs in "cellular automata." Von Neumann's (1966) original cellular automaton consisted of a large number of identical "cells" connected in a uniform pattern. The term "cell" was chosen by Von Neumann and others as the indivisible subunit in "cellular automata" based on biological "cells" as indivisible subunits of life. Much like atoms once indivisible, are now recognized to be composed of electrons, protons, neutrons, quarks, etc., it is now apparent that biological cells are complex entities whose actions depend on collective functions of intracellular structures including the cytoskeleton. Nevertheless, "cellular" in cellular automaton jargon means an indivisible grain, a discrete subunit with a finite number of states. The essential features of cellular automata are 1) at a given time, each cell is in one of a number of *states.* 2) The cells are organized according to a fixed *geometry.* 3) Each cell communicates only with other cells in its *neighborhood*; the size and shape of the neighborhood are the same for all cells. Depending on geometry, the number of neighbors may be 4 (rectangular), 6 (hexagonal), 8 (rectangular with corners) or more neighbors per subunit or cell. 4) There is a universal *clock.* Each cell may change to a new state at each tick of the clock depending on its present state, and the present states of its neighbors. The rules for changing state are called the *transition rules* of the cellular automata. At each clock tick (or "generation") the behavior of each cell depends only on the states of its neighbors and its own state. In cellular automaton, simple neighbor rules can lead to complex, dynamic patterns.

Cellular automaton may be considered similar to lattice models such as a two dimensional Ising generator. Based on magnetic spin states of components within a lattice, Ising generators evolve to stable patterns in which states of opposite spin align in one direction, and like spins align in another direction. A generalized two dimensional Ising generator is shown in Figure 1.11. Cellular automaton models in microtubules (Chapter 8) evolve to similar states in which opposite states align in one direction, and similar states align in another direction (Figure 1.12).

Von Neumann studied how cellular automata could perform useful computations. He assumed a large number of cells start in a *quiescent*, or inactive state and that input was encoded by placing a number of contiguous cells in a specific pattern. By then running the clock through a sequence of generations, an output can be obtained by the patterns of states at a later time. A cellular automaton is said to be *universally computing* if, for any solvable problem, there is an initial configuration of the cellular automaton which evolves to a configuration containing the solution. As far as implementing such computing capabilities, access to every cell must be established to set its initial state and read its final state. Von Neumann discovered a universally constructing cellular automaton in which an initial configuration of a small number of cells (the "constructor") can set initial states of *distant* cells to the pattern required to solve any problem. The constructor communicates with distant cells through intermediate cells according to transition rules. If a cellular automaton is universally constructing, it can be "programmed" to solve any problem, even if only a few cells can communicate with the outside world. Several universally constructing cellular automata have been devised in simulation; "constructors" as patterns of cytoskeletal subunit conformation would be useful mechanisms for biological computation.

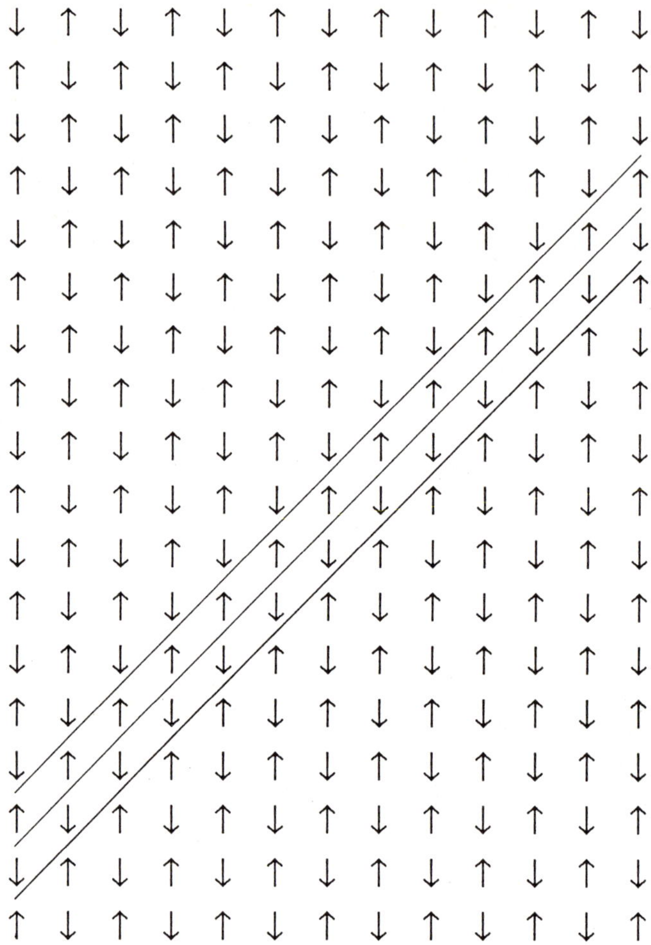

Figure 1.11: Two dimensional Ising generator evolves to stable state of opposite spin states aligned horizontally, and like spin states aligned diagonally. This stable configuration is similar to MT automaton simulation (Figure 1.12). Computer generation by Conrad Schneiker.

Cellular automata are frequent topics of Scientific American's *Mathematical Games* columns. Written by Martin Gardner and, more recently, A. K. Dewdney these columns have intermittently focused on the game of "Life," a cellular automaton invented by Cambridge mathematician John Conway in 1968 (Gardner, 1970). "Life" is played on a large two-dimensional grid of square cells. Each cell has eight neighbors, four at the edges and four

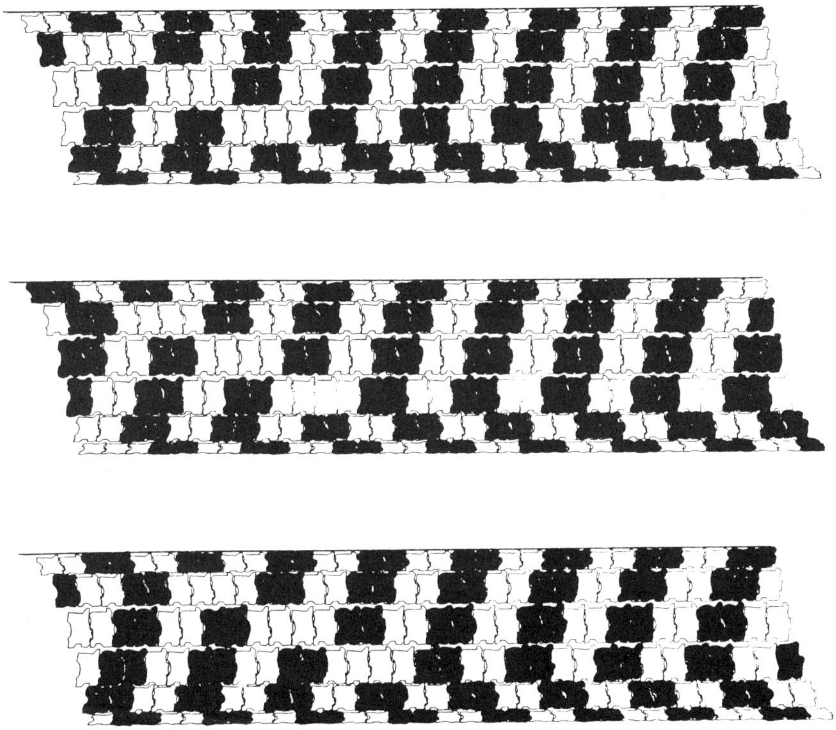

Figure 1.12: Cellular automaton model in microtubules (Chapter 8) reaches stable state in which opposite states are aligned along long axis of MT, and like states aligned along rows. A "kink-like" pattern is seen moving through the structure. By Paul Jablonka.

at the corners, and exists in one of two states: "dead" or "alive." At each generation, cells may die or come alive, their fate determined by the number of living neighbors. For example a living cell with fewer than two living neighbors, or more than three, will not survive (due to lack of sustenance or overcrowding, respectively). A dead cell will be born in a subsequent generation if it has exactly three living neighbors (or "parents"). Conway's game was named "Life" because the cells could be either dead or alive, however the behavior of the patterns of "living" cells included some "life-like" behaviors. These included movement through the grid and oscillatory patterns which came to be called blinkers, beacons, gliders, and beehives. Repeating von Neumann, though in a much simpler format, Dewdney (1985) showed that a computer could exist within the game of Life.

Carter Bays has extended the game of "Life" to three dimensions (Dewdney, 1987). In his version, each cell is a cube with 26 neighbors, but the neighbor rules are essentially the same as in Conway's two-dimensional "Life." A variety of interesting behaviors ensue in Bay's "Life," dependent on initial patterns. For example he observed a 10 cube "glider" traveling through space like a "sofa in free fall." Another 7 cube form (a "greeter") dies unless it is in the presence of another structure. Gliders which pass near greeters are grabbed and held until rescued by a second glider which collides with the repressive greeter. Other stable patterns emerge which Bays has called arcades, stairs, helices, and space-time barriers. Life and other cellular automata have enraptured computer buffs who can now create their own realms. Beyond that, cellular automata have serious scientific and mathematical implications.

Stephen Wolfram (1984) has viewed cellular automata as systems of simple components capable of complex collective effects such as the simulation of partial differential equations and deterministic chaos. He has described four general behaviors for cellular automata patterns. 1) They disappear with time, 2) they evolve to a fixed finite size, 3) they grow indefinitely at a fixed speed, 4) they grow and contract irregularly. Type three, which grow indefinitely at a fixed speed, are often found to be self similar in scale; parts of such patterns when magnified are indistinguishable from the whole. Thus these cellular automata patterns are characterized by a fractal dimension.

Wolfram notes that the mechanisms for information processing in natural systems appear more similar to those in cellular automata which are highly parallel than to conventional serial processing computers. The "results" are given by the configuration obtained; the "medium is the message." Further, "it is common in nature to find systems whose complexity is generated by the cooperative effect of many simple identical components." Cellular automata are sensitive to initial conditions and their behavior is characterized by the stability or predictability of their behavior under small perturbations in initial configurations. With a given set of rules, changes in a single initial site value can lead to markedly different patterns. Such perturbations have characteristic effects on Wolfram's four classes of cellular automata: 1) no change in final state, 2) changes only in a finite region, 3) changes over an ever increasing region, 4) irregular change. Thus at least some cellular

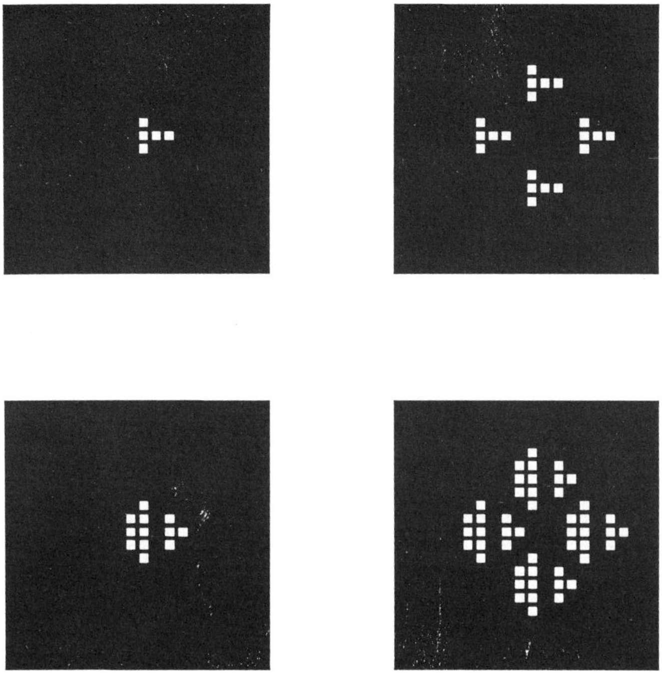

Figure 1.13: Self replicating automata described by Edward Fredkin (Dewdney, 1985). "Off" states are shown as all black. Computer generation by Conrad Schneiker.

automata patterns are nonlinear and deterministic.

Cellular automata can be ascribed to exist within a variety of environments. Perhaps the most extreme view is that the universe is a cellular automaton. MIT's Edward Fredkin has been contending that the universe may work according to the same principles as a cellular automaton (Wright, 1985). He believes the basic material of which everything is made of can be considered as *information* rather than mass and energy. Working at the interface of physics and computer science, Fredkin has become intrigued with the relations between cellular automata and nature. With the right rules, a cellular automaton can simulate the formation of a snowflake, mollusc shell, or galaxy. Fredkin's view is to apply cellular automata to fundamental levels of physics and the rules needed to model the motion of molecules, atoms, electrons, and quarks. With sufficient information to model these particles, an automaton may be designed that describes the physical world with perfect precision. At that level, says Fredkin, the universe is a cellular automaton

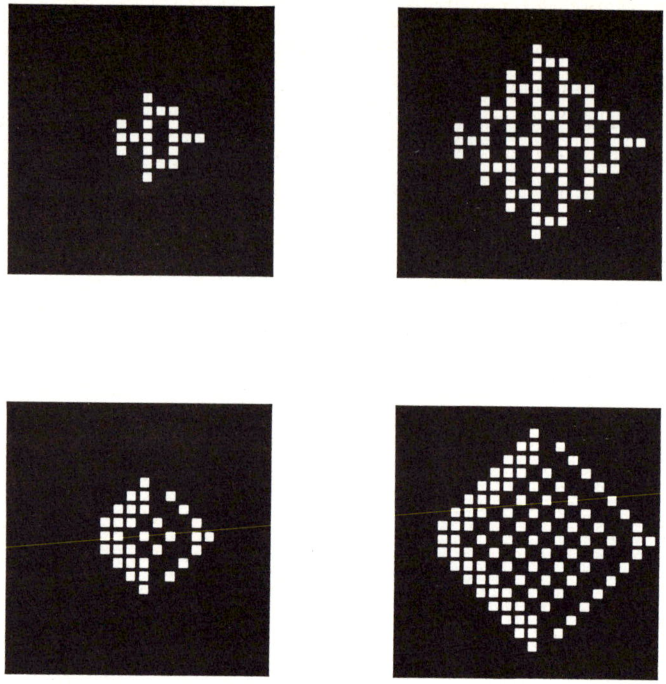

Figure 1.14: Self replicating automata described by Edward Fredkin (Dewdney, 1985). Computer generation by Conrad Schneiker.

in three dimensions: a lattice of interacting logic units, each one deciding billions of times per second whether it will be "off or on" at the next instant. Fredkin sees this information as the fabric of reality, the stuff from which matter and energy are made. He argues that cellular automata can represent the universe as usefully as can differential equations, the prevalent mathematical alternative. The cellular automaton view is by far the simpler. A child can understand the rules governing a cellular automaton and with pencil, paper and enough time can predict the course of an automaton including charting the growth of a snowflake, the ripples of a pond or a sound wave. Cellular automata are the language of pure information and may be involved in biological information processing as well as future computer devices. Forrest Carter (1984) of the Naval Research Lab and James Milch (1986) of Eastman Kodak have both proposed the construction of molecular automata, extolling the virtues of the cellular automaton concept applied to problems of interfacing to molecular scale devices. With a large cellular automaton molecular computer, communication to the "macro" world need only interface with a "constructor," a small portion which can take advantage of the entire cellular automaton capacity.

Subunits of cytoskeletal microtubules may be particularly suitable for "cellular" automaton behavior and information processing in the nanoscale. Cytoskeletal automata and their dynamic consequences may be an important substrate of biological computing ranging from the actions of single cells to brain/mind consciousness. Will they pave the way to Ultimate Computing?

# Chapter 2

# Brain/Mind/Computer

## 2.1 Metaphors of Consciousness

Systems for information processing are evolving within both biological life forms and computer technologies. The most highly evolved information processing system currently appears to be human consciousness which resides in the human brain. The scientific relationships between consciousness and structural brain activities remain obscure and are often referred to as the brain/mind "duality." To explain this duality, humans have historically perceived their minds in the context of predominant cultural themes, particularly information technologies. Author Julian Jaynes (1976) has chronicled how the metaphors of the mind are the world it perceives. The trail of brain/mind metaphors perhaps began during the Greeks' Golden Age. According to Plato, Socrates said:

> Imagine ... that our minds contain a block of wax ... and say that whenever we wish to remember something we hear or conceive in our own minds, we hold this wax under the perceptions or ideas and imprint them on it as we might stamp the impression of a seal ring.

The Greeks traveled about in freedom (while their slaves did the work) and consciousness was perceived by free men as a free entity. Heraclitus described consciousness as an "enormous space whose boundaries could never be found out." Later, Augustine of Carthage described "the mountains and hills of my high imagination," "the plains and caves and caverns of my memory" with "spacious chambers wonderfully furnished with innumerable stores."

The geological discoveries in the 19th century revealed a record of the past written in layers of the earth's crust. Consciousness became viewed as layers recording an individual's past in deeper and deeper layers until the original record could no longer be read. This emphasis on the unconscious mind grew until the late 19th century when most psychologists thought that consciousness was but a small part of the mind. As chemistry superceded geology in scientific esteem, consciousness became viewed as a compound structure that could be analyzed in a laboratory into precise elements of sensations and feelings. When steam engines became commonplace, the *sub*conscious was perceived as a boiler of straining energy demanding release, and when repressed, pushing up and out into neurotic behavior. In the early part of the 20th century, mind metaphors continued to encompass technologies for information processing such as telephone switching circuits, tape recorders, clocks, holograms, and computers.

The computer is the most recent brain/mind metaphor and has evolved qualitatively beyond its predecessors (as has human consciousness). Computer technology has approached, and in some cases surpassed, some aspects of human brain function such as "brute force" calculations. Computers may also be used to simulate dynamical systems (including the brain), thus providing a metaphorical medium. In efforts to construct computing machines capable of independent logic and decision making, artificial intelligence (AI) researchers have examined what is known about the workings of the brain and mind. Accordingly, they have been led away from classical "serial" computers towards massively parallel systems with high degrees of lateral interconnection. Because the brain at first glance is a parallel aggregation of billions of neurons with tens of thousands of connections per neuron, AI researchers of the connectionist school have viewed and modeled the brain as "neural networks" which may be simulated on conventional computers. These neural net models, to be discussed later in this book, are based on relatively simple assumptions regarding interneuronal synapses as switches between neurons. Dynamic patterns of neural net activity can simulate systems capable of learning, independent recognition, different "mental" states, and with some imagination, rudimentary consciousness. The general architecture of parallel computers is similar to neurons within the brain, and can take advantage of simultaneous processing with lateral resolution of conflicting concepts. Despite these apparent similarities, the brain's complexity and the dynamic vastness of human consciousness remain unassailable by current technology. The mind remains enigmatic to brain and computer.

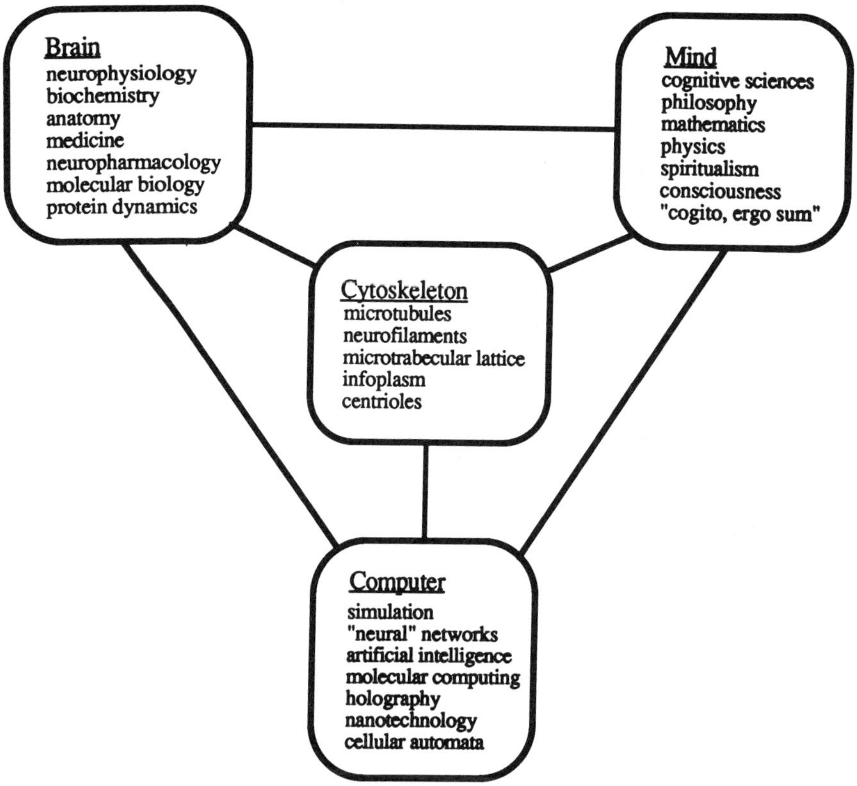


Figure 2.1: The Brain/Mind/Computer metaphorical triangle. Is the cytoskeleton the key to understanding?

Brain, mind, and computer are mutually metaphorical; each is related to the other in ways that are not clearly understood. This impasse, the "brain/mind/computer triangle," is based on an incorrect assumption. The irreducible substrate of information processing within the brain has been assumed to be the notoriously slow interneuronal synapse. Consequently, synapses have been compared to simple switches, and the brain has been compared to a computer composed of a collection of synaptic switches. Because each neuron within the brain has up to several hundred thousand synapses, it must "integrate" information from among these synapses to regulate its output. Neurons utilize a variety of analog functions including dendritic morphology, slow wave membrane properties, and cytoskeletal activities which determine their responses within neural networks, and which alter synaptic efficacy as apparent mechanisms of learning. Thus *each of the billions of neurons in the brain is a computer*. Similarly, single cell organisms which have no synapses and are independent agents perform complex tasks involving rudimentary decision making, behavior, and organization. Thus, the basic irreducible substrate of information should reside *within* biological cells, and the brain may then be viewed as an organized assembly of *billions of computers* in which collective emergent properties may be specifically related to consciousness. The hierarchy of brain organization may thus have a secret basement—a new "dimension." Advances in intracellular imaging and molecular biology have illustrated the complex dynamic organization of intracellular cytoplasm. Specifically a dense, parallel, highly interconnected solid state network of dynamic protein polymers, the "cytoskeleton," is a medium which appears to be ideally suited for information processing, and which is actively involved in virtually all cell functions. Appreciation of this "cytoskeletal dimension" may be the key to the brain/mind/computer triangle (Figure 2.1).

## 2.2 Historical Perspectives—Consciousness as ...

Many disciplines have concerned themselves with attempts to understand consciousness. Like the proverbial group of blind men trying to describe an elephant, each discipline's perception is highly dependent on its orientation and particular elephant part it happens to contact. The blind men succeed, largely because they have the elephant surrounded.

Some feel the mind is too complicated to be described by the human brain. Perhaps the mystery of the mind is a necessary barrier to man's "roboticization"? As philosopher Richard Rorty has said: "the ineffability of the mental serves the same cultural function as the ineffability of the divine—it vaguely suggests that science does not have the last word" (Jaynes, 1976). Despite these worries, a progression of theories and metaphors of the mind have evolved and are reviewed historically in Julian Jaynes' book, *The Origin of Consciousness and the Breakdown of the Bicameral Mind.* Jaynes describes eight solutions to the brain/mind problem developed through the 20th century. They describe consciousness as a property of matter, of protoplasm, of learning, as a metaphysical imposition, a helpless spectator, an emergent property of evolution, behavior, and as activity within the brain's reticular activating system. These are reviewed with modifications and additions relevant to computer technology and the cytoskeletal dimension.

### 2.2.1 Consciousness as Particle/Wave Physics

Great discoveries in 19th century particle physics dissolved the solidity of matter into mere mathematical relationships in space. Thoughts, feelings, introspection, and mind-environment interactions were related to the brain as waves to electrons. In the 20th century, Nobel biochemist Albert Szent-Gyorgyi (1960) wrote *Introduction to a Submolecular Biology* in which he perceived the essence of life and consciousness to exist in coordinated electron movement within semiconductive proteins. Others, including Russian physicists Pullman and Pullman (1963) compared life and consciousness with the mobility of electrons within resonant bond orbitals. Scottish biologist A. G. Cairns-Smith (1985) has theorized that life developed from crystals of clay. The molecular lattice structure in clay allows for shifting neighbor relationships and processing of information which Cairns-Smith has likened to genetic development. These views equate life's basic processes with those of atoms and sub-atomic particles. Information *is* represented as dynamic electron patterns within computers, and life and consciousness are certain to be related to fundamental particle activities. The questions are how, where and at what level of organization?

### 2.2.2 Consciousness as a Property of Protoplasm

Believing that consciousness is a fundamental property of all living things, some 19th century biologists saw its essence in the irritability of the smallest one cell organisms. Popular books of this era included, *The Animal Mind* by M. F. Washburn, and *The Psychic Life of Microorganisms* by Alfred Bonet. Observation of an amoeba hunting food or responding to various stimuli, or paramecium avoiding obstacles or conjugating led to application of human psychology to such behavior. These concepts were accepted by Charles Darwin and E. B. Titchner, who saw such rudimentary consciousness related to man through the course of evolution.

Circumstantial support for this thesis is found in the inhibitory effects of general anesthetic gases on protoplasmic streaming in slime molds, and anesthetic inhibition of amoeboid and paramecium motility. This suggests a common link between these primitive organism activities and brain activities related to consciousness in that all are reversibly sensitive to the same anesthetic gas molecules at comparable concentrations. Protoplasmic streaming, amoeboid movement and paramecium motility all depend on dynamic activities of cytoskeletal structures including "computer-like" cytoplasmic microtubules, actin sol-gel transitions, and ciliary appendages (Chapter 5). The cytoskeletal link among anesthetic sensitive processes could be a clue to the brain/mind/computer triangle.

Jaynes objects to protoplasmic consciousness, suggesting that humans may be projecting their own mind functions onto protozoan behavior which he believes to reside entirely in physical chemistry rather than introspective psychology. However, introspective psychology itself is in all probability a function of physical chemistry at some level. If an amoeba, or slime mold, or paramecium are not conscious, at what point in the evolutionary hierarchy does consciousness emerge? Stanford University Professor Karl Pribram (1966), known for his conceptualization of mind functions as "holographic," recalls being confronted with this issue during a lecture at the Montreal Neurological Institute in the late 1950's. Famed neuroscientist Wilder Penfield asked Pribram whether the difference between man and the non-human primates was quantitative or qualitative. Pribram replied that the difference was quantitative but to such an extent that qualitative changes emerged. He cited the relatively new computer technology as an example: vast increases in the capacity of memory and central processors had changed computational power not only quantitatively but qualitatively. Penfield argued for a more fundamental distinction to distinguish man. Pribram countered by saying that, although the only difference in brain structure between man and other animals is quantitative, changes in organization, chemical composition, developmental sequence, and in time and duration of critical periods had led to *collective emergence* of *qualitatively* distinctive properties. The *quantitative* common link of consciousness may be the cytoskeleton within cells ranging from single cell organisms, viruses, (and perhaps more simple "life" forms such as "prions," or independent protein structures) to neurons within the human brain. The qualitative differences appear to lie in nonlinear collective properties related through evolution to structural complexity.

### 2.2.3 Consciousness as Learning

Proponents of this view believed that consciousness began at some specific time after life evolved and was directly related to learning. The rationale was: if an animal modifies its behavior on the basis of experience, it must be having an experience and therefore must be conscious. By equating learning, experience and consciousness, this school viewed associative processes as the essential element of consciousness.

Models of associative memory in neural net computer simulations may be bolstered by the historical glorification of learning per se, and learning *is* important for biological success. Structural correlates of learning in mammalian brain (discussed in Chapter 4) appear to involve strengthening of specific synapses brought about by a dynamic reorganization of the neuronal cytoskeleton. However, as Jaynes observes, learning and consciousness are separate problems. AI systems can learn, but they clearly are not conscious. Information may be perceived in human consciousness, exist in short term memory, but fail to be stored in long term memory—hence no "learning" occurs. Certain drugs, including some anesthetics and tranquilizers, specifically block long term memory storage and retrieval in *conscious* patients. Thus consciousness constitutes more than learning.

### 2.2.4 Consciousness as a Metaphysical Imposition

Assessment of the evolutionary link, but intellectual chasm, between civilized man and apes resulted in a metaphysical view: consciousness could not have evolved merely by natural selection from assemblages of molecules and cells. Something must have been added from outside of the closed system to account for an entity so different as human consciousness. This school was founded by Alfred R. Wallace, co-discoverer of the theory of natural selection with Charles Darwin. Wallace believed that some metaphysical force had directed evolution at three different points: the beginning of life, the beginning of consciousness, and the beginning of civilized culture. Because Wallace sought evidence for a metaphysical force among vitalists, spiritualists, and seances, he was discredited and Darwin became known as the discoverer of evolution.

Some so called vitalists and spiritualists attempted to apply particle/wave physics to what was then known about cell biology in their search for consciousness. Like Wallace, they were vilified because they had no proof and the scientific establishment felt that to explain consciousness by metaphysical imposition was outside the realm of science.

Modern bioelectromagnetic field theories pertaining to embryology and consciousness have been proposed by many authors but remain undocumented. Dynamic nanoscale activities within a cytoskeletal information system could provide such a field yet be beyond detection by current technologies. Future nanotechnology may permit detection of these fields, if they exist. The metaphysical imposition theory, its "vitalist" and particle/wave physics counterparts remain speculation, but the degree to which they irritate the scientific establishment is noteworthy. Perhaps it is because they blur the distinction between science, philosophy and religion. This may presage violent opposition to the future development of artificial consciousness.

### 2.2.5 The Helpless Spectator Theory

A materialistic view of the origin of consciousness arose in response to the metaphysical imposition theory. The helpless spectator theory suggests that life is like a roller coaster ride and that consciousness does nothing at all, being an epiphenomenon to important biological activities. As a helpless spectator of cosmic events, consciousness was described as

> the heat given off by wires, colors laid on the surface of a mosaic, the movement of a train going along tracks that have determined its destiny, the melody that floats from a harp but cannot pluck its strings, the foam raging from a river that cannot change its course, the shadow of a pedestrian (Jaynes, 1976).

Giving up on free will, T. H. Huxley bleakly summarized "we are conscious automata." (The negative connotation of automata as helpless spectators prevails in the context of robots and machines, however should not be confused with the notion of *cellular automata* which may independently process information and deterministically compute, and which have been likened to biological processes.) The helpless spectator theory was rejected by William James who found inconceivable the notion that consciousness should have nothing to do with the business it so faithfully attends. He asked, "why is consciousness more intense when action is most hesitant, why are we least conscious when doing something most habitual?"

### 2.2.6 Emergent Evolution

In this view, consciousness was rescued from the undignified position of a helpless spectator by reconciling the metaphysical imposition view with collective emergent properties. One metaphor used was: as the property of wetness cannot be derived from the properties of hydrogen and oxygen atoms alone, so consciousness emerged at some point in evolution in a way underivable from its constituent parts. John Stewart Mill and others suggested that as properties of matter emerged from an unspecified forerunner, properties of complex compounds emerged from conjunction of simpler compounds, and properties distinctive of living things emerged from the conjunction of these complex compounds, and finally consciousness emerged from these living things (Jaynes, 1976). Thus a scaffolding of new conjunctions were thought to result in previously unseen relationships bringing new emergent phenomena. Coalescing as something genuinely new at a critical stage of evolution, consciousness assumed guidance over the course of events in the brain, and causal efficacy in bodily behavior. In some ways this view is like the "Indian rope trick" in which the Fakir tosses a rope into the air where it mysteriously stays, he then climbs up the rope, pulls it behind him and disappears. Evolutionary processes may have provided for the development and existence of consciousness which then assumed control and guidance of biological systems. The conditions leading to the appearance of consciousness may be viewed as a nonlinear emergence from evolutionary events.

The emergent evolution theory liberated biologists and neuroscientists from their burden of needing to base all of their results on known physical properties. The mind could thus be dealt with in a subjective sense, allowing psychiatry and Freudian theory to become acceptable without a concrete basis for concepts such as ego, id, and superego. Significant questions which then arose included: When did consciousness emerge? Where? In what species? And what was it? The brain/mind duality still existed and in fact the mind was dealt with only in broad and nebulous generalities.

### 2.2.7 Behaviorism

The problem of consciousness could be solved by ignoring it. Behaviorists traced their roots to the so called epicurians of the 18th century and before that to attempts to generalize plant tropisms to the actions of animals and man. Behaviorists explained all cognitive processes on reflexes and conditioned responses which were comparable across wide varieties of organisms. Thus human behavior, no matter how noble or furtive, could be explained on reflex responses to given situations or needs to satisfy bodily functions. Behaviorism did lend itself to experimentation very readily and was a boon to the credibility of scientists studying the mind. Behaviorist laboratories flourished in universities throughout the world; rat mazes and conditioned responses became the operant paradigms. Behaviorists were able to attract university positions, grant money, and behaviorist laboratories dominated neuroscience for a significant period of time. The failing of behaviorism is that it is a method rather than a theory and is patently hypocritical in denying or ignoring consciousness. Behaviorism did, however, purge psychology to place it squarely in the mainstream of academic science.

### 2.2.8 Consciousness as Dynamic Activities of the Brain's Reticular Activating System

As technology accelerated, brain and mind theorists eventually turned back to the brain—the three and a half pound lump of pinkish-gray "wonder tissue." Mathematician-philosopher Rene Descartes, who had defined the brain/mind duality by his statement "cogito, ergo sum," (I think, therefore I am) chose the brain's pineal gland as the site of consciousness. His choice was partly based on the fact that the pineal gland is a midline, single structure. Thus, unlike nearly all other brain regions it had no duplicate, and was therefore thought to be essential. Descartes' proposal was readily refuted by neurophysiologists but did start the search for a single site of brain consciousness. Since many researchers viewed the brain/mind as a hierarchical arrangement of information processing, the logical conclusion was that at a certain site or region all information was recognized and assimilated by the "Mind's Eye," the site of consciousness, the Grandfather neuron, or some manifestation of a hierarchical apex. Recently, most of the brain has been found to be involved with wide ("distributed") parallel files of information, memory, and cognition. Historically, however, many workers focused on the reticular activating system (RAS) as the neural substrate of consciousness.

Maintenance of consciousness depends to an important extent upon the RAS, an organized tangle of tiny interconnecting neurons extending from the top of the spinal cord up through the brain stem into the thalamus and hypothalamus. The RAS integrates collaterals from sensory and motor nerves, has direct lines to half a dozen major areas of the cortex and probably all of the nuclei of the brainstem, and sends fibers down the spinal cord where it influences the peripheral sensory and motor systems. Its function is to sensitize or awaken selective neurons and nervous centers and desensitize others such that it can regulate the activity and wakefulness of the entire brain. Anesthetic induction drugs such as sodium thiopental are thought to exert their effects largely on the RAS. Destructive lesions of this area also produce permanent sleep and coma, and stimulating the RAS electrically can wake up a sleeping animal. It can also regulate the activity of most other parts of the brain through its own internal electrical excitability and neurochemistry.

Although the RAS regulates wakefulness, the riddles of consciousness

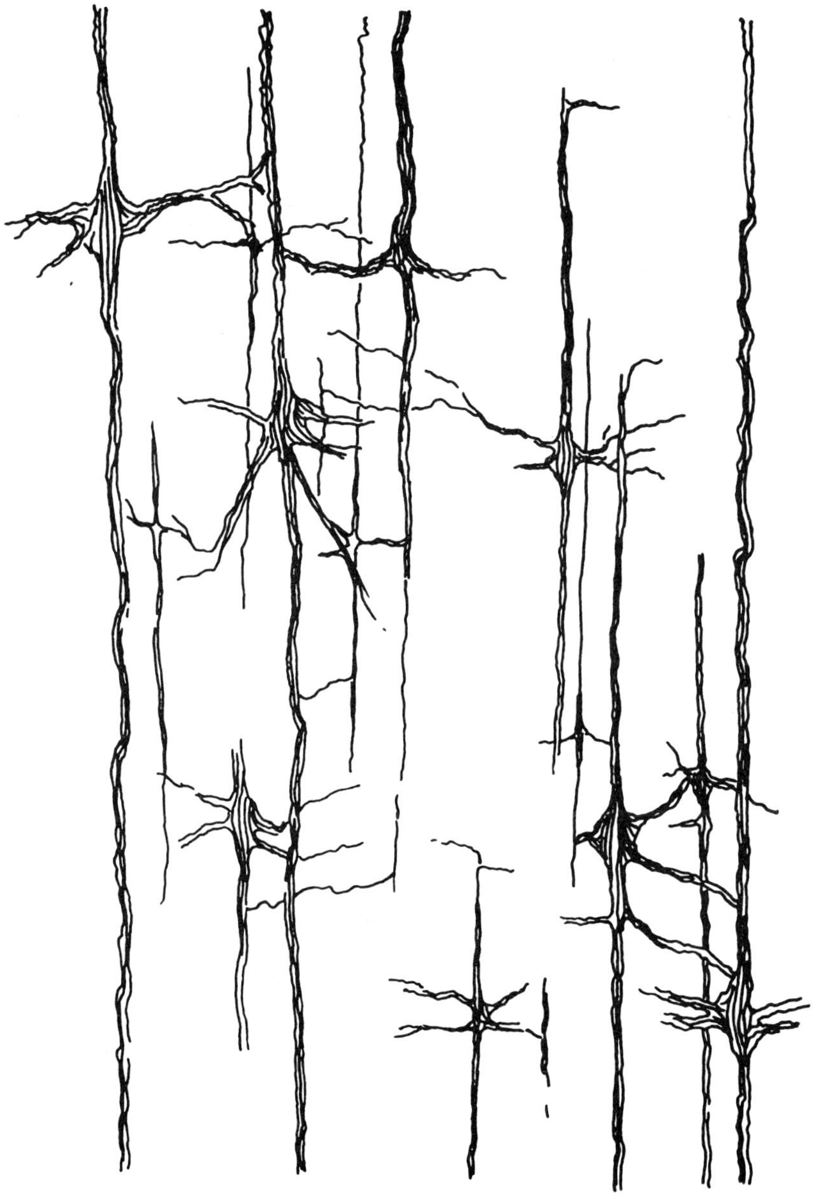

Figure 2.2: A group of associated neurons such as brain cortical pyramidal cells. The long axons and dendrites extending from cell bodies are on the order of a few microns (thousand nanometers) thick. Some lateral synaptic connections are evident. By Jamie Bowman Hameroff.

are unsolved. For example, high level integration and associative functions implicit to cognition occur predominantly in the cortex, structurally more evolved in advanced animals such as man. Conversely, the RAS is one of the oldest evolutionary parts of the nervous system and is relatively unchanged when compared among lower animals and humans. Jaynes observes that even if we had a complete wiring diagram, if we were aware of every transmitter in the nervous system, understood all of the billions or trillions of synapses, we could still not discern that a specific brain contained a consciousness like our own.

### 2.2.9 Neural Net Connectionism

This movement has been fortified by computer scientists' efforts to mimic the brain by constructing artificial intelligence systems. Based on approximation of cortical neurons as linear threshold units, a large number of "neural net" models have been constructed and simulated on computers. A key concept in relating neural network dynamics to the facts of psychology is the *cell assembly* introduced by Donald Hebb in 1949. In his view, *cell assemblies* are specially organized reverberatory circuits that constitute elements of thought. An individual neuron may participate in many of them just as an individual member of society participates in many social assemblies. By allowing strengthening or reinforcement of repeatedly used connections ("synaptic plasticity"), recognition, learning and problem solving become manifest in lowered thresholds of specific loops. By assigning energy levels to various patterns ("landscapes") within the net, mathematical solutions can also be imposed. The "intelligence" or capabilities of a given neural net model depends on the richness of its interconnections and nonlinear feedback. Neural net connectionist theory may help to advance robotic and computer systems for artificial intelligence, and may provide significant insight into brain function. These theories have provided evidence that dynamic activities within a given network can at least mimic some aspects of brain activities.

Shortcomings of early neural net models are that they have been based on hypothetical neurons with huge assumptions about neural function. Each neuron has been considered an on/off gate or switch, and interneuron synapses viewed as variable weight interconnections. More recent models incorporate axonal impulses, synaptic delays, dendritic analog functions and spatial coherence. In their most elegant form, neural net theories provide possible representations of mental objects ("consciousness"?) in the transient instan-

Figure 2.3: Closer view of neurons with intraneuronal cytoskeleton visible. Dendrites, ascending from below into cell bodies, have numerous dendritic spines. Several synapses are depicted. By Jamie Bowman Hameroff.

taneous patterns of network activity. "Temporally stable cooperative coupling" among sets of neurons are suggested to manifest thoughts and images by the work of Hebb, Kohonen, Edelman, Thom, von der Malsburg, Hopfield, Pellionisz, Llinas, Changeux, and others. Some of their work suggests the brain forms sets of "prerepresentations" of what is expected from which sensory input induces "selection" of its reality candidate. In this context, Changeux (1985) has defined consciousness as "a kind of global regulatory system dealing with mental objects and computation using those objects." Neural net models and associative memories have significantly advanced understanding of collective neural capabilities. Their essential features (parallel processors with lateral variable connections) may also be operant *within* neurons in the cytoskeleton.

### 2.2.10 Holography

Holograms may be formed from interference patterns generated from two or more coherent wave sources. Initially a laboratory curiosity, holography became important with the advent of lasers as coherent light sources in the early 1960's. A hologram recorded in a photographic plate appears to be mere garble until reilluminated with one of the original coherent wave sources and thereupon projects three dimensional spatial images. Two properties of holograms have attracted interest and comparisons with consciousness and mental imagery. One is that holograms have an enormous capacity for information storage, although coherent reference waves are necessary for recall. The second is that much information in a given hologram is contained in any one small portion of it, although with reduced resolution and signal to noise ratio. Recently, dynamic real-time holography has been developed with the use of photorefractive crystals (Gower, 1985).

Denis Gabor (1948), who received the 1969 Nobel prize for his invention of holography, remarked:

> for some years now this property of holograms has attracted the interest of neurophysiologists who were puzzled by the difficulty in locating the 'engram' in the human or animal memory. As is now well known, especially since the famous experiments of Lashley, large parts of the brain can sometimes be destroyed without wiping out a learned pattern of behavior. This has led to speculation that the brain may contain a holographic mechanism.

Gabor was skeptical, however, about the "existence of waves or tuned resonators in the brain." The mantle of holographic brain theory was taken up by Stanford's Karl Pribram (1986) who has contended that the brain perceives sensory information by analysis of the interference of neural firing frequencies. What results within the brain, according to Pribram, is a holographic domain in which space and time are enfolded. Consequently transformations into ordinary domains can be achieved from any part of the encoded records. This is the property of distributedness of information which characterizes holograms and brain functions. Holographic brain models have been based on coherent wave interference at the level of neuronal activities, particularly dendritic-dendritic interactions. Verification of coherency among neurons has been lacking, and the neural level hologram remains merely an interesting mathematical model. However, the cytoskeleton *within* neurons (and all cells) may be well suited for holographic mechanisms due to its spatial coherence (i.e. 8 nanometer periodicity) and potential temporal coupling (coherent nanosecond oscillations, see Chapter 6). Thus intracellular cytoplasm surrounding the cytoskeleton may be the substrate for holographic consciousness.

The holographic concept of consciousness has psycho-physical implications (Weber, 1975). Physicist David Bohm has remarked that our perceptions of reality are conditioned on lenses (eyes, cameras, microscopes, etc.) which focus, objectify, form boundaries, and particularize. Lensless holograms are distributed, lack boundaries and are "holistic." Bohm suggests that the reality of the universe is mathematically similar to a hologram (the "implicate" domain) which deals with frequency domain, and fluctuating waveform properties as opposed to our lens conditioned Euclidean-Newtonian impressions. This approach seems consistent with modern physics views on wave/particle duality and the uncertainty principle. Experiences reported by mystics, schizophrenics and hallucinogenic drug experimenters describe loss of spatial and temporal boundaries, and a holographic ("fractal") characteristic of the whole being represented in every part. One may argue whether mystical/schizoid/drug induced perceptions are aberrations or clarified reality, but certain properties of holograms (distributedness of information, vast storage capacity, three dimensional spatial imagery) bear some resemblance to consciousness.

### 2.2.11 Cytoskeletal Basis of Consciousness

Classical approaches to understanding the brain/mind have assumed a hierarchy of organization in which interneuronal synapses are the indivisible

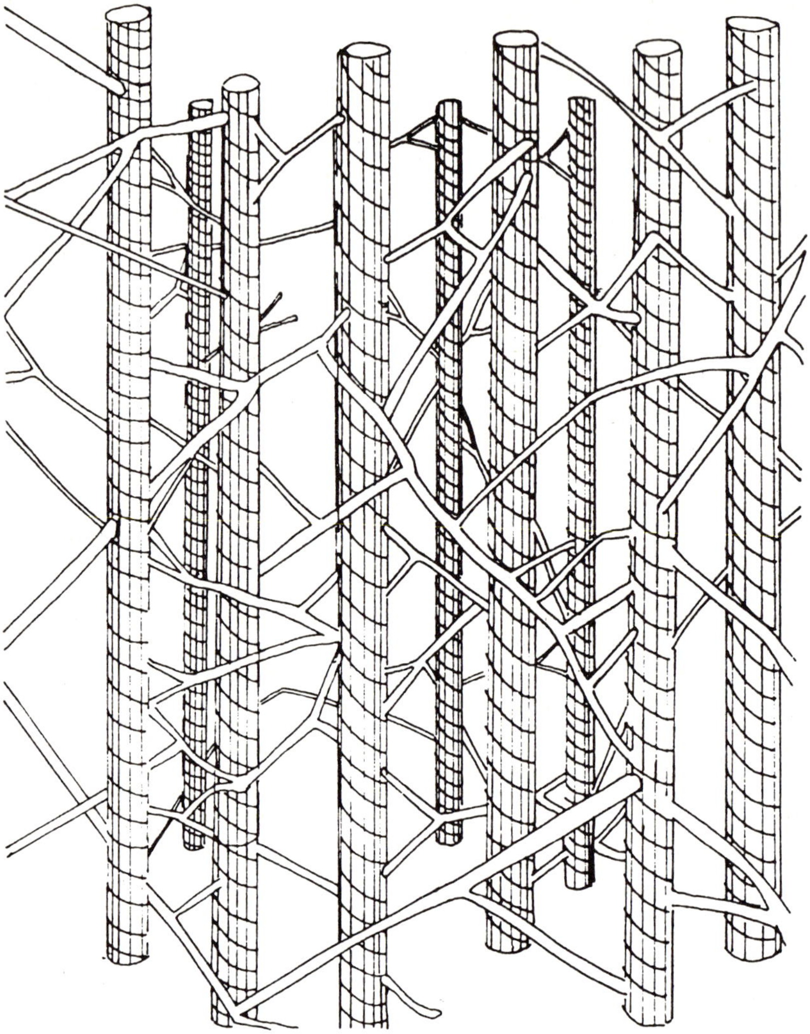

Figure 2.4: Interior of neuron showing cytoskeletal network. Straight cylinders are microtubules, 25 nanometers in diameter. Branching interconnections are microtrabecular lattice filaments. Neurofilaments are not shown. By Jamie Bowman Hameroff.

substrates of information transfer (Figure 2.2). However, neurons appear far too complex to be simple digital switches or gates and must contain intrinsic information processing systems.

The neuron is often and mistakenly described as a simple device that compares a weighted sum of dendritic "analog" input signals with some threshold level above which an output "digital" pulse is transmitted along an axon. The structure of neurons, in which vast arborizations of dendritic fibers may accept some 100,000 synaptic inputs per neuron, indicates a very complex system. While some have viewed the dendritic tree as being a passive transmitter of impulses, data suggest action potentials in large dendrites, slow depolarization waves in others, and the possibility of elementary logic operations at branching points (Scott, 1977). Each dendritic branch point may act as a logical "OR" gate if a pulse on either daughter branch can supply sufficient charge to excite a pulse on the parent; otherwise it may act as a logical "AND." Dendrites may also be transmitting presynaptic information to other neurons by graded potentials at dendrodendritic synapses, a process called "whispering together" (Adey, 1977). The extent and significance of "electrotonic" current pathways among dendritic membrane glycoproteins, membrane patches and dendritic spines are unknown but could also be important. The net summation of dendritic potentials in cerebral cortex is recorded at the scalp as electroencephalography (EEG). On the axonal or output side the parent fibers do not necessarily excite all daughters at each branching point, and branch points of some axons contain regions where high frequency filtering, alternate firing and other forms of information processing can occur (Scott, 1977). Branch point conductance may be influenced by small changes in local geometry and electrical coupling, thus providing intraneural regulation of neural transmission. Composition of membranes, spatial distribution of glycoproteins, ion channels, myelin gaps ("nodes of Ranvier") and clustering of receptors have also been viewed as information coding. All potential neural information processing modes (synaptic plasticity, axon and dendrite morphology, membrane protein distribution) depend on the dynamic cytoskeletal functions of axoplasmic transport and trophism (Chapter 5).

The neuron's complexity may indeed be more like a computer than a single gate. But what are the gates? What is the indivisible substrate of neuronal function? Where is the neuron's neuron? How does an amoeba or paramecium perform complex tasks without benefit of a synapse, neuron, or brain? Relatively recently, with perfection of electron microscopic fixation, immunofluorescence and other techniques, the interior of living cells has been revealed. It has been shown to possess complex, highly parallel interconnected networks of cytoskeletal protein lattices which connect to and regulate membranes and all other cellular components. The structure of these protein polymers, their dynamic activities and the lack of a clear alternative understanding of cognition have led to theoretical consideration of dynamic cytoskeletal activities as functional information processing modes such as cellular automata and holograms.

This cytoskeletal view, developed in detail later in this book, is consistent with many of the earlier schools of understanding consciousness. From the cytoskeletal perspective, consciousness *is* a property of protoplasm (specifically related to cytoskeletal proteins) but the vertebrate variety of consciousness is a nonlinear collective effect—an emergent evolution—of that which exists in simple organisms. An early nonlinear jump in the evolution of consciousness may have occurred with the introduction of cytoskeletal centrioles and microtubules, and the concomitant transformation of prokaryotic cells to eukaryotic cells about one billion years ago (Chapter 3). Perhaps, as cytoskeletal networks and higher organizational levels such as neural networks reached sufficient complexity in the brains of mammals, collective properties emerged nonlinearly due to cooperativity, resonance, phase transitions, and coherent phenomena allowing for automata, holography, or some other mechanism of information processing. Neural network theory, parallelism, connectionism, and the AI approach to consciousness have provided enlightenment regarding the brain/mind. Many of these approaches may be applied to the cytoskeleton, a fractal subdimension of neural networks (Figures 2.3 and 2.4). Levels of neural network connections such as the reticular activating system or hippocampal circuits may depend on intracellular cytoskeletal dynamics for their regulation. For example, learning in neural net theories is based on Donald Hebb's (1949) suggestion that circuits of connected neurons develop more conductive synapses which facilitate activation of that circuit. Firing along given patterns following a specific stimulus is thought to represent a specific concept, thought, or memory—information. Learning is then thought to occur by reinforcement or strengthening of synaptic connections, or formation of new synapses along given neural net patterns. Synaptic strength and neuronal connection depend on intracellular cytoskeletal rearrangements and axoplasmic transport to maintain synaptic efficacy and to form new synapses and dendritic spines. Ingredients for synapses, membrane proteins, receptors, ion channels, enzymes, and apparatus for neurotransmitter releasing mechanisms are synthesized in cell bodies and transported to synapses along relatively great lengths by axoplasmic transport, a coordinated effort of cytoskeletal microtubules. Synaptic modulation may represent a mechanism by which separate dynamic layers in an information hierarchy communicate and "tune" each other.

A dozen models of cytoskeletal information processing have been published and will be reviewed in Chapter 8. Nanoscale activities in cytoskeletal lattices may offer a future bridge between consciousness and emerging nanotechnology. As complex and highly connected as neuronal branches are (i.e. dendritic trees) the cytoskeleton within all neurons may be a *forest within those trees.*

# Chapter 3

# Origin and Evolution of Life

## 3.1 Soup vs Mud, Chicken vs Egg

What is life? Living organisms have certain properties that are nearly synonymous with the trait of being alive—organization, growth, reproduction, dynamic purposeful activities and (at least in higher organisms) intelligence and consciousness. Life forms that we have come to know are all based on the same type of genetic blueprints (DNA, RNA) and building blocks (proteins), suggesting a common ancestry. That ancestry, life's emergence, is generally viewed as a rearrangement of cosmic matter originally produced in the "Big Bang" which is presumed to have given birth to the universe some 14 billion years ago. Life's molecular emergence can be viewed in the context of two basic questions concerning place of origin ("soup vs mud") and molecular cause and effect ("chicken vs egg").

In the 1920's Russian biochemist A. I. Oparin (1938) and British biologist J. B. S. Haldane (1947) described their concept of a "primordial soup" of organic molecules existing in the earth's oceans a mere 4 billion years ago. Their soup was thought to be a product of geochemical processes and energy sources acting in an atmosphere of unoxidized gases such as methane, ammonia and hydrogen, similar to what exists currently on Jupiter. This primordial atmosphere was the view of eminent chemist Harold Urey (1939), whose graduate student Stanley L. Miller carried out a key experiment in the early 1950's. Miller created a closed environment containing such a primitive atmosphere and passed electric sparks simulating lightning through it. He detected organic molecules relevant to living processes. Fifteen percent of the original methane carbon was found in molecules which included four amino acids, the building blocks of proteins. Miller also found precursors of DNA and ribose sugars from which RNA is formed. Because the most central molecules of life are identical in all organisms on earth, Miller's primordial soup has been considered to represent *the* conditions from which life emerged. Other research has questioned the hydrogen rich atmosphere upon which Urey and Miller based their experiment and still other work has shown that at least some organic precursors of life can be generated in many types of atmospheres.

There are other candidates for the site of life's origin. Conditions above the thermal vents recently discovered on the ocean floor are believed conducive to the formation of organic compounds, leading some to propose these spots, rather than the traditionally imagined ponds or tidal pools, as the cradle of life ("deep soup"). Thermal vents are home to strange and exotic life forms such as giant tube worms which thrive in the great pressures of the deep ocean. Organic compounds have also been found commonly in intrastellar dust, in comets, and in meteorites that fall to earth. Many believe the supply of organic precursors to life was augmented from space while few admit to believing that primitive *cells* were transplanted to earth from space.

An alternative explanation has been advanced by A. Graham Cairns-Smith (1982) of Glasgow University, who suggests that early organisms utilized pre-existing information templates in the form of wet clay crystals ("mud"). Crystalline inorganic materials appear to have many "life-like" properties such as the ability to store and replicate information in the form of crystal defects, dislocations, twin boundaries, and substitutions. Clay minerals like kaolinite crystallize at ordinary temperatures from aqueous solutions of common rock. Their catalytic surfaces and complex morphology suggested to Cairns–Smith an environment not unlike living material. He observed that defects in crystals could supply multiple, stable alternative configurations which can store and process information much like modern computers. Crystal defects which can move are very similar to primitive cellular automata, dynamic patterns occurring in lattice neighborhoods capable of computing. Cairns-Smith reasoned that certain clays proliferated with their replicating defects (representing information) acting as primordial genetic information and proving useful in alignment of amino acids and protein synthesis. As more efficient organic synthesis developed, Cairns-Smith argues that clay machinery became expendable and was jettisoned in favor of a new biotechnology—DNA and RNA.

Whether or not Cairns-Smith's clay theory is correct, he demonstrates the capacity for information storage in crystal defects. Perfectly ordered crystals which are repetitive and homogeneous have no capacity for information storage but are also extremely rare or do not exist at all. Real crystals have defect structures superimposed. Simply to be finite—to have a shape and size—is a defect, but many other features are almost invariably present. Units are often missing or are replaced by others, and sections of the crystal structure may be misaligned in various ways. While such features can be very small in scale, they provide real crystals with a large potential capacity for information. Certain classes of crystals might have defect structures that replicate as the crystal grows by having the right combination of structural characteristics, growth patterns and cleavage properties. Cairns-Smith (1982) concludes by posing a challenge to discover crystal genes of various materials. He asks:

> ... Imagine doing experiments with crystals that could evolve, setting them problems—applying selection pressures—and seeing how they cope. This would be an interesting thing to do any way whatever the crystals are made of. We would soon find out whether mineral versions of replicating systems are plausible although we might lose interest in our ultimate ancestors once we had in our hands the first organisms of another kind: the first organisms of our own contriving.

The implications of Cairns-Smith's ideas include the possibility of alternative life forms from propagating crystalline structures and a suggestion that DNA and RNA are not necessarily the only carriers of genetic information. This is in concert with a demystification of life in general. At an international meeting on the origins of life (Eckholm, 1986), Dr. Cyril Ponnamperuma of the University of Maryland suggested "the division between life and nonlife is perhaps an artificial one." He views the animate and inanimate as lying on a continuum both over evolutionary time and among currently existing systems. On such a scale prions, proteinoids, and some viruses would lie near the middle as might some ancient unknown protocell that became the ancestor of life on earth. To speak of advanced chemistry rather than divine creation is certain to disturb religious fundamentalists. Equating life with oscillations in crystals does have an almost biblical resonance, and narrows the conceptual gap between life molecules and technological devices.

Regardless of the precise environment in which life-related molecules emerged, other major questions include whether the carriers of genetic information, DNA and RNA, preceded proteins whose amino acid sequences they determine, or whether proteins, including enzymes and structural elements seemingly necessary for genetic replication, came first. Thus a chicken (DNA, RNA) vs egg (protein) conundrum regarding life's origins has developed. A primary information flow from nucleic acid to protein (chicken before egg) was a "central dogma" in molecular biology. Fox and Dose (1972) challenged this conviction by proposing an alternative evolutionary continuum from the beginning of the material cosmos to the first organisms. In their "egg before chicken" view the sequence of life's organization was from interstellar gases, to amino acids, to polymers, to organized microsystems. As evidence they cite the self organization of amino acids and proteins into "proteinoid" microspheres which can establish communication links and perform other "life-like" functions. Evidence for a "chicken before egg" view has been found by chemist Leslie Orgel and colleagues (Schwarz and Orgel,1985) of the Salk Institute in La Jolla, California. They recently discovered a 15 nucleotide long DNA-like molecule that had formed spontaneously from much simpler carbon compounds and zinc in the absence of living cells or protein enzymes. Other work has suggested that RNA can function enzymatically to facilitate reactions and the bulk of recent findings leans towards the primacy of nucleic acids.

Manfred Eigen (1971) views this cause/effect problem as a "closed loop" whose original starting point is unimportant. What is important, in Eigen's view, is how molecular self-organization occurs from random events and feedback which lead to macroscopic functional organization, self-reproduction, selection, and evolution: "hypercyles." Eventually, according to Eigen, such systems can escape the prerequisites of their origin and change the environment to their own advantage.

A view of primary nucleic acid (*chicken*) organization in a primordial aqueous environment (*soup*) is summarized and elaborated in the writings of biologist Lynn Margulis and Dorion Sagan (1986). They view as logical the facts that RNA and DNA spontaneously formed in the shallow seas of early earth and also became able to self replicate perfect copies of themselves. They liken RNA molecules to half of an open zipper. With the proper complementary ingredients, the missing half forms by using the existing RNA as a template.

Margulis and Sagan (1986) note:

> An RNA molecule can do more than copy itself. The sequence of its nucleotides can also serve as a signal for a neighboring strand of RNA to attach the amino acids in its environment, thus

> forming a portion of a protein which will in turn accelerate the matching of other RNA molecules producing more RNA, more protein like fragments, and so on.

This suggests that at a critical level of evolution, nonlinear accelerations occurred due to the level of associative inter-relationships among evolving molecules. This can help explain how biological systems can produce "order from chaos," and thus apparently violate the second law of thermodynamics which states that ordered systems must dissipate towards disorder.

Margulis and Sagan describe the following scenario for the development of life on earth. RNA formation in the primordial soup led to the evolution of double stranded DNA eons later. This in turn allowed the full variety of life—as manifest in the richness of structures and functions of proteins and other macromolecules. Survivability was enhanced by enclosure of dynamic molecules inside membranes, apparently formed when phospholipid hydrocarbons aligned and, because they were charged on one end, formed spherical droplets which sequestered biomolecules. With the advent of ion channels and other membrane proteins came regulatory voltages and a discrete microcosm: the "prokaryotic" bacterial cell.

## 3.2 Prokaryote to Eukaryote—Symbiotic Jump

Proliferation of prokaryotes literally changed the face of the earth. According to the Margulis/Sagan scenario, collective teams of bacteria gathered nutrients, disposed of toxins, recycled organic matter by turning waste into food and stabilized the atmosphere. Prokaryotic bacteria produced ammonia which adjusted the acidity of oceans and lagoons and increased the earth's temperature through a "greenhouse" effect similar to that of carbon dioxide (which lets in more solar radiation than can escape). About two billion years ago purple and green photosynthetic bacteria began using water to manufacture hydrogen rich compounds, giving off oxygen which was poisonous to most ("anaerobic") prokaryotes. This "toxic waste crisis" pressured adaptations including motility systems to escape oxygen exposure, detoxification, and eventually oxygen breathing. The resultant early "aerobic" prokaryotic bacteria flourished for a few hundred million years, but as atmospheric oxygen increased, aerobic and anaerobic bacteria begat a new improved form of life, the eukaryotic or nucleated cell.

We are all eukaryotes, as are all animals and nearly all plants existing on earth today. Eukaryotic cells differ from their prokaryotic ancestors by having organized cell interiors (cytoplasm or protoplasm) including separate membrane enclosed compartments (nuclei) which contain, among other structures, chromosomes: DNA libraries and their supportive proteins. Eukaryotic cytoplasm usually contains mitochondria, chemical energy factories which utilize oxygen to generate ATP to fuel cellular activities (respiration) and, within green plants, chloroplasts which convert solar energy to chemical energy foodstuffs (photosynthesis).

Eukaryotic cells are enormously sophisticated compared to their prokaryotic predecessors. Fossil records indicate that eukaryotes appeared abruptly, with no apparent intermediate form which would indicate progressive genetic mutation from prokaryotes. This evolutionary gap, which separates bacteria and blue-green algae from all other present day cellular life forms, is a mysterious dichotomy, an evolutionary chasm. Explanations based on symbiosis—a mutually beneficial association—were advanced by Marishkowski in 1905 and Wallen in 1922 (Margulis and Sagan, 1986). They proposed that eukaryotic cells resulted from a symbiotic association of two types of prokaryotes—a primitive "monera" and a more advanced cocci-type bacteria. Ingestion of the cocci by the monera is thought to have led to a stable symbiosis in which the more evolved cocci became the nuclear material and the monera became the cytoplasm. Marishkowski proposed other examples of symbiosis such as the emergence of green plants from a union of colorless nucleated cells and minute cyanophycae which became chloroplasts specialized for photosynthesis. This proposed union is similar to the symbiosis of green algae and fungi to form lichen, and of chloroplasts in metazoa such as hydra. Wallen proposed that mitochondria originated as symbiotic bacteria which entered, and became indispensably entrenched within, animal cells.

A more complete "endosymbiotic" theory of eukaryote origin was introduced by biologist Lynn Sagan (later Lynn Margulis) in 1967. She suggested that prokaryotic cells (specifically anaerobic heterotrophic bacteria) underwent a series of three symbiotic events leading to the first eukaryotes. During the period of adaptation to oxygen breathing an aerobic heterotroph was engulfed by an anaerobic heterotroph. The aerobic bacteria became the ancestor of the mitochondria, converting oxygen to ATP and remained as an intracellular organelle. The next symbiotic event, according to Sagan-Margulis, was the ingestion of a spirochete—a motile organism which traveled by whip-like beating of its tail-like flagellum composed of cytoskeletal proteins. Ingestion of flagellae and their intracellular anchors, basal bodies, are thought to have led to cilia, centrioles, and microtubules—cytoskeletal structural and organizational elements which brought the capabilities for cell movement, cytoplasmic organization, and (apparently) information processing (Figure 3.1). Multiple cilia attached to cell membranes and extending outward enabled single cell organisms such as paramecium to swim about in their aqueous medium, greatly expanding their ability to find food, avoid predators, and increase their horizons. In other stationary organisms cilia could flow the environmental medium past the organism, achieving the same results. Within the cytoplasm, cytoskeletal structures such as centrioles, basal bodies and microtubules organized, oriented, and transported organelles and materials. The eukaryotic cytoskeleton took on functions akin to mechanical scaffolding, conveyor lattice, and the cell's own nervous system.

Basal bodies, cilia, flagella, and centrioles are assemblies of microtubules, themselves complex cylindrical assemblies of protein subunits, and are ubiquitous throughout eukaryotic biology. In these organelles, nine pairs or triplets of microtubules are arranged in a super-cylinder, which may have an additional microtubule pair in its center (9+2 or 9+0 arrangements, Figure 3.2). Involvement of these structures in nearly *all* instances of dynamic cell activities (mitosis, growth and differentiation, locomotion, food ingestion or phagocytosis, cytoplasmic movement etc.) greatly accelerated the capabilities of eukaryotic cells. Utilizing the chemical energy from mitochondrial ATP, these cytoskeletal elements appear to have provided not only stable structure and motility, but also a sophisticated "computer-like" information processing system.

Eukaryotic microbial technology was as different from the basic bacterium as a main frame computer to an abacus. The eukaryotes flourished, evolved and solved environmental problems by mixing and merging. Forming new collectives, they eventually found their way from water to land and air and branched into the myriad forms of plant and animal life that have since populated the biosphere. The human brain and nervous system are recent innovations; Homo sapiens apparently appeared about 50 thousand years ago.

The endosymbiotic theory can explain the nonlinear jump in evolution that occurred with the advent of eukaryotes, but does have its detractors. For example, Hyman Hartman (1975) of MIT has noted that, while mitochondria and chloroplasts are agreed to have originated as free living prokaryotic cells, there is some question as to the pedigree of basal bodies and centrioles. Sagan-Margulis (1967) had claimed that:

> upon entry into a host, such a symbiot may lose from none to all

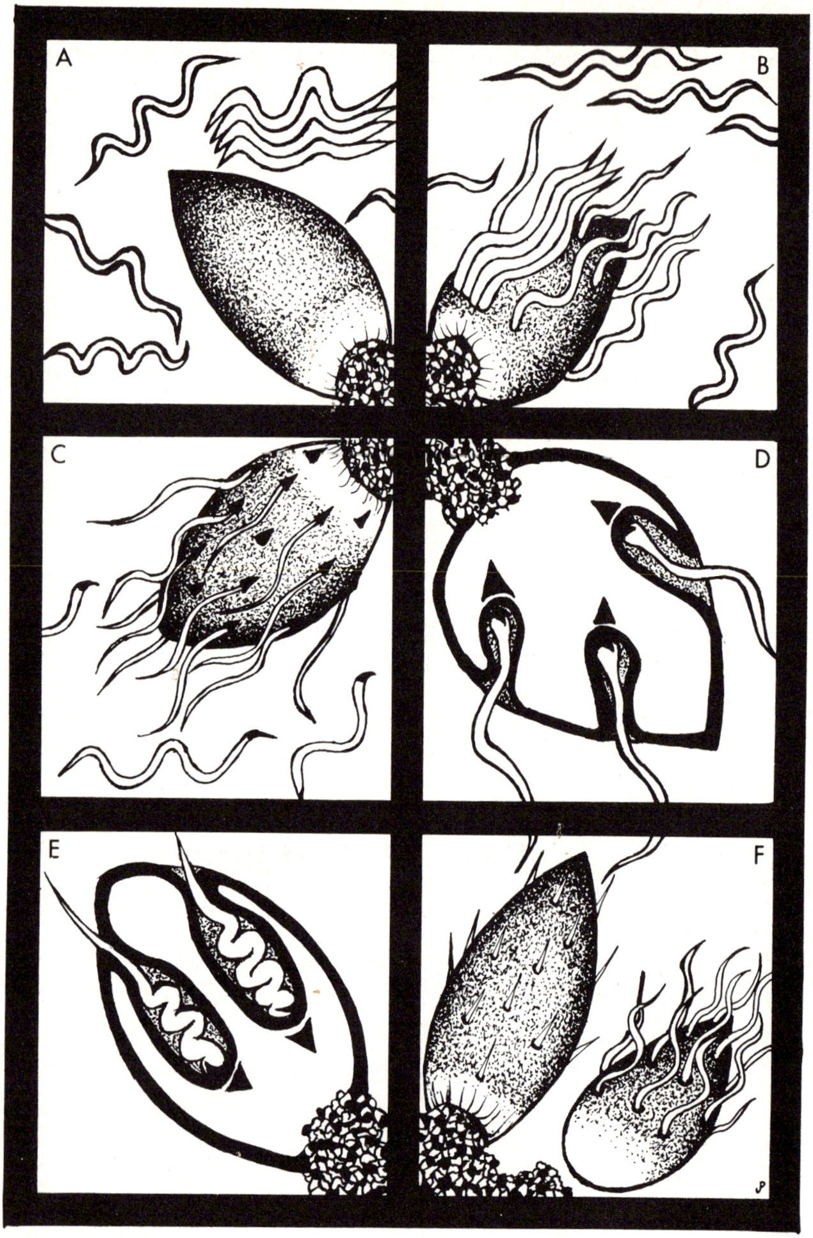


Figure 3.1: Symbiotic ingestion of motile spirochetes by a primitive bacteria, resulting in the first "eukaryotic" cells. The spirochete's filamentous proteins became, according to Sagan-Margulis (1967), the centrioles and cytoskeleton providing movement and organization of cytoplasm. By Paul Jablonka

> of its synthetic capabilities except the ability to replicate its own DNA and synthesize complementary RNA from that DNA—the sine qua non of any organism.

This implies that symbiotic organelles which originated as separate organisms must retain their nucleotide synthesis capabilities. Mitochondria and chloroplasts have been shown to have their own DNA, however DNA has not been isolated with centrioles or basal bodies. These cytoskeletal organelles in fact, routinely self replicate *without* DNA, although Hartman has shown that RNA may exist in association with basal bodies. Perhaps centriolar DNA became lost in the evolutionary shuffle, or the cytoskeleton possesses other mechanisms of information transfer. Evidence of cytoplasmic information being transmitted over hundreds of generations of paramecium without genetic involvement (Aufderheide, Frankel and Williams, 1977) suggests that centrioles and other cytoskeletal elements may have a degree of independence (Figure 5.27). Real time information processing is in the cytoskeletal province, so DNA replication may *not* be the "sine qua non" of living organisms. Dynamic, collective activities of centrioles, microtubules, and other cytoskeletal proteins may manifest biological intelligence and be closer to life's essence than are genetic mechanisms. Ambiguous life forms may be particularly important in the future.

## 3.3 Centrioles—Evolution's Hijackers

Appearance of centrioles and related cytoskeletal structures as motile intelligent organizers on the evolutionary scene one billion years ago may have been the key to success for eukaryotes, and initiation of the lineage that has led to human consciousness.

Centrioles are the specific apparatus within living cells which trigger and guide major reorganizations of cellular structure occurring during mitosis, growth and differentiation. Centrioles are composed of two similar cylinders (each of which is also referred to as a "centriole") whose diameters are 0.2 microns (200 nanometers). Each cylinder possesses a nine fold radial symmetry and consists of microtubule triplets longitudinally fused. A cartwheel filamentous structure (or "pinwheel") appears to hold together the end of each centriole cylinder. One centriole begets another by replication which occurs at right angles to the long axis of the cylinders. This perpendicular replication which initiates mitosis is counter-intuitive compared to longitudinal replication or fission, and remains one of the mysteries surrounding

Figure 3.2: Centrioles are cylinders of 9 triplets of microtubules, shown here in cross-section. Pairs of centrioles associate at right angles to each other, shown here at bottom. By Paul Jablonka.

centrioles.

Centrioles reside within a portion of the cytoplasm known as the centrosome (or centrosphere) adjacent to the cell nucleus outer membrane. By triggering and guiding polymerization of microtubules and other cytoskeletal elements, centrioles temporally and spatially organize cytoplasm ("microtubule organizing center"—MTOC, Figure 3.3 and Chapter 5). For example centrioles and microtubules (mitotic spindles) separate chromosomes and establish daughter cell architecture (Figure 3.4)—features which had enormous advantages in accelerating eukaryotic evolution.

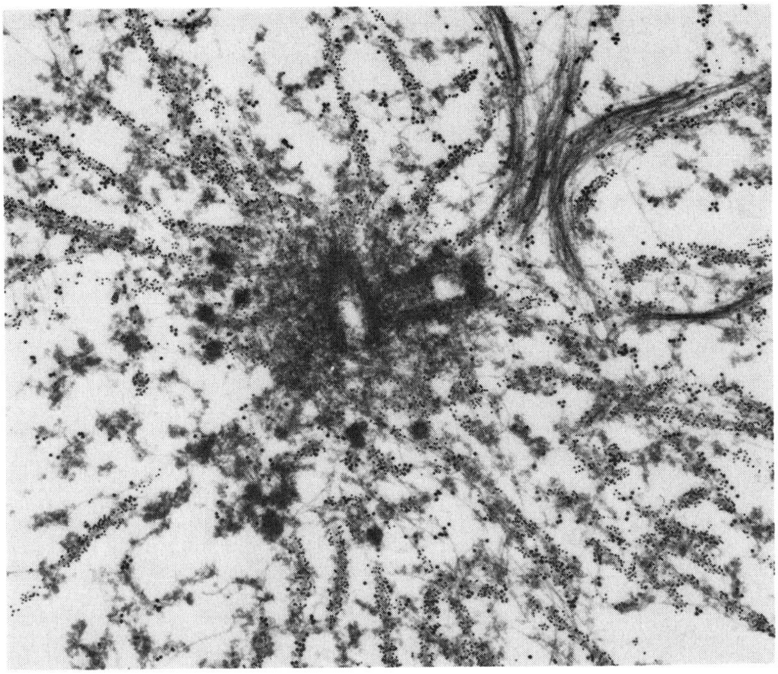

Figure 3.3: Spindle pole centrioles in PtK2 kidney mitotic cell. Perpendicular centrioles are seen in the dense pericentriolar material from which MT radiate, dotted by immunogold. A filamentous network is seen to the right of, and above, the microtubule organizing center (MTOC). Scale: 3.3 millimeters on micrograph = 100 nanometers. With permission from Geuens, Gundersen, Nuydens, Cornelissen, Bulinski and DeBrabander (1986), courtesy of Marc DeBrabander and Janssen Pharmaceutica Research Laboratories

Centriole-like basal bodies, acting near cell membranes, induce formation of cilia as appendages which protrude from outer surfaces of cells. Cilia have structures virtually identical to centrioles except being membrane covered and, in the case of motor cilia, having a central microtubule pair and contractile interconnections which act to bend and wave cilia in a variety of control functions. These range from propulsion of single celled paramecium to expulsion of dust and particles from human airways. Similarly, sensory cilia permit communications with external environments across a wide biological range from single cell organisms to the inner ears of human beings, transducing mechanical sound into the nervous system. Generations of motile and sensory cilia allowed eukaryotic cells like paramecium to roam about vaster quantities of their aqueous environment, serving to maximize their food supply as well as perceive sensory information from their environment. This resulted in complex activities involving logic and information processing which led 19th century scientists to ascribe rudimentary consciousness to such organisms.

Another mystery surrounding centrioles is their command of orientation in space and ability to convey that information to other cytoskeletal structures. Navigation and gravity sensation have been suggested to represent a "gyroscopic" function of centrioles (Bornens, 1976) which have also been described as perfectly designed signal detectors (Albrecht-Buehler, 1981). These and other models of information processing and intelligence in centrioles and the cytoskeleton will be covered in Chapter 8.

Relevant to evolution is that centrioles provided eukaryotes with a sophisticated cellular information processing and communication system. The consequences of such a system on biology is perhaps analogous to the potential impact of computers on societies. The mystery and aesthetic elegance of centrioles, as well as the fact that in certain instances they appear superfluous, have created an enigmatic aura about this marvelous organelle. In the forward to Wheatley's (1982) book *Centrioles: The Central Enigma in Cell Biology,* biochemist B. R. Brinkley states:

> Before Gallileo's telescope challenged their views, early scholars argued that the earth was the center of the universe around which revolved the sun. Following their discovery by light microscopists, centrioles were given an equally permanent role in the cytoplasm of eukaryotic cells. This minute organelle was thought to be the center of the cytoplasmic universe ... .

Centrioles' structural beauty, unfathomable geometry, intricate behavior, navigational command, and apparent origin as invader from the prokaryotic kingdom add to their mystique. Writing in Wheatley's book, Patelca states: "biologists have long been haunted by the possibility that the primary significance of centrioles has escaped them."

A possible conclusion is that centrioles are intelligent nano-engines who "jumped ship" from a previous species to symbiotically upscale their lifestyle. By so doing, they have coopted biology and, in concert with other dynamic cytoskeletal structures, pushed intelligence to its current stage of evolution. The next symbiotic event may have equally profound implications. Nanoscale technologies may directly interact with biomolecular intelligence.

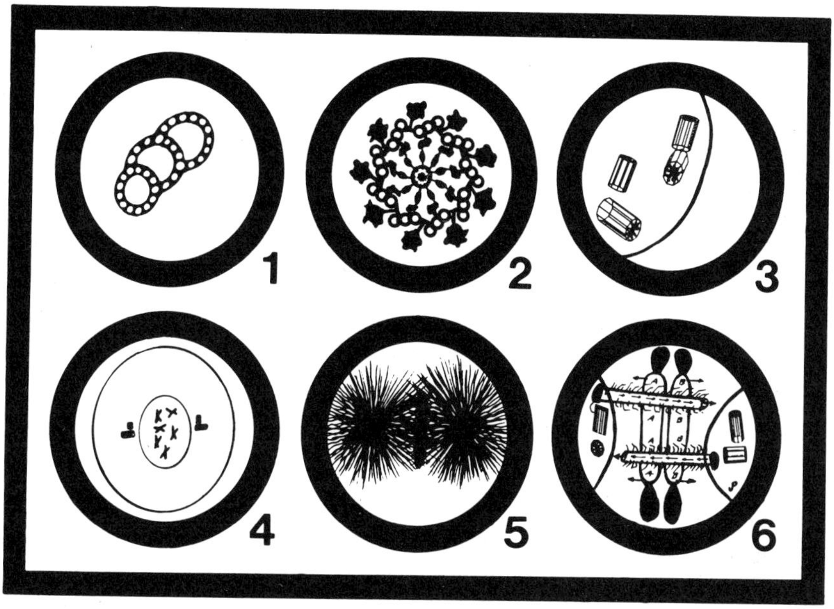


Figure 3.4: Centrioles in cell division. 1) Cross-section of centriole microtubule triplet. 2) Cross section of a centriole with 9 microtubule triplets, 9 satellite bodies, and central "pinwheel" structure. 3) Centriole pairs near cell nucleus, prior to cell division. 4) Centriole pairs have separated and migrated; chromosomes ready for separation. 5) Mitotic spindles, composed of microtubules, have formed from centriole centers (MTOCs) with chromosomes in middle. 6) Centriole anchored microtubules separating a pair of duplicate chromosomes. By Paul Jablonka.

## 3.4 Biotech Evolution—The Next Symbiosis

There are several indications that the evolution of technology will force another nonlinear acceleration in biological evolution which has dealt with crises such as toxic oxygen two billion years ago, utilized new energy sources, inhabited new environments, developed new forms, and spawned technologies which themselves have evolved. Many observers have been alarmed by technological evolution. Nineteenth century scientist/satirist Samuel Butler (Margulis and Sagan, 1986) considered the possibility of machines suppressing humans and assuming supremacy of earth:

> man will become to the machine what the horse and the dog are to man—he may continue to exist, even improve, and will probably be better off in a state of domestication under the beneficent rule of the machines than he is in his present wild state. After all we treat our horses, dogs, cattle, and sheep on the whole with great kindness. We give whatever experience teaches us to be best for them. In like manner it is reasonable to suppose that machines will treat us kindly for their existence is as dependent upon ours as ours is upon lower animals.

Of course we eat some animals, and experiment upon others. It seems every new technology is a double-edged sword with capacity for good or evil—the basic "Frankenstein" scenario. But reliance on new technology is probably necessary and inevitable for adaptation and survival in an ever crowding and progressively toxic world. Margulis and Sagan (1986) cite general systems theorist John Platt who is a student of evolutionary acceleration and believes that life on earth may be nearing an enormously important turning point. The global computing and communication that has emerged following World War Two has become, according to Platt: "a collective social nervous system for managing millions of our problems, and its importance for the long range future may be as great as that of the first learning nervous system."

New technologies may help biology to deal directly with current and future crises. In their book, *Microcosmos: 4 Billion Years of Evolution from our Microbial Ancestors,* Margulis and Sagan (1986) describe some surprising possibilities. They feel that current man is little more than communities of bacteria, modular manifestations of the nucleated cell, and that new "artificial" life forms will emerge from symbiotic fusion of biology and technology. They see this happening along three lines: genetic biotechnology, computer robotics, and biochips.

> ... one day soon entire suits of genes, proteins and hormones may be dovetailed in the laboratory to create new species of microbes. As we gain a greater understanding of embryology and immunology we will surely clone cells into progressively larger and more complex organisms sure to intervene in our own evolution.

As computer robotics evolve smallward to become nanotechnology, collective interactions with genetic biotechnology and natural biochips could precipitate the next evolutionary phase transition: mind/tech symbiosis.

Margulis and Sagan (1986):

> Robotics and bacterium [may become] ... ultimately united in biochips based not on silicon, but on complex organic compounds. ... Manufactured molecules would exchange energy with their surroundings ... to turn it into information [and] open doors to 'cybersymbiosis,' the comingling of human and manufactured parts in new life forms and ultimately enable us to remake our species. ... Homo sapiens might survive only as a rudimentary organ, a delicately dissected nervous system attached to electronically driven plastic arms.

Hans Moravec (1986) of Carnegie–Mellon University's Robotics Laboratory and author of *Mind Children* has his own vision of mind/tech symbiosis in which ultra-precise robotic brain surgeons transfer the software of human consciousness to a supercomputer. He describes advantages of existing in silicon or gallium arsenide with robotic bodies. These include being impervious to harsh environments, electronic transportation across galaxies and immunity to disease. Max Headroom is a hypothetical television personality whose consciousness exists solely within computers and electronic equipment. The mind content of a head injured motocyclist ("Max. Headroom 2.3m" was his last image before the crash) is somehow transferred, collected, and actively existing in electronic circuitry. Somewhat of a video cult figure, Max Headroom may be the first of a breed of technocognitive entities.

Comingling of mind and technology would be a neat trick, fraught with potential benefits and dangers! Certainly it would depend on an understanding of the mechanism of consciousness which is not currently available. Perhaps imminently available nanosensors will be able to interact dynamically at the level of cytoskeletal protein lattices within all living cells. This interaction may lead to the next symbiosis, one which will have as profound effects on biology as did the conversion from prokaryote to eukaryote. If nanotechnology and biology become symbiotic, consciousness can be a commodity.

# Chapter 4

# From Brain to Cytoskeleton

## 4.1 Nervous System Evolution

The German philosopher Nietzsche wrote:

> "Then you must be a scientist whose field is the leech" said Zarathustra, "and you must pursue the leech to its last rock bottom, you conscientious man!" "Oh Zarathustra!" answered the man, "that would be an enormity, how could I take up such a huge task? What I am the master and connoisseur of is the *brain* of the leech: that is my field and it is a whole universe."

The human brain appears to have evolved from predecessors of earthworms and leeches whose development was a milestone in eukaryotic evolution (Somjen, 1983). These organisms' nervous systems probably consisted of a chain of organized clumps of nerve cells called ganglia, or perhaps two chains of symmetrically paired ganglia with an enlarged head ganglion at the front end. The polarity and preferred axis of orientation which defined these basic nervous systems are related to polarity and asymmetry *within* their component nerve cells, or neurons, each a "universe" of its own. As will be described in the next chapters, neuronal orientation and asymmetry are determined by the cytoskeleton which, in many ways, is the nervous system *within* all higher plant and animal cells.

Over the course of evolution the primitive leech's head ganglion began to dominate other members of its chain, performing "decisions" which required cooperation of the entire assembly. Each segmental ganglion still retained some autonomy of action and, when cut into pieces, such a creature may have been able to regenerate complete new individual organisms like its current descendants. Pairs of leech ganglion chains resemble sympathetic ganglion chains of vertebrates which retain a measure of autonomy. For example, man's autonomic nervous system can efficiently regulate heart, intestine, blood vessels and other organ systems even when disconnected from the brain and spinal cord.

Transition from a segmented organism to a nervous system like our own probably occurred due to fusion of the paired chains of ganglia into a tube-like structure of nervous tissue. Paired nerve roots then emerged from the primitive central system similar to the spinal roots of today's vertebrates. These roots connected the central nervous system with the peripheral sensory organs, muscles and glands. Eventually, the head end of the neural tube increased in size and importance until it dominated most nervous system functions, a process termed "encephalization" by famed English neurologist Hughlings Jackson (Somjen, 1983). Encephalization, which occurred over eons and may be continuing presently within man, reflects development of a hierarchical organization in an otherwise parallel, distributed system. Brain components which are more highly organized and capable of more complex functions are generally newer on the evolutionary scale (i.e. "neocortex"). A collective hierarchy of parallel information processing systems based on functional organization which includes subcellular elements (cytoskeleton and cytoplasmic ground substance) is shown in Table 4.1.

## 4.2 Nervous System Organization

Brain activities have been intensively studied by various disciplines for many years. In the following sections, essential elements of brain organization are presented. The purpose is to provide sufficiently comprehensive background in order to justify the contention that the cytoskeleton is an underlying medium of information processing within brain neurons.

### 4.2.1 Architecture

The central nervous system (CNS) of vertebrates including man is organized in an ascending hierarchy of parallel structures—spinal cord, brain stem, and brain. The peripheral nervous system consists of peripheral nerves and the ganglia of the autonomic nervous system. Human brains contain about a hundred billion neurons. Evolution has caused a "cephalic" shift of importance, relative size, and control towards the higher centers or neocortex, which in man is larger and more complex than in other mammals. There

| Brain/Mind |
|---|
| consciousness, "self," "Mind's Eye," attention |
| |
| Brain Systems, Homunculi, "Centers" |
| functionally related neurons, anatomical regions, assemblies of networks, reverberation |
| |
| Neural Synaptic Networks, Cartels, Modules |
| cooperativity due to dense interconnectedness, parallelism, associative memory, learning, synaptic plasticity |
| |
| Neuron |
| multiple synaptic inputs and outputs, dendritic processing, synaptic plasticity, axoplasmic transport |
| |
| Cytoskeleton |
| centrioles, microtubules, filaments, synaptic morphology, spatiotemporal cellular organization, cellular automata, coherent oscillations |
| |
| Cytoplasmic Ground Substance ("Infoplasm") |
| sol-gel states, geodesic actin, tensegrity structures, ordered water, dissipative patterns, holographic interference |

Table 4.1: Collective hierarchy of parallel information processing systems.

are generalized similarities in structure, composition and functioning of central nervous systems in all vertebrates. Neurons within all nervous systems are themselves organized by their component cytoskeletons.

Two types of cells make up nervous tissue: neurons and satellite cells. In the central nervous system, the satellite cells are called neuroglia and in the periphery, Schwann cells. These satellite cells wrap layers and layers of myelin sheeting around neurons forming what is generally considered to be merely insulation which increases the velocity of propagating signals.

The parts of a neuron are the *dendrites*, the *cell body* (or perikaryon) and the *axon* which is also referred to as the nerve "fiber." Dendrites and the cell body generally receive incoming signals and the cell body transforms these into an outgoing signal carried by the axon (Figure 4.1). The "white matter" of the central nervous system consists of fiber tracts and their myelinating glial cells while the "gray matter" refers to clumps of cell bodies and dendrites known as "nuclei." A schematic diagram of brain functional organization is shown in Figure 4.2.

### 4.2.2 Neuronal Signaling

Signals which transmit information among nerve cells consist of electrical potential changes produced by ionic currents flowing across their surface membranes. The currents are carried by ions such as sodium, potassium, calcium, and chloride and occur due to the opening and closing of membrane protein ion channels. The nerve maintains an electrical polarization across its membrane by actively pumping sodium ions out, and potassium ions in. Thus when the ion channels are opened (by voltage change, neurotransmitters, drugs, etc.) sodium and potassium rapidly flow through the channel creating a depolarization. Depending on the spatial location and temporal sequence of channels, activation can result in waves used as signals.

Neurons carry only two obvious types of signals: localized "gated" potentials which are older on the evolutionary scale and *analog*, and propagating "all or none" action potentials which are newer and *digital*. Localized gated potentials can spread only one to two millimeters, are attenuated and distorted by local resistivity, and are essential where spatial and temporal summation ("integration") is required. This occurs at sensory nerve endings ("receptor potentials"), neuronal synaptic junctions where both excitatory and inhibitory potentials are integrated ("synaptic potentials"), and as slow waves arising as rhythmic depolarizations in dendrites. Gated potentials in dendrites are also integrated at the cell body to initiate (when appropriate) propagating action potentials along axons. A primary role for localized dendritic potentials in cerebral neuron information processing has been emphasized by several authors including Alwyn Scott (1977) and Ross Adey (1966) who feel dendritic slow wave potentials allow cerebral neurons to "whisper together."

Action potentials ("nerve impulses") propagate as membrane depolarization waves *along axons*. They occur due to sequential opening of membrane channels which allow passive diffusion of ions. Gaps in myelinization ("nodes of Ranvier") along axons contain abundant ion channels so that impulses propagate rapidly between nodes where they are slowed and susceptible to modulation ("saltatory conduction"). With this exception, action potentials occur on an "all or none" basis (i.e. digital) from integration of dendritic input (i.e. analog) at the cell body region of the neuron. The frequency of firing is related to the stimulus intensity; a sensory nerve responding to muscle stretch fires at a rate proportional to the degree of stretch. Action potential velocity is fixed for given axons dependent on axon diameter, degree of myelinization, and distribution of ion channels. Typical action potential velocities of about 100 meters per second allow effective communication within relatively large nervous systems.

### 4.2.3 Interneuronal Synapses

Action potentials and axons terminate at synaptic connections with other neurons or effector cells such as muscle or gland. Final branch portions of axons are thin with swollen synaptic terminals known as boutons. Some axons may have multiple boutons, each one forming a synapse. Generally, synapses form between the axon terminal and another neuron's dendrite, although axon-cell body, axon-axon, and dendrite-dendrite synapses also occur. Many, even most, dendritic synapses occur on dendritic "spines"—knobby dendritic protuberances.

Two modes of synaptic signaling have been recognized: electrical and chemical. At electrical synapses, currents generated by an impulse of the presynaptic nerve terminal spread directly to the next neuron through a low resistance pathway (which may be networks of extracellular protein filaments known as "synapsin"). The sites for electrical communication between cells have been identified in electron micrographs as gap junctions in which the usual intercellular space of several tens of nanometers is reduced to about two nanometers. There appear to be a huge number of electrical synapses in mammalian brain (estimated to be as high as 80 percent of all synapses), but because of difficulties in isolation, characterization, and inability to study them by pharmacological manipulation, their significance remains unknown. Chemical synapses have been extensively studied.

At chemical synapses the fluid gap between presynaptic and postsynaptic membranes prevents a direct spread of current and the lack of an electrical connection between neurons. The synaptic bouton contains large quantities of spherical vesicles (about 50 nanometers diameter) which contain neurotransmitter molecules. Acetylcholine, norepinephrine, serotonin, dopamine, gamma amino butyric acid (GABA) and various peptides and amines have been identified as neurotransmitters. Some are excitatory while others (i.e.

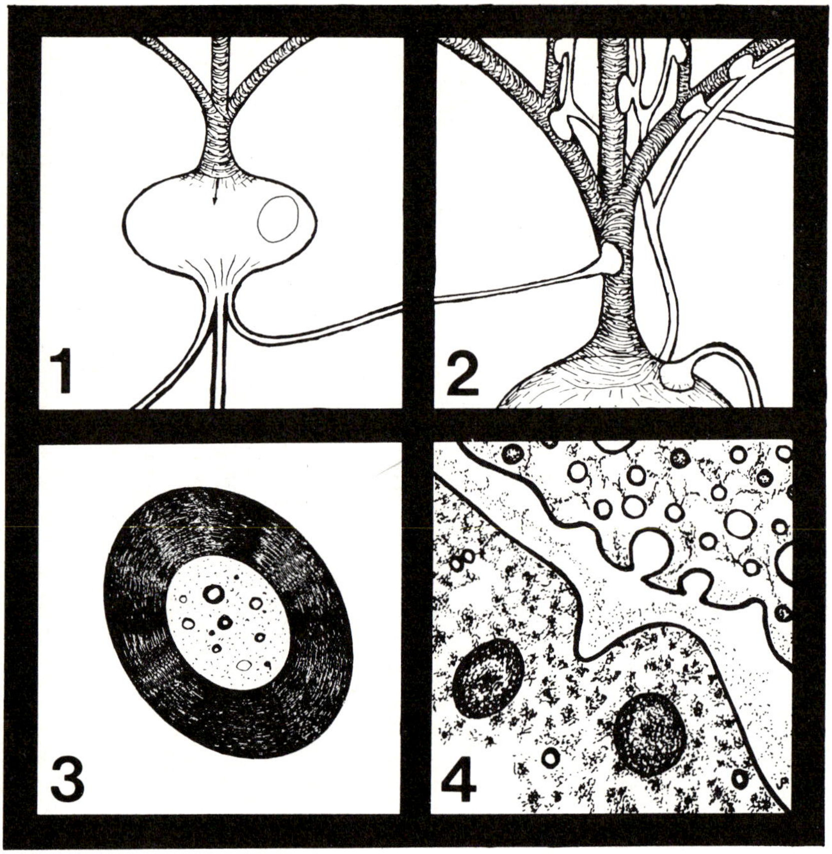


Figure 4.1: Neuronal organization: 1) Branching dendrites (top) entering cell body or perikaryon; branching axons exiting at bottom. 2) Axons forming synapses on dendrites and cell body. 3) Axon surrounded by 100 layers of myelin which increases conduction velocity; structures visible in axon include mitochondria, neurotransmitter vesicles, and microtubules. 4) Synaptic cleft; neurotransmitter vesicles (top right) fuse with membrane as they are released. By Paul Jablonka.

GABA) are inhibitory.

The sequence of events liberating neurotransmitter molecules from nerve endings is remarkably uniform at all synapses. Transmitter molecules are stored inside presynaptic nerve terminals in small vesicles that are analogous to the secretory granules of gland cells. The amount of neurotransmit-

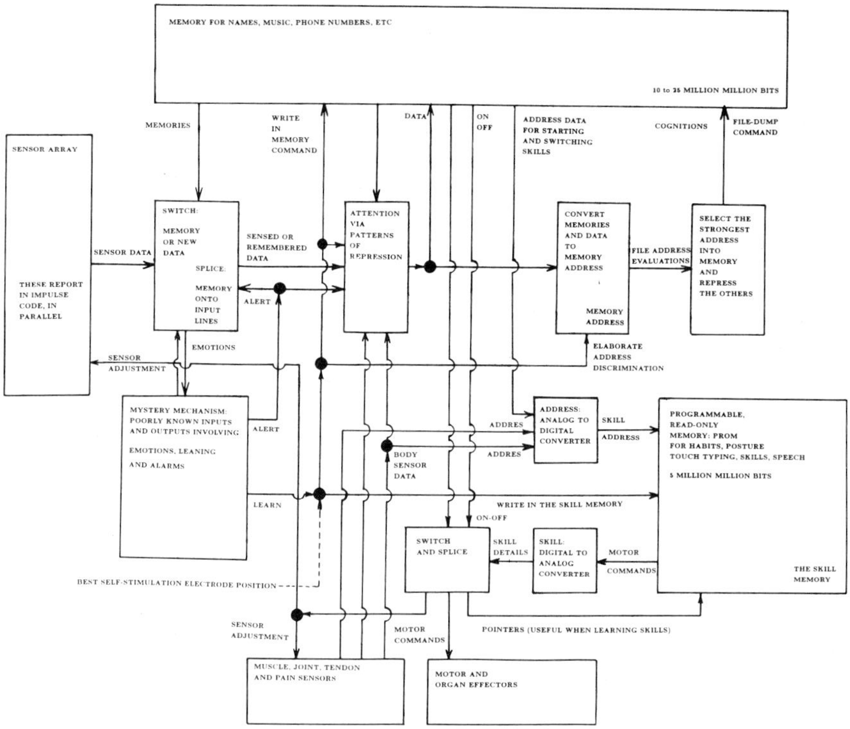


Figure 4.2: Schematic diagram of brain functional components. By Paul Jablonka.

ter in one vesicle is considered a "quantum" and in neuromuscular synapses each quantum consists of one thousand to five thousand molecules of acetylcholine. Each action potential reaching the presynaptic terminal releases a number of vesicle quanta ranging from a few to several hundred. The coupling mechanism between the action potential and vesicle release involves both calcium and the cytoskeleton. As an action potential impulse arrives at the presynaptic nerve terminal, calcium ions enter the cytoplasm through the membrane by way of voltage gated channels. In presynaptic nerve terminals, inward ionic current of the action potential is thus carried partially by sodium and partially by calcium. Free calcium in the cytoplasm of the terminal causes vesicles to fuse with the surface membrane and to expel their contents. The mechanisms by which calcium triggers the release of vesicles from the presynaptic terminal is not clearly understood, however cytoskeletal proteins including contractile actin, myosin, and other filamentous proteins are involved. Calcium mediates dynamic contractile activities in flagella and skeletal muscle and appears to trigger cytoskeletal expulsion of neurotransmitters from nerve terminals.

After release, transmitter molecules diffuse across the synaptic cleft and bind reversibly with the postsynaptic membrane receptors. The distance between the two membranes is sufficiently small such that the diffusion takes about a millisecond—a relatively slow event compared to switching in semiconductors. Whenever a transmitter molecule binds to the postsynaptic membrane, it causes a small voltage change in the postsynaptic membrane called the miniature endplate potential which can be either excitatory or inhibitory depending on the neurotransmitter molecule and postsynaptic receptor. In the resting state there are spontaneous releases of individual vesicles causing a background rate of miniature end plate potentials which are below the threshold for depolarization of the postsynaptic cell.

Specific binding of neurotransmitter molecules to post synaptic receptors changes the membrane permeability to specific ions which produce localized receptor potentials which are either excitatory or inhibitory. The post synaptic membrane "integrates" the local receptor potentials spatially and temporally such that when a "threshold" is exceeded, signals propagate in the post synaptic dendrite, cell body, or axon.

Whether the response is excitatory or inhibitory depends on the species of ion channel carrying the synaptic current. For example, in the neuromuscular junction, acetylcholine increases post synaptic permeability to sodium and potassium, leading to an excitatory, depolarizing action potential. Post synaptic acetycholine activated channels open for one to two milliseconds and allow a net entry of about $2 \times 10^4$ ions. Other synaptic channels stay open for tens of milliseconds and pass $10^5$ ions or more. Still other post synaptic receptors couple directly to cytoskeletal changes and do not involve ionic conductance at all. At inhibitory synapses, GABA may increase permeability to chloride ion, driving the membrane potential away from threshold. Inhibition may also occur pre-synaptically in which case release of excitatory neurotransmitter is prevented. By combining multiple inputs, synapses "compute" to determine their output.

Nerve cells influence each other by either excitation, producing impulses in another cell, or by inhibition, tending to prevent impulses from arising in an adjacent cell. Lateral inhibition also occurs; activity in a group of active axons inhibits firing in nearby fibers—an apparent sharpening or focusing mechanism. A neuron receives many excitatory and inhibitory inputs from other cells (convergence) and in turn supplies many others (divergence). The process whereby neurons combine together all of their incoming signals is known as integration. Thus *each cell* must integrate a multitude of synaptic inputs (up to 200,000 synapses per neuron) to determine its own output. Additional levels of processing at dendritic branch points, dendritic spines, active nodes between myelin sheaths, and changes in synaptic efficacy illustrate the complexity at the level of individual neurons. Rather than a simple switch, each neuron is more like a computer. The intraneuronal cytoskeleton is the nervous system within the nervous system.

## 4.3 Representation of Information

The central enigma in brain science is how information is represented within nervous systems. Understanding mechanisms of the vast capacities for storage, retrieval, and processing of information within the brain would be of enormous benefit not only to neuroscience, but to workers in computer science, particularly artificial intelligence (AI). Indeed, vast strides have been made in AI by utilizing "good guesses" about brain information processing and, conversely, understanding of brain functions and capabilities is being advanced by AI related theory including neural nets and parallel connectionism.

The underlying assumption about brain function and the comparative basis for AI considers parallel networks of connected units in which neurons and their synaptic connections are the fundamental substrates. However individual neurons perform a significant amount of analog processing both at the level of dendrites and within their cytoskeleton. For example, modification of synaptic transmission threshold, the cornerstone of neural net learning models, is regulated by the cytoskeleton and its cytoplasmic connections. Viewing neurons as fundamental digital substrates (switches or gates in a computer) may be overlooking an important dimension available for the organization of intelligence.

### 4.3.1 Integration—Sherrington's Reflex Centers

Processing the dynamic excitatory and inhibitory patterns of activity within masses of neurons ("reflex centers") was described as "integration" by the famed neuroscientist C. S. Sherrington during the 1930's.

The brain is continually faced with the task of making decisions on the basis of information about the outside world provided by sensory end-organs and information stored in memory. At any one instant, incoming signals from diverse sources in the periphery excite the brain. The mechanisms by which the various types of information are taken into account and assigned priorities is called "integration" which is carried out at all levels of brain organization. In a global example of integration, an animal confronted by danger integrates input towards a binary output decision: "fight or flight." Our higher centers continually receive information arising in a great variety of sources on the surface of the body and in the internal organs. A typical central neuron faces a task similar to that of the brain as a whole. It is a target of converging excitatory and inhibitory signals that it transforms ("integrates") into its own impulses. The general principles of integration were discovered in the early 20th century by Sherrington (1933, 1947) who recorded tension in skeletal muscle by the stretch reflex before electrical recording from individual cells was possible. Integration appears to occur at all levels of nervous systems and among various types of organisms: crustacean, fish, and mammals. Sherrington proposed and cited evidence for integration by groups of neurons which he termed neural masses or reflex centers and suggested that they correlated with anatomically identifiable "nuclei."

A nucleus is a compact region of *gray matter* of relatively homogeneous neural architecture and recognizable boundaries which contains a high density of neuronal cell bodies and synapses. (White matter connotes a high density of cable-like axon fibers.) A reflex center is an assembly of neurons performing a specific function. A nucleus is purely morphological or structural while a center is functional. Nuclei *may* coincide with centers, but often do not (Freeman, 1972).

The concept of neural centers may convey an erroneous impression of anatomically specific function, but remains as a vestigial reference to Sherrington's concept of the nervous system and now denotes groups of neurons whose destruction leads to loss of specific function and/or the stimulation of which evoke a certain behavioral or physiological function. Brain functions are clearly not divided among centers in the same way as the work of a large organization or factory is divided among its various offices and workshops. The relation between anatomic regions devoted to specific functions, and the brain-wide distribution of information is perplexing and complicated. For example, the satiety center is located in the hypothalamus; if this general region is stimulated in an animal having a meal, the animal will stop eating as though it has had enough. If the same structure is destroyed, the animal eats too much and gets fat as though it is never satisfied. Thus clearly the satiety center neurons are essentially related to evoking the sensation of fullness or satiety. Feeding behavior, however, is regulated by a much wider range of many neuronal circuits in different regions. The satiety center integrates multiple inputs to a binary output: eat or don't eat. Body representations such as motor and sensory homunculi and other concrete evidence of anatomical localization of neuronal function may also integrate wide sources of distributed input to representations of anatomical sensation and action. Anatomical hardware such as satiety centers, motor and sensory homunculi appear to be evolutionary adaptations necessary for larger and more complex nervous systems.

Sherrington is a key figure in the history of neuroscience. His concept of integration by reflex centers illuminated possible modes of information processing by neural structures. It is now appreciated that information transfer functions occur at all levels of nervous system organization and include functions now used in computers such as summation, ramp triggers, analog/digital conversion, and logic.

### 4.3.2 Pulse Logic

Electrical signaling in the nervous system was, according to legend, discovered during a demonstration given by Luigi Galvani to a class in the late 18th century. While Master Galvani was dissecting a frog, an electrostatic generator (which discharged electric sparks) was being played with by a bored student. Each time he drew a spark, the frog's leg (held in a forceps by Galvani) twitched. Galvani pursued this curious observation and published his treatise on *Animal Electricity* in 1791. Galvani felt that animal electricity was the material of excitation of nerves and of the contraction of muscles. Only after his death did his followers demonstrate that animals indeed generated their own electrical currents, rather than merely responding to applied electricity (Freeman, 1972).

The development of the microelectrode and appropriate electronic support equipment after 1940 revealed discrete electrical nerve impulses to be ubiquitous throughout the central and peripheral nervous systems. Electrophysiological techniques afforded an approach to complex problems by recording the activity of individual cells or small groups of cells. For example, the neural events involved in perception of touch were studied by recording signals from a neuron that terminates in the skin and whose function—touch sensation—was unambiguous. These signals consist of brief electrical pulses about 0.1 volt in amplitude, they last for about one millisecond, and they move along nerves at a speed up to 120 meters per second. Electrophysiological pulses within "Sherringtonian centers" had been correlated with appropriate forms of behavior, for example bursts of firings of neurons in medullary reticular formation were related to breathing. Neural activity was conceptually limited to actuation of information stored in specific neural regions. Newer approaches introduced the possibilities of encoding by dynamic neural firing patterns.

In 1943 McCulloch and Pitts proposed that neurons might be approximately described by the following assumptions: 1) the output activity of a neuron is an "all or none" process, 2) a certain number of input synapses must be excited within a brief period to excite the neuron's output, and 3) the only significant delay within the nervous system is synaptic delay. These "McCulloch-Pitts neurons" connected in networks could, in theory, perform any computation. Behaviorally significant information ("psychons") were conceived as action potential patterns in single neurons much like mass discharge in "centers" had been previously considered. A finer structure of neural information processing was perceived.

All or none action potentials which were the central element of pulse logic are relatively new over the course of evolution compared to graded wave-like events which occur in more primitive nerve nets. In the olfactory bulb, granule cells have no axons and do not generate extracellularly detectable action potentials at all. Thus the nerve impulse is not the only basis for transmission either between neurons or within neurons. What *is* the basic substrate of information transfer and processing within the nervous system? McCulloch (1943) considered:

> ... Why have I chosen to quantize in nervous impulses? ... If we think (of the brain) in terms of its ultimate particles, one might split this at the level of the atoms or one might split this at the level of the neurons and so on. The question is at what level can one split the behavior so as to define a set of units in terms of which to work? And obviously the nervous impulse at the level of the neuron is a fairly nice unit for working ... . But what I'm looking for is something that will perform a logical task. I'd like a thing that has a grain and I'd like to take that grain as my unit.

The grain which would explain information processing within *all* eukaryotic systems is the dynamic activity of subunits within cytoskeletal proteins.

McCulloch, Pitts and others had guessed that information was coded in the pattern and sequence of nerve impulses somewhat like Morse code in telegraph lines. However, Adrian showed in 1947 that the frequency of firing in a nerve cell is a quantitative measure of the intensity of the stimulus. When skin is pressed, the stronger the pressure applied to the skin, the higher the frequency and the better maintained the firing of the cell. Frequency coding appears limited to information about the intensity of a stimulus and impulses in a given cell appear identical with those in other nerve cells. The significance and meaning are quite specific for each cell; for example skin sensory neurons indicate that a particular part of the skin has been pressed. Although there appears to be no reason why a great deal of information could not be conveyed by any predetermined signal including a code made up of different frequencies, the frequency or pattern of discharges do not appear to stand on their own as qualitative information. Following this recognition, the meaning of a signal became attributed to origins and destinations of the nerve fibers which convey it: "connectionism." The importance of connections is exemplified by sensations of light produced by nonvisual stimulation of the eyeball, or the phantom limb phenomenon in which an amputee may have sensation of a limb long since removed. In each case, mental representation is determined by the location of stimulation.

McCulloch and Pitts elaborated their model by adding a term which included the possibility that a firing decision might depend on inputs of times more remote than the synaptic delay period (dendritic memory) and by considering circular neural networks: closed pathways with logical feedback and reverberation. Thus pulse logic evolved into connectionism and neural networks as media of neural information representation.

### 4.3.3 Connectionism and Neural Networks

The form and structure of neurons and the observation of neuroanatomy was made available to optical microscopy by Italian anatomist Camillo Golgi in 1875. He found a method by which, seemingly at random, a very few neurons in a brain region became stained in their entirety, with all their branches becoming evident. The cytoplasm of selected neurons take up a brightly colored stain and are thus exposed against the tangled morass of less visible cells. Golgi's contemporary, Spaniard Santiago Ramon y Cajal (1955) used the Golgi stain to investigate nearly every part of mammalian nervous systems. His neuroanatomy texts are still classic and he resolved the question of whether nerves were separate entities like all other cells, or part of a continual network. He also demonstrated that the complex connections among neurons are not random, but highly selective and specific.

Subsequent generations of neuroanatomists and neuroembryologists including Roger Sperry (1969, 1980) have emphasized the meticulous detail with which neural connections are formed, and initially supported the concept of the nervous system as a pulse logic device, superseding the older concept of the brain as a switchboard of reflex centers. Adrian's discovery that neural pulse coding was limited to a frequency/intensity coupling shifted emphasis to connectionism per se. Because the electric signals the brain uses to communicate among cells were seen as *stereotyped*, or nearly identical, they were viewed as symbols which do not themselves resemble the external world they represent. The consensus of opinion regarding brain functions shifted to a concept in which the shape of a neuron and its fiber origins and destinations determine mental representation as part of a neural network. The meaning of stereotyped signals was thought to be derived from specific connections of neurons.

The high degree of precision with which nerve cells are connected to each other and to different tissues in the periphery became emphasized in the connectionist concept. Orderliness of connections formed during development became viewed as essential for integrating mechanisms and representation of information in some way. The nervous system appeared to be constructed as if each neuron had built into it an awareness of its proper place in the system. The question of mental representation refocused on the embryological development during which synaptic connections were formed. During development, the neuron grows towards its target, ignores some cells, selects others, and makes permanent contact—not just anywhere on a cell but with a specified part of it. Further, neurons behave as if they were aware when they have received an appropriate synaptic connection. When they lose their synapses they respond in various ways. For example, neurons or muscle fibers disconnected from their neuronal contacts may die, but first develop "super-sensitivity" to their chemical neurotransmitter by means of an abundance of new synaptic membrane receptor proteins. Cell death or dysfunction induced by denervation occurs due to a loss of morphologicial "trophism," a neural function which conveys structural and functional material and information by cytoskeletal axoplasmic transport. Atrophy, dystrophy and spasticity of muscles and limbs which occur after strokes and other nervous system insults are examples of the loss of normal trophism. Microtubules and other cytoskeletal proteins responsible for trophism and axoplasmic transport also allow growth and extension of neuronal axon growth cones, dendrites and dendritic spines and thus play a key role in neural connections. Super-sensitivity, spasticity, atrophy and dystrophy are examples of "synaptic plasticity," changes in connections or connection strength among neurons which are relevant to brain and bodily function.

Association of learning with ongoing alteration of synaptic function was considered by several late 19th century writers and was popularized due to the Pavlovian and behaviorist influences of conditioned responses. Pavlov (1928) proposed that conditioned reflexes are established by forming new connections between cortical neurons that receive a conditioned stimulus (one accompanied by a reward or punishment) and those that receive an unconditioned stimulus. Once a new pathway was established the unconditioned stimulus would acquire the same power of evoking the response that only the conditioned stimulus originally possessed. Pavlov's idea of a new connection became fused with Donald Hebb's (1949) concept of plastic changes in synaptic efficacy to correlate with learning. Because it was believed that new fibers and therefore new synaptic connections, could not grow in adults, long term facilitation of anatomically preformed, initially nonfunctional connections became the likely alternative. This implied that at birth there existed a vast number of redundant and ineffective synaptic conditions which became "selected" during the individual's lifetime of experience. An alternative view is that, at birth, excitations can pass between any two points of the CNS through a random network of connections. As maturation, experience, and learning occurred, synaptic activity gradually sculpted usable patterns by suppressing unwanted interconnections.

Thus the connectionist brain/mind became viewed as one of two types of systems: a blank slate ("tabula rasa") in which acquired learning and internal organization result from direct environmental imprinting, or a "selectionist" network chosen from a far vaster potential network. Selectionists believe that the brain/mind spontaneously generates variable patterns of connections during childhood periods of development referred to as "transient redundancy," or from variable patterns of activity called prerepresentations in the adult. Environmental interactions merely select or selectively stabilize preexisting patterns of connections and/or neural firings which fit with the external input. Selectionists further believe that, as a correlate of learning, connections between neurons are eliminated (pruning) and/or the number of accessible firing patterns is reduced. Supporting a selectionist viewpoint is the observation that the number of neurons and apparent synapses *decreases* during certain important stages of development in children. However, this reduction could be masking an *increase* in complexity among dendritic arborizations, spines, synapses, and cytoskeleton. The selectionist view is also susceptible to the argument that *new* knowledge would appear difficult to incorporate.

On the assumption that the basic mode of learning and consciousness within the brain is based on synaptic connections among neurons (connectionist view) several attempts to model learning at the level of large assemblies of interconnected neurons have been made. Hebb pioneered this field by proposing that learning occurred by strengthening of specific synaptic connections within a neuronal network. This led to a concept of functional groups of neurons connected by variable synapses. These functional groups as anatomical brain regions have been described by various authors as networks, assemblies, cartels, modules or crystals. These models are aided by the mathematics of statistical mechanics and have been rejuvenated due to the work of Hopfield (1982), Grossberg (1978), Kohonen (1984) and others who drew analogies between neural networks within the brain and properties of computers leading to applications for artificial intelligence. They emphasized that computational properties useful to biological organisms or to the construction of computers can emerge as *collective* properties of systems having a large number of simple equivalent components or neurons with a high degree of interconnection. Neural networks started as models of how the brain works and have now engendered chips and computers constructed with neural net connectionist architectures utilizing hundreds of computing units and linking them with many thousands of connections. Hopfield (1982) remarks that neural net chips can provide finely grained and massively parallel computing with:

> a brainlike tolerance for fuzzy facts and vague instructions. Some of the general properties you get in these systems are strikingly like ... properties we see in neurobiology .... You don't have to build them in; they're just there .... 

Neural networks had formally appeared in Rosenblatt's (1962) "perceptron" model of the 1950's and 1960's. Perceptrons created enthusiasm, but failed to reach their potential due to limitations of the model and its mathematics. AI experts Marvin Minsky and Seymour Papert (1972) wrote a harshly critical review which discouraged neural net research until Hopfield's resurgence in the 1980's. Hopfield introduced an energy function so that information in a neural net circuit would settle into a series of stable energy states much like rain water falling on mountains flows through valleys into lakes and rivers. Depending on the rainfall, an information state (i.e. memory, conscious image, thought) would be a given watershed pattern. Hopfield's neural nets are loosely based on aspects of neurobiology but readily adapted to integrated circuits. The collective properties of his model produce a content addressable memory (described by a phase space flow of the state of the system) which correctly yields an entire memory from any sub-part of sufficient size. The algorithm for the time evolution of the state of the system is based on asynchronous parallel processing. Additional emergent collective properties include some capacity for generalization, familiarity recognition, categorization, error correction, time sequence retention, and insensitivity to failure of individual components. Hopfield nets and similar models are best categorized with the "tabula rasa" view of learning in which the initial state is taken as a flat energy landscape which becomes progressively contoured, eroded and complicated by direct interactions with the environment.

A selectionist approach to neural net theory has been taken by Jean Pierre Changeux, who pioneered description of allosteric interactions among proteins. Turning to the brain/mind, Changeux and colleagues (1984, 1985) have proposed a model of learning by selection based on the most recent advances in the statistical mechanics of disordered systems, namely the theory of spin glasses. Spin glasses are materials which are not crystalline, yet whose atoms possess a high degree of similar neighbor relationships and a finite number (i.e. 2) of magnetic spin states influenced by their neighbors. Aggregates of "like spin" states beget similar states among neighbors. Consequently the spin states of atoms in a spin glass can be viewed as a network (or cellular automaton) much like a collection of neurons in a nervous system. Changeux also uses terms from mathematical chaos theory like basins and attractors to describe the states to which the spin glass model evolves. Unlike the blank slate approach, the brain's initial state is viewed by Changeux as a complex energy landscape with an exuberance of valleys typical of spin glasses. Each valley corresponds to a particular set of active neurons and plays the role of a prerepresentation. An input pattern sets an initial configuration which converges towards a valley whose entry threshold is lowered by synaptic modification. Starting from a hierarchical distribution of valleys, the "lowest" valleys (sea level fjords) would correspond to maximal comprehension, ultimate answer, best correlation. The learning process is viewed as smoothening, gardening, and evolutionary pruning as already stored information influences the prerepresentations available for the next learning event. Changeux's spin glass model of neural nets is elegant, and successfully presents a hierarchical pattern of static information sorting. It's shortcomings are that it is unidirectional and fails to describe dynamic, "real time" information processing.

Another selective connectionist network model of learning is that of

George Reeke and Gerald Edelman (1984) of Rockefeller University. They describe two parallel recognition automaton networks which communicate laterally. Automata are dynamic patterns of neighbor interactions capable of information processing (Chapter 1). The two parallel recognition automata which Edelman and Reeke devised have distinct and complementary personalities. They are named Darwin and Wallace after the co-developers of the theory of evolution, and utilize different approaches to the problem of recognition. "Darwin" is highly analytical, keyed to recognizing edges, dimensions, orientation, intensity, color, etc. "Wallace" is more "gestalt" and attempts to merely categorize objects into preconceived classifications. As in all parallel processing systems, output of the individual processors must be reconciled if they are not identical. Lateral communicating networks between Darwin and Wallace resolve conflicting output and form an associative memory. Because they operate on an unchanging connectionist network, Darwin and Wallace are considered "selectionist." Similar recognition automata may be operating in dynamic cytoskeletal networks *within* neurons.

### 4.3.4 Distributedness

Neuroanatomical appreciation of synaptic structure gave credence to connectionist theory, and permitted Hebb to extend the theory of synaptically regulated, discretely assembled neural nets to memory and behavior in humans. However, the only experimental *data* came from 30 years of work by Karl Lashley (1929, 1950) who had shown that memory and perception appear to be *distributed throughout* the brain. Lashley's quest was the site of representation—the "engram"—of information, memory and learned behavior in the brains of laboratory animals.

Lashley's experiments in search of the engram consisted of ablation of specific parts of animal brains with careful testing for retention of habits learned before the ablation, ability to relearn, and ability to learn new tasks. The overall conclusion was that memory function is disturbed in proportion to the amount of cortex destroyed, irrespective of which part of the cortex had been removed. As far as learning, Lashley felt that all parts of the cortex were equipotent except for the specific sensory receiving areas involved in that particular learning modality. Lashley's findings were a disappointment to connectionists and engendered a bleak, hopeless outlook. In his famous life's work, *In Search of the Engram,* Lashley (1950) wrote that after 30 years of searching for the location of the engram in the brain he was convinced that "learning is just not possible. Nevertheless, in spite of all the evidence against it, learning does sometimes occur." Lashley rallied from despair to propose an alternative to localized storage of memory traces attached to spatially fixed reflex pathways. Agreeing that recall involves reactivating a previous pattern of neuronal excitation, he insisted that the pattern representing any one memory could be evoked not just in one specific set of neurons, but in many sets, in many places, and perhaps anywhere in regions of the brain having to do with memory function. This distribution of information in memory was explained by Lashley as excitatory patterns in cortex related to the spread of interference patterns on the surface of a liquid disturbed at several points at once. (His theory of interference patterns later inspired comparisons between the brain, storage of information and laser holography).

Lashley (1929):

> Nerve impulses are transmitted ... through definite intercellular connections, yet all behavior seems to be determined by masses of excitation ... within general fields of activity without regard to particular nerve cells. It is the pattern and not the element that counts.

Hebb (1949) commented:

> Lashley has concluded that a learned discrimination is not based on the excitation of any particular nerve cells. It is supposed to be determined solely by the pattern or shape of the sensory excitation ... this suggests that the 'nmemonic trace', the neural change that is induced by experience and constitutes memory, is not a change of structure ... [it] is a lasting pattern of reverberatory activity without fixed locus like some cloud formation or an eddy in a mill pond.

"Engram" was originally described as the brain's "nmemonic" trace of an elementary idea as if it were the atom of mental content. An earlier proposal was that one engram was stored in the firing pattern of one cortical neuron: a "psychon" within a "grandfather" neuron at the apex of a neuronal hierarchy. Pulse logic saw "psychons," (or engrams) in the discharge patterns of neurons and connectionist neural nets equated engrams with specific circuits of synaptically connected neurons. Activity within those circuits was thought to represent consciousness and memory that would occupy specific locations within the brain. Memory was viewed as libraries, filing cabinets, digital computers, and junk boxes in which information was stored in particular places and retrieval involved finding where it was stored. Lashley's experiments suggested that information is not stored anywhere in particular, but rather is stored everywhere; memory was distributed. One prevalent interpretation was that information was stored in the relationships among units and that each unit participated in the encoding of many, many memories. Information could thus be distributed over large spatial areas. With distributed memory, individual traces from within a complex of traces can be found much like the way filters can extract individual frequency components from complex acoustic waveforms. Filters are able to detect the presence of specific frequencies even when they are completely intertwined with others. Consequently, a filtered, distributed memory system can operate as a storage and retrieval device. If memories are not independent of one another, the storage of one memory can affect another. Previously stored information tends to evoke an original pattern of activity even though the inputs to the system may differ in many details. This description is similar to the hologram concept in which coherent reference waves are necessary to retrieve information from an interference pattern.

Lashley (1950):

> It is not possible to demonstrate the isolated localization of a memory trace anywhere within the nervous system. Limited regions may be essential for learning or retention of a particular activity, but within such regions the parts are functionally equivalent. The engram is represented throughout the area. All of the cells of the brain must be in almost constant activity either firing or actively inhibited. Every instance of recall records the activity of literally millions of neurons. The same neurons which retain the memory traces of one experience must also participate in countless other activities.
>
> Recall involves the "synergic" action or *some sort of resonance* among a very large number of neurons ...

Hebb's view of engram representation was a closed loop of neurons firing in a confined region. Lashley's studies led him to suggest an open, parallel, distributed network covering wide regions of brain. Hebb's linkage of learning and synaptic efficacy transcended this conflict because it related to both concepts, which may also operate within cytoskeletal networks.

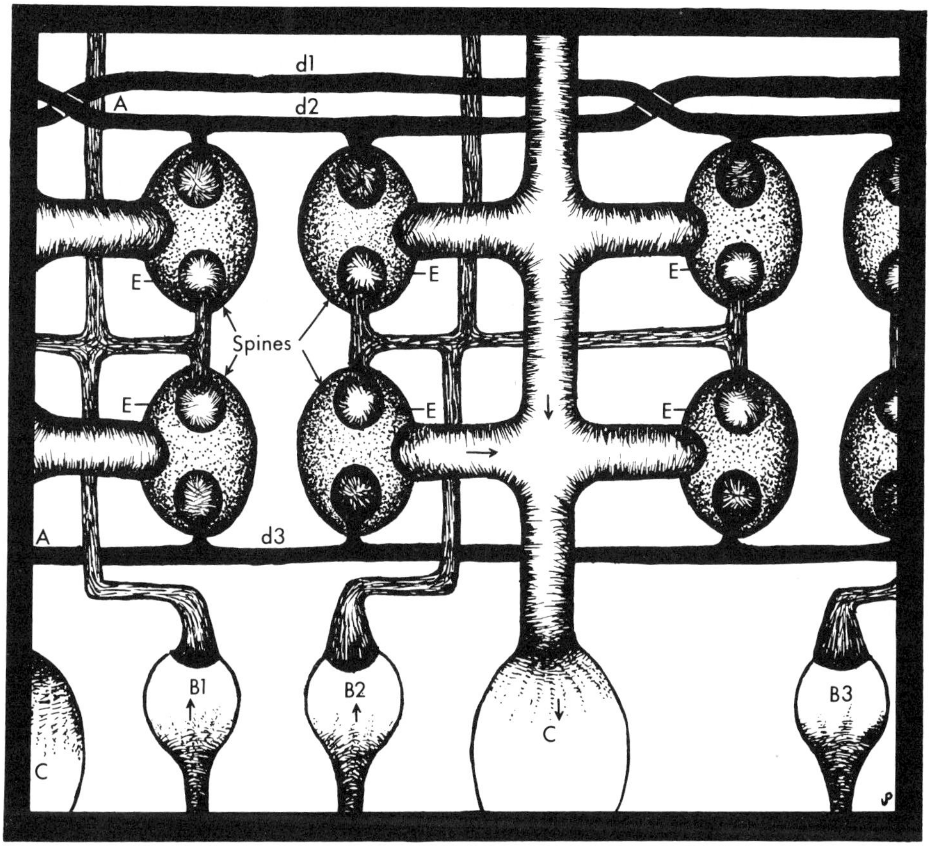


Figure 4.3: Brain memory bank based on "conventional" notions of dendritic spine synapses. Input lines ($B_{1-3}$) are axons which branch and synapse on dendritic spines (E) connected to output lines (C). "File dump lines" (A, $d_{1-3}$) direct spine activities to "copy" or "dump." Such a configuration allows each spine to hold about 3 bits of information, remarkably low when compared to the capacities of biomolecules such as DNA or microtubules. Cytoskeletal activities *within* dendritic spines may contribute. By Paul Jablonka.

### 4.3.5 Synaptic Mechanisms of Learning and Memory

Memory processes have traditionally been divided into two classes; short term and long term. Short term memory, or working memory, is apparently how we remember telephone numbers from the time we look them up in the directory until we dial them. Long term memory or reference memory is used in recording information for long term reference. Short term or working memory is thought to be of relatively small capacity with a maximum of 5 to 9 items at any one time. It is labile and easily disrupted if attention is diverted and it automatically erases within minutes. Continual verbal rehearsal counteracts erasure and re-enters the contents as long term memory. Once laid down, long term memory can endure for a long time—perhaps an individual's lifetime. Long term memory has a large capacity, and difficulty in recalling specific items arise not because memory traces fade, but rather because their address is lost. Well practiced access routes to long term memory items include "nmemonics," easily remembered reference clues. Items which have been stored and then not used will become increasingly difficult to recall. Once they are retrieved, such seemingly forgotten memories become again accessible. Conventional wisdom has held that items become stored in long term memory only after they have been first held in short term memory. This is a process known as consolidation which is thought to require a finite period of time on the order of 45 seconds (Cherkin and Harroun, 1971). Some regard the two processes as independent parallel functions. If brain activity is markedly altered by seizures, trauma resulting in unconsciousness, or by general anesthesia the phenomenon of amnesia may occur in which the subject cannot remember events that occurred immediately before the disruption. Events farther in the past are not forgotten by such interventions. One explanation is that short term memory depends on some dynamic process such as the continued circulation of impulses in a pattern which, when disrupted, is erased. According to this scheme, long term memories are stored by an enduring change in the neurons or in connections between them which would be unaffected by such upheavals. An analogy to computer jargon would suggest that a software program becomes "hardwired." Other models of memory process describe more of a continuum in the consolidation process divided into three or five stages, or entirely different mechanisms with varying methods of entry, storage capacities, lifespan and accessibility. Still other models regard all memory functions as the same basic mechanism differing only in secondary characteristics.

Learning is linked to memory; learning a complex task probably involves generating a pattern suitable for memory storage. For example, the image of a person's face may be recalled by exciting neurons in a pattern similar to the one generated when that face was actually perceived. The ability to swing a tennis racket implies the existence of a program or pattern for activating muscles in the proper sequence and degree. Acquiring such neuronal programs has been attributed to changes in the functioning of synapses between specific neurons (Figure 4.3). Structural changes which lead to alteration in the function and sensitivity of interneuronal synaptic connections are the cornerstone of current concepts of learning and memory. Classifications of the types of plastic changes that could occur in synapses include habituation, long term potentiation, and heterosynaptic potentiation.

Habituation to a response is the simplest form of learning. When an animal hears a new sound it may respond by perking its ears showing some form of attention. If that sound is repeated continuously, the animal learns that the sound or stimulus is neither threatening nor interesting and becomes "habituated" to it. Habituation is different from other forms of decreased response such as synaptic fatigue or desensitization and is specific for the stimulus and its intensity. If the habituating stimulus is withheld for a period of time and presented again, the response reappears ("dishabituation"). Habituation to noxious stimuli does not occur and when non-noxious and noxious stimuli are paired there is no habituation to either of the two. The mechanism of habituation has been studied in detail in the marine organism aplysia, or sea slug. The sea slug has a "gill withdrawal reflex" which is convenient for study. When the skin of the slug's syphon is stimulated, the animal withdraws its gill. This response shows habituation, dishabituation and other features typical of more complex mammalian responses. The neuronal network mediating this reflex response has been extensively mapped and the participating synapses studied electrically. The habituation of gill withdrawal might have been the result of many processes including synaptic inhibition, but in fact has been shown to be the result of decreased output of excitatory neurotransmitter at the presynaptic axon terminals mediating the withdrawal reflex. Neuroscientist Eric Kandel (1976) of Columbia University has shown that the habituation is different from ordinary fatigue, does not involve exhaustion of available transmitter, and is related to decreased flux of calcium ions through presynaptic calcium channels. Thus a behavioral response has been elegantly related to molecular level events. The activity of presynaptic calcium channels and other neural proteins have been viewed as "allosteric"—an ability to spatially and temporally integrate multiple converging signals to a specific protein state, or "conformation."

The second form of neuronal plasticity is long term potentiation (LTP). In LTP, repeated use of a synapse makes transmission through that synapse increasingly easy. In the synapses of mammalian brain hippocampus, the effect endures for many hours and LTP has been classically related to learning and memory. If two pathways share an interneuron, then LTP can enhance transmission through the pathway not originally excited, a form of associative memory and recall. The duration of LTP effect of many hours to days corresponds with the morphological turnover and trophic maintenance of synaptic membrane proteins by axoplasmic transport, a function of the neuronal cytoskeleton.

The third form of neuronal plasticity is heterosynaptic potentiation in which activity on one synapse changes efficiency of transmission in another on the same postsynaptic membrane. This may occur either through change in sensitivity of the post synaptic neuron to the transmitter, or by change in the amount of transmitter released. There is some evidence that LTP may be due to increased numbers of receptors in post synaptic membranes and a similar mechanism could occur in heterosynaptic potentiation in which more then one neuron is involved. Habituation, long term potentiation, and heterosynaptic potentiation can account for synaptic plasticity and some aspects of learning and memory. Brain processes related to representation, memory, learning, and consciousness thus focus on molecular level alterations in synaptic membrane proteins which are regulated by the neuronal cytoskeleton.

There is some evidence for direct cytoskeletal involvement in cognitive processes. Activities of cytoskeletal microtubules and turnover of microtubule subunits ("tubulin") have been shown to be increased in the brain during specific times of learning, memory and experience. Mileusnic, Rose, and Tillson (1980) have utilized a learning model in baby chicks who can be readily trained not to peck at a bright bead coated with an unpleasant tasting substance. These authors have studied some of the neurochemical correlates of this "passive avoidance learning" and point out that tubulin, the major constituent of microtubules, is present in large amounts in the developing brain of young chicks. Significant amounts of tubulin are associated with synaptic membranes leading the authors to conclude that any model of learning and memory which postulates modulation of synaptic structure, as consistent with Hebb's postulates, must involve tubulin in learning. Other work from their laboratory and others have shown that both incorporation of precursor amino acids and total quantity of tubulin may be enhanced by experience and learning during early development.

John Cronly-Dillon and co-workers (1974) of Britain's Manchester University have found that when baby rats first open their eyes, genes in visual cortex suddenly begin producing vast quantities of tubulin which presumably form microtubules involved in establishing new synaptic connections. When the rats are 35 days old, the critical phase for learning is over and tubulin production is drastically reduced. Their conclusion is that tubulin turnover and microtubule activity are involved in synaptic plasticity aspects of learning and memory.

Another dynamic mode of synaptic plasticity focusing on dendritic spines has been proposed by Francis Crick (1982). He suggested that dendritic spines can "twitch" and change their shape, thereby altering their synaptic thresholds by mechanical changes. The placement and architecture of dendritic spines are determined by microtubules, but spines themselves are comprised mostly of contractile actin (Matus, Ackermann, Pehling, Byers, and Fujiwara, 1982). Dynamic spine plasticity, orchestrated by dendritic MT and cytoskeleton, may be an important mechanism of short term memory, and a link between the cytoskeleton and synaptic level neural networks. Another link is axoplasmic transport which maintains and supplies the form and functions of dendritic spines and all neuronal synapses and structures.

### 4.3.6 Axoplasmic Transport

Synaptic membrane proteins including ion channels and receptors, cytoskeletal protein structures which expel neurotransmitter vesicles, organelles including mitochondria, and enzymes required for the synthesis and metabolism of transmitters are manufactured only in the cell bodies of neurons where biochemical machinery exists for protein synthesis and assembly (Golgi apparatus and ribosomes). These materials or their precursors are then moved through the axon (or dendrite) to the nerve terminal by a cytoskeletal mechanism similar to a conveyer belt or bucket brigade. Time lapse photography of neurons in cell culture show mitochondria (large organelles which produce chemical energy in the form of ATP) floating down axons like barges on a river. Recent technology such as video enhanced contrast microscopy (Allen, 1987) has shown vesicles zipping along microtubules on the surfaces of axoplasm extruded from squid neurons. Transmitters are synthesized throughout the entire neuron as the enzymes which catalyze transmitter formation move along the cytoskeletal apparatus from cell body to terminal. The highest enzymatic activity and transmitter concentrations are reached in the terminal boutons. Thus the plasticity of a synapse, its efficacy, readiness and threshold which appear to regulate learning and memory in neural nets over time all depend on axoplasmic transport.

Several separate and independent axoplasmic transport processes have been identified by following the movement of various tracers. The fastest move at a rate of 400 millimeters per day (about 500 nanometers per second), the slowest barely one millimeter per day. The mechanical parts of the system are microtubules and contractile proteins attached to specific sites on microtubule walls. These contractile proteins (dynein or kinesin) utilize chemical energy in the form of ATP hydrolysis to contract in orchestrated sequences of bucket brigade activity. What is *not* understood is the mechanism by which microtubules orchestrate the cooperative sequential activities of the attached contractile proteins. The main stream of axoplasmic transport can be stopped by drugs such as colchicine, which depolymerizes microtubules. Axoplasmic transport generally flows from the cell body toward the tips of fibers. In motor nerves, axoplasmic transport flows in the same direction as the impulse traffic; in primary sensory neurons it flows in the opposite direction to the sensory impulses. In dendrites, the main flow is also from the cell body to the periphery. These are all examples of anterograde axoplasmic flow. There is also simultaneous transport in the opposite direction toward the cell body called retrograde axoplasmic flow which apparently brings feedback information to the cell machinery to regulate the production of transmitter enzymes and other materials. It might also return worn or broken down cell constituents to be recycled. These trophic feedback mechanisms create dynamic neurons capable of changing shape and function as an adaptation to ongoing experience without excessive loss of old information. Synapses, dendritic spines, dendritic branch patterns and membrane proteins are continually changing, yet the memories they contain are somehow maintained by the ever present and ever-changing cytoskeleton.

### 4.3.7 Parallelism, Collective Cooperativity, and the Grain of the Engram

Neuron to neuron synapses operate on the order of milliseconds whereas modern high speed computers operate with semiconductor switches which function on the order of nanoseconds. Yet the brain is able, in a few hundred milliseconds, to perform processing feats that are impossible to emulate in hundreds of minutes of computer time. The assumption is that the brain accomplishes this feat through the simultaneous operation of many, many *parallel* components.

Parallel, distributed models of collective mind organization have been proposed by two noteworthy authors coming from different orientations. Michael Gazzaniga (1985) is one of the first researchers to work with "split brain" patients in whom a severed corpus collosum has separated right and left hemispheres. He contends that minds consist of large collections of smaller semiautonomous parts with limited communication among them. Gazzaniga has developed convincing evidence that our minds are "modular"; they are organized into relatively independent functioning units that work in parallel. The mind does not operate in a single way to solve problems but has many identifiably different units that contribute to our conscious structure in ways that can sometimes be isolated by clever experiments. Specific modules might be devoted to face and visual image recognition, language interpretation, and what Gazzaniga calls an "inference engine." Located in the brain's dominant hemisphere close to the language interpreter module, the inference engine is thought to "coordinate" consciousness. These modules may be compared to those described by Pribram or the "cartels" described by Freeman and represent a level of brain organizational hierarchy just below that of the entire brain. Gazzaniga relates free will to a basic feature of brain organization, and suggests that the particular belief in free will itself follows from the modular theory of mind. He observes that we are continually interpreting behaviors produced by independent brain modules as behaviors that are produced by the "self." We conclude that we are acting freely whereas at the root of it we don't really know why we do almost anything. This notion may be compared to the parallel connectionist concept (i.e. "Darwin" and "Wallace") which requires a vote or caucus to determine a summary output. Gazzaniga also states that basic cognitive phenomena such as acquiring and holding social beliefs are just as much a product of human brain organization as our behaviorist desires to eat, sleep, and have sex. He argues that we are "hardwired to have beliefs."

A comparable conclusion has been reached by Marvin Minsky (1986). In *The Society of Mind,* the patriarch of AI discusses the mind as a vast number of "agents." Information is represented by "frames" which are multiple connected knowledge nodes. Minsky throws a much more complex grid of agents over the mind than Gazzaniga's modules, but both describe levels of organization between the neuronal synaptic level and the whole brain which are more or less representative of neural network theory. Gazzaniga argues that the brain is more a social entity than a psychological one. Rather than being an indivisible whole as was once believed, it is a vast confederacy of relatively independent modules, each of which processes information and activates its own thoughts and actions. But what are the modules? Where is the *grain* of consciousness? Isn't the mind more than an array of squabbling modules?

John O'Keefe and Andrew Speakman (1986) at the University College of London have completed a series of experiments on the activity of rat hippocampal neurons while the animals were performing spatial working memory tasks. These and other results suggest a hippocampal cognitive map in which the representation of place in an environment is distributed across the surface of the hippocampus. O'Keefe and Speakman find that the grain of the representation is *at or below* the single cell level. That is, each cell in a small cluster participates in the representation of different patches of environment. Conversely, any cluster of about 8 to 10 cells appears adequate to provide a coarse representation of an entire environment. The addition of more cells to the network increases the resolution, or grain of the representation, but does not alter it. The representation of an environment is thus distributed and the "graininess" is at a level *below that* of individual nerves and synapses. E. R. John and collegues (1986) from New York University have used metabolic mapping of memory traces in cats to show that information is extensively distributed, requiring that "(nerve) cells participate in multiple memories." They implicate:

> ... cooperative processes in which the nonrandom behavior of huge ensembles of neural elements mediate the integration and processing of information and the retrieval of memories. Memory and awareness in complex neural systems may depend on presently unrecognized properties of the system as a whole.

Walter Freeman (1972, 1983) of the University of California at Berkeley agrees that nervous systems are more than the sum of their parts. According to Freeman, this occurs because interconnections of numbers of neurons give rise to collective properties belonging to the neural populations in general rather than to specific individual neurons. Such *collective* properties related to mental processes are thought by Freeman to be generated by the interconnectedness of numbers of neurons of at least ten thousand or more.

Freeman and others who advocate cooperative collective aspects of mental processes cite the electroencephalogram (EEG) as supportive evidence. EEG is a continuous electromagnetic wave which pervades the brain and is composed of frequencies from one hertz (= Hz = cycles per second) to a few thousand Hz, but concentrated in the range between 3 and 50 Hz (Chapter 7). EEG has been used for half a century to diagnose brain disease, but not until the 1960's was the source of the "brain waves" clarified. EEG arises *not* from the sum of propagating action potentials in axons, but rather from the slow, graded potentials produced by dendrites and cell bodies. Local potentials combine to form regional and brain-wide patterns. Individual neurons and their dendrites slip in and out of phase with the surrounding EEG field. Studies by Adey (1977), John (1980) and many other scientists have correlated mental activities in animals and humans with EEG pattern changes. Because the EEG is produced by large numbers of neurons, these correlations may suggest that collective neuronal effects are the basis for mental activities. Freeman proposes that functionally significant EEG wave phenomena occur within neural masses in which there exist feedback connection of one neuron with many others in the same mass. Collective waves of graded potentials are Freeman's candidate for the grain of consciousness. Localized wave activity within neural masses or "cartels" would be related to EEG waves in the same way that atmospheric temperature and pressure waves generate cloud patterns—observable side effects which may yield information about the internal dynamics.

Opinions regarding the significance of local collective EEG wave fields vary from superfluous epiphenomena, to information transmitters, to the substance of consciousness itself. E. R. John (1984) is perhaps the strongest proponent; he points out that individual neurons are sensitive to the fields they generate. Adey (1984) and colleagues have applied "EEG-like" fields to the brains of experimental animals and found they produce behavioral effects. John proposes that a specific electromagnetic field is evoked by sensory stimuli and "resonates" with similar patterns stored in memory. New patterns bring new resonances which lead, according to John, to the stream of conscious experience. Wave patterns modify neuronal structure, forming memories to be evoked by later resonances.

E. R. John (1980) asserts that:

> Consciousness is a property of these improbable distributions of energy in space and time, just as gravity is a property of matter.

## 4.4 Toward Molecular Consciousness

How is electrical wave energy coupled to neuronal structure, and what neuronal structures are most suitable for coupling and representation of cognitive content? Simultaneous recognition (and cooperative coupling) by large numbers of neural elements requires rapid changes in chemical state of widely distributed macromolecules. Likely candidates are the "allosteric" proteins which can transduce regulatory signals (binding of molecules, ions/acidity, voltage fields etc.) to undergo functional conformational changes. Hyden (1977) initially proposed that proteins rapidly change their conformation in response to weak, oscillating electric fields. W. Ross Adey (1977) has elaborated on the coupling of neural protein conformation and function to EEG waves. He has suggested that webs of hydrated glycoproteins (extending from neural membranes into the extracellular space), membrane proteins, and the cytoskeleton are primed to undergo rapid conformational changes in response to localized and selective spatiotemporal EEG patterns, as well as biochemical signals. Transduction of electromagnetic energy into conformational states by widely distributed proteins can cooperatively represent dynamic information.

The common thread of biological intelligence, the "grain of the engram," may be found within cooperative dynamics of a molecular network whose structure and functions appear perfectly adapted to information processing: the cytoskeleton.

# Chapter 5

# Cytoskeleton/Cytocomputer

Living organisms are collective assemblies of cells which contain collective assemblies of organized material called protoplasm. In turn, protoplasm consists of membranes, organelles, nuclei and the bulk interior medium of living cells: *cytoplasm*. Dynamic rearrangements of cytoplasm within eukaryotic cells account for their changing shape, repositioning of internal organelles, and in many cases, movement from one place to another. We now know that the *cytoskeleton*, a dynamic network of filamentous proteins, is responsible for cytoplasmic organization (Figures 5.1 thru 5.3).

## 5.1 The Nature of Cytoplasm

The nature of cytoplasm has been scientifically studied for at least a century and a half. That history was described by Beth Burnside (1974) in a landmark meeting devoted to the cytoskeleton at the New York Academy of Sciences.

An early French observer of cellular material, Felix DuJardin proposed in 1835 that all cells were composed of a motile material called "sarcode" that had both structural and contractile properties. In 1861, Austrian E. Brucke linked the mechanical and physiological properties of cells to a fundamental organization or architecture of cytoplasm as distinguished from purely chemical or physical properties. Early Dutch microscopist van Leeuwenhoek observed that red blood cells became deformed passing through capillaries and then sprang back into shape, demonstrating the elasticity of cytoplasm. The variety of elaborate forms that cells assume and maintain also require some cytoplasmic rigidity, properties which led nineteenth century biologists

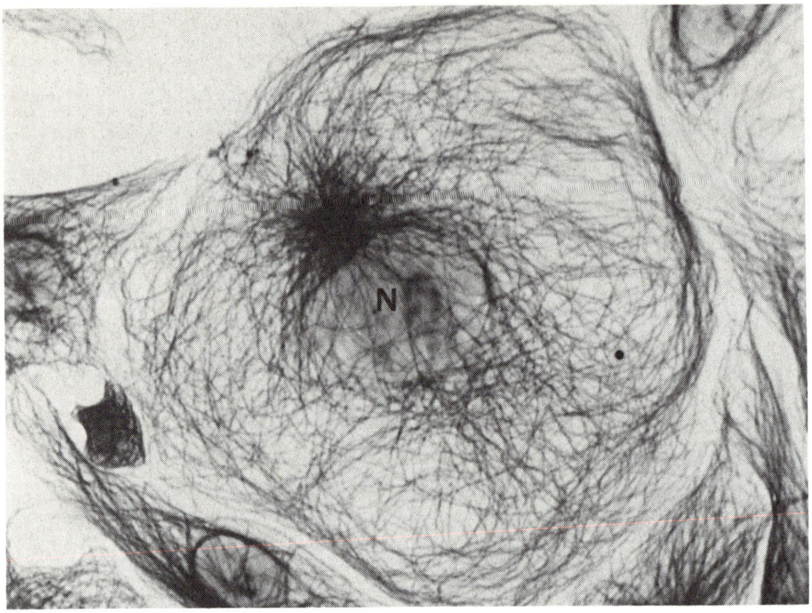


Figure 5.1: Cytoskeletal network in a rat kangaroo kidney cell (ptK2) illustrated by tubulin antibody and light microscopy. Microtubules emanate from a dense MTOC region near the nucleus (N). With permission from Marc DeBrabander (1982).

to conclude that cytoplasm is not merely a liquid nor an emulsion nor an aqueous suspension of life bearing granules. Rather, cytoplasmic architecture and contractility could be explained by the proposal of a mesh-like "reticular" or "fibrous" substructure whose interstices were filled with fluid. To some early biologists, the structure of cytoplasm therefore consisted of a continuous reticular network of delicate fibrils extending through the cell (reticular theory). Others claimed that the fibrils forming the cytoplasmic substratum were unbranched and discontinuous (fibrillar theory). Both of these theories were initially supported, but later deflated, by microscope preparation techniques of fixation and staining which arose between 1870 and 1890. Fibrillar and reticular frameworks appeared everywhere in many types of fixed cells, particularly in muscle, nerve, cartilage and epithelial cells.

Fischer and Hardy (1899) showed that the new fixatives and stains *induced* artificial coagulation of gelatin, egg albumin, and other proteins. The coagulation gave rise to beautiful reticular and fibrillar formations and even

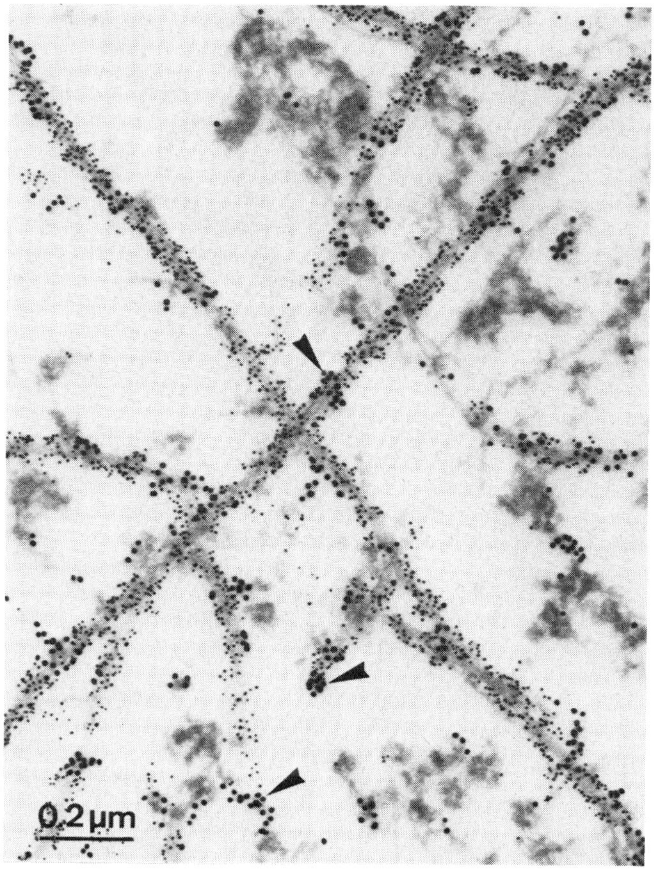


Figure 5.2: Microtubules from PtK2 cells double labeled with antibody and immunogold under high magnification. Large circles (10 nanometer gold particles, arrows) tag glutamated tubulin subunits, small circles (5 nanometer gold particles) label tyrosinated tubulin subunits. With permission from Geuens, Gundersen, Nuydens, Cormellisen, Bulinski, and DeBrabander (1986), courtesy of Marc DeBrabander and Janssen Pharmaceutica Research Laboratories.

produced strikingly real, but phoney mitotic spindles! These studies discouraged support for the fibrillar and reticular theories to the extent that many cytologists denied that meshworks seen in fixed material had any validity

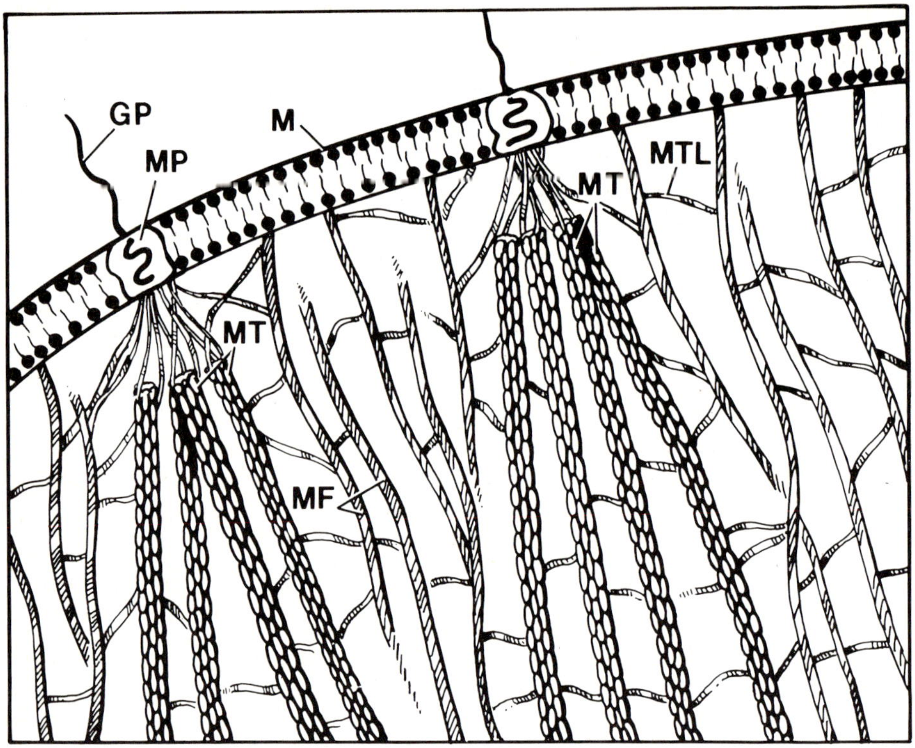


Figure 5.3: Schematic of cellular cytoskeleton/membrane. M: cell membrane, MP: membrane protein, GP: glycoprotein extending into extra-cellular space, MT: microtubules, MF: microfilaments (actin filaments or intermediate filaments), MTL: microtrabecular lattice. Cytoskeletal proteins which connect MT and membrane proteins include spectrin, fodrin, ankyrin, and others. By Fred Anderson.

in the living cell. Butschli added his alveolar foam theory of cytoplasmic structure, claiming that the reticular appearance seen in fixed materials and sometimes in living cells resulted from a honeycomb of vacuoles of one substance crowded together in a continuous phase of another. This theory died because it failed to account for the fixation observations and because vacuolated cytoplasm was uncommon. Meanwhile, observations of true fibrillar formation in living cells were accumulating. The use of polarized light microscopy revealed "birefringent" submicroscopic rods in the cytoplasm of muscle, nerve, sperm, and spindles of dividing cells. Biologists began to consider cytoplasm as a "gel," in which rod-shaped filaments formed cross linkages. Gel aptly describes the mechanical properties of cytoplasm: an elastic intermeshing of linear crystalline units giving elasticity and rigidity to a fluid while allowing it to flow.

The protoplasmic rods revealed by birefringence in polarized microscopy were thought to be proteins, or linear aggregates of proteins, which were held together by hydrogen bonding. Seifriz concluded that the fibrillar "artifacts" produced in cells or protein solutions upon fixation were significant. He suggested that the relatively coarse microscopic threads in fixed materials may be artifactually produced aggregates of important *submicroscopic* threads, probably linear proteins. To account for the elastic and tensile properties of cytoplasm, Seifriz proposed the "brush heap theory" of interlacing fibers. He and others later suggested that cytoplasmic fibers could exist in the brush heap form or could form lateral associations by hydrogen bonding into paracrystalline aggregates, and that switching between the two states accounted for cytoplasmic behavior observed in living cells. We now know that subunits of the protein actin can polymerize in a variety of forms including interlacing gels or filamentous bundles. Cytoplasmic "gel" states occur due to crosslinking of actin filaments and other cytoskeletal proteins, and can be converted to more aqueous "sol" states by calcium ions and other factors.

Development of the electron microscope through the 1960's initially did *not* illuminate the substructure of cytoplasm. Portions of cells which were optically empty by light microscopy persisted in being empty in electron micrographs. The cell was perceived by many to be a "bag of watery enzymes." However, in some cases the fibrillar structures seen with light microscopy appeared as fine tubular filaments with the electron microscope. They comprised the internal structure of cilia, flagella, centrioles and basal bodies, and were prominent in the mitotic apparatus of dividing cells. Tubular filaments were also frequently encountered in protozoa, nucleated red blood cells and the dendrites of neurons. Ironically, the fixative then used in electron microscopy, osmium tetroxide, had been *dissolving* filamentous elements so that their presence was observed only sporadically. Like the 19th century fixative induced coagulation which induced superfluous fibrillar structures, the use of osmium tetroxide delayed recognition of the cytoskeleton. Later, with the advent of glutaraldehyde fixation, delicate tubular structures were found to be present in virtually all cell types and they came to be called microtubules. Bundles of microtubules seen with the electron microscope corresponded to birefringent fibers seen in living cells. Their ubiquity suggested that they were the fibrillar substructure of cytoplasm previously predicted on theoretical grounds. After some skepticism, microtubules became generally accepted as "household organelles" in nearly all cells studied. Subsequent characterization of actin, intermediate filaments, the microtrabecular lattice or "ground substance," and the structure of centrioles led to the recognition that cells were comprised of dynamic networks of connecting filaments and brought the cytoskeleton out of the closet.

## 5.2 Microtubules

Soifer (1986):

> When microtubules are required by a cell for a particular function, microtubules assemble in the appropriate part of the cell, with the necessary orientation. As microtubules are no longer needed, they depolymerize.

Essential functions within living cells ranging from single cell amoeba and paramecium to neurons within earthworms and Nobel scholars are performed by similar cytoskeletal structures. The most visible and widely studied cytoskeletal elements are microtubules (MT), slender cylinders intimately involved in important cell functions. As DNA is the common genetic data base containing hereditary information, *microtubules are real time executives of dynamic activities within living cells.*

### 5.2.1 Microtubule Structure and Function

Microtubules (MT) are hollow cylinders about 25 nanometers in diameter whose walls are polymerized arrays of protein subunits. Their lengths may range from tens of nanometers during early assembly, to possibly meters in nerve axons within large animals. The protein subunits assemble in longitudinal strings called protofilaments; thirteen parallel protofilaments laterally align to form the hollow "tubules." The subunits are "barbell shaped" proteins ("dimers") which in turn consist of two globular proteins ("monomers") known as alpha and beta tubulin (Figure 5.4). Alpha and beta tubulin monomers are similar molecules with identical orientation within protofilaments and tubule walls. Each monomer consists of about 500 amino acids, weighs about 55 kilodaltons (Chapter 6), and has a local polarity or charge orientation. MT which grow from cell centers have a plus end (beta tubulin)

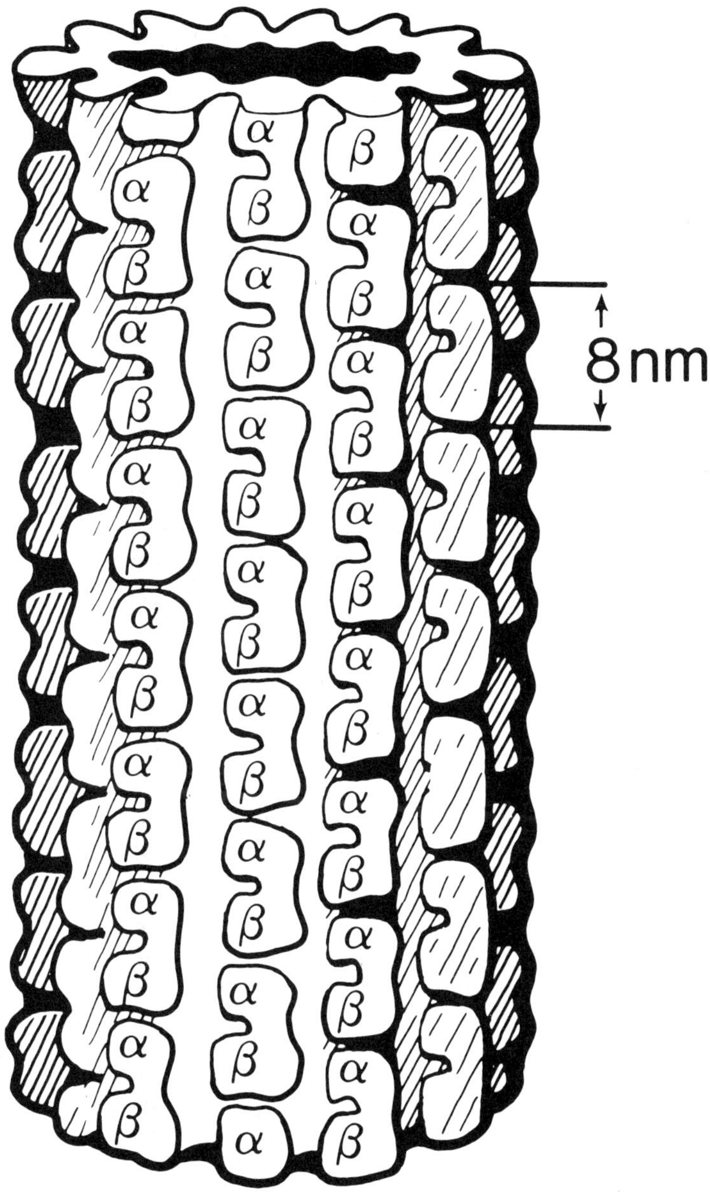


Figure 5.4: Microtubules are cylinders whose walls are 13 protofilaments, each a string of 8 nm tubulin dimers. Alpha and beta tubulin monomers form the dimers. Each dimer has 6 neighbors. By Fred Anderson.

which extends outward from the cell center ("centrosome") into the cell periphery. The minus end (alpha tubulin) remains anchored to a microtubule organizing center (MTOC) within the centrosome. Each dimer, as well as each MT, appears to have an electrical polarity or dipole, with the negative end oriented towards the alpha monomer and cell center, and the positive end towards the beta monomer and cell periphery.

The dimers are held together by relatively weak Van der Waals hydrophobic forces such as dipole coupling (Chapter 6). Dimer neighbors form hexagonal lattices with a "leftward" tilt and several helical patterns may be discerned in the relations among dimers. The crystal-like symmetry packing of tubulin in microtubules has been evaluated by Djuro Koruga (1986) of the Molecular Machines Research Unit at the University of Belgrade in Yugoslavia (Chapter 8).

MT from different life forms have marked similarities, but subtle differences. Comparison of MT from nerve cells of earthworms and mammals shows that the more primitive worm MT are more variable in geometric structure with MT ranging from 9 to 11 protofilaments, whereas mammalian MT generally have 13. Tubulins from among different species including mammals and plants bind to common antibodies and tubulins from different species may coassemble into hybrid MT. Despite these common traits, the diversity of tubulin gene expression has proved far greater than imagined years ago. Analysis of tubulin by amino acid sequencing and advanced electrophoretic techniques have shown that multiple, different alpha and beta tubulins exist concurrently, with the greatest diversity shown by beta tubulin. For example, Lee (1986) and colleagues at St. Louis University have shown that as many as 11 different tubulin forms exist in rat thyroid microtubules and 17 different forms exist in rat brain microtubules. Thus alpha and beta tubulin are families of "isozymes," each of which may have specific functions or binding of microtubule associated proteins (MAPs). Another tubulin variable, detyrosination, occurs in the cytoplasm subsequent to DNA transcription. Detyrosination is the removal of the terminal amino acid, tyrosine, from the polypeptide chain which comprises beta tubulin. Removal of tyrosine exposes an acidic amino acid, glutamate. Local factors in the cytoplasm independent of genetic programming determine whether or not individual tubulin subunits are "tyrosinated" or "glutamated." Marc DeBrabander (1986) and collaborators at Janssen Pharmaceutica in Belgium have been able to identify specific tubulin subunits within assembled microtubules which are either tyrosinated or glutamated. Their elegant studies show heterogeneous patterns of tyrosinated and glutamated tubulin which

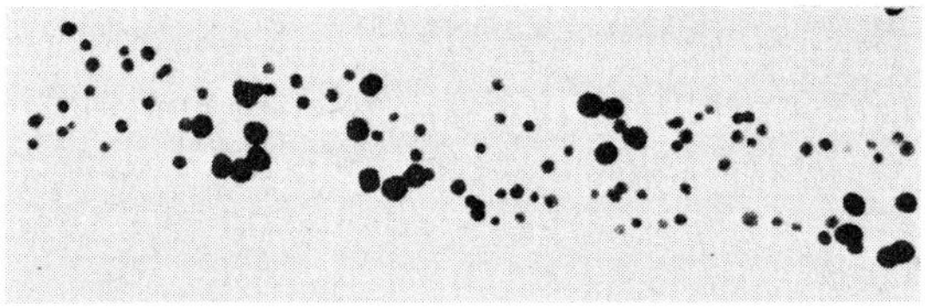

Figure 5.5: Microtubule double labeled with immunogold tubulin antibody. Large circles, 10 nanometer gold particles, label glutamated tubulin; small circles, 5 nanometer gold particles, label tyrosinated tubulin. With permission from Geuens et al (1986), courtesy of Marc DeBrabander and Janssen Pharmaceutica Research Laboratories.

could indicate an information representation coupled to specific MT functions by the action of MAPs (Figure 5.5).

Since early electron microscopy studies, microtubules have been invariably described as being surrounded by a "clear zone" which gives the impression of a "halo" around them when they are viewed in cross section. A 5–10 nanometer distance from the surface of MT is free of cytoplasmic ground substance or any other material normally seen elsewhere throughout the cell. These clear spaces were initially thought to be electron microscopic artifacts, or layerings of less dense filamentous proteins on the surface of the microtubules. However, newer fixation and staining techniques and freeze etching have confirmed that the space immediately surrounding microtubules are seldom encroached upon by other organelles or cytoplasmic ground substance. Stebbings and Hunt (1982) have studied the "clear zone" and point out that the surface of microtubules is strongly "anionic" since tubulin is an acidic protein due to its high content of acidic amino acids such as glutamate and aspartate. These amino acids give up positively charged hydrogen ions to solution, leaving MT with excess electrons. Stebbings and Hunt propose that anionic, or electronegative *fields* at MT surfaces can explain the clear zones as well as the staining of MT by positively charged dyes, binding to MT of positively charged proteins, cations such as calcium, metals and other compounds. Electronegative fields surrounding MT may act as excitable ionic charge layers ("Debye layers") which are also thought to occur immediately adjacent to cell membranes (Green and Triffet, 1985). Excitable "clear zone" charge layers next to MT could facilitate collective communicative mechanisms within the cytoskeleton (Figure 6 and 8).

The question of the hollow core within MT is even more mysterious; it too appears devoid of ground substance. It is unknown whether the interiors of MT are also electronegative zones, or perhaps positive ones which would create voltage gradients across MT walls. Del Giudice and colleagues (1986) at the University of Milan have even suggested electromagnetic focusing and "superconductivity" within microtubule cores (Chapters 6 and 8). Insulated from "aqueous" surroundings and held together by water-excluding hydrophobic forces, MT and the rest of the cytoskeleton comprise a "solid state" network within living cells.

What do microtubules do? For openers they *are* the cyto*skeleton*, being the most rigid structures in most cells. To establish the pattern of the cytoskeleton and the form and function of living cells, MT assemble from subunits at the proper time, place, and direction. They are often anchored and guided by MT organizing centers (MTOC) containing centrioles. Once in place, they participate in movement of cytoplasm, organelles and materials, growth, reproduction, synaptic plasticity, and nearly all examples of dynamic cytoplasmic activity. The mechanisms for MT organization are unknown, but several theories of MT based information processing have been proposed and will be described in Chapter 8.

Many MT functions involve control and signaling of the activities of attached proteins including those which interconnect MT in assemblies like centrioles, cilia and flagella. Active sliding by motile bridges attached along MT are involved in cell shape determination, and extension of cytoplasmic projections like neuron growth cones, dendritic spines and amoeba lamellipodia. These extensions contain actin filament bundles without MT, however MT generally establish the architecture which orients these extensions, provide their raw materials, anchor them and integrate their functions with the cell. In cytoplasmic transport, components moving along parallel arrays of cytoplasmic microtubules deliver cellular materials wherever they are needed. In lengthy asymmetrical cytoplasmic processes like nerve axons, specialized mechanisms of axoplasmic transport have evolved to transport cell constituents at rates up to five thousand nanometers per second! MT orchestrate the motile force supplied by "dynein" mechanical arms to move organelles which include chromosomes, nuclei, mitochondria, neurotransmitter synaptic vesicles, liposomes, phagosomes, granules, ribosomes, and other proteins. MT assembly determines the form and function of biological systems.

MT also participate in sensory perception of the cell's external environments. Many sensory receptors are modified cilia, assemblies of microtubules similar in structure to centrioles. Insect mechanoreceptors and sensory cilia

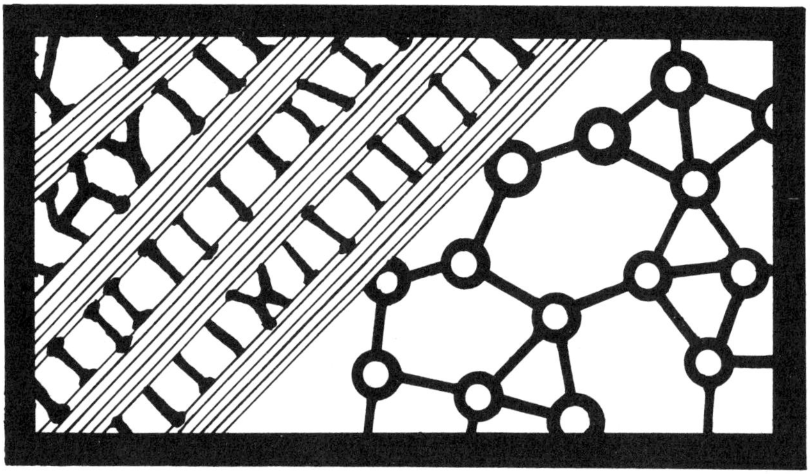

Figure 5.6: Microtubule arrays interconnected by MAP bridges. Hollow circles are MT arrays in cross section. By Paul Jablonka.

in our ears appear to utilize MT to transduce mechanical force to the base of each cilium and into the nervous system. Atema (1973) has proposed that propagation of conformational changes along sensory cilia convey information to the cell as a whole. Moran and Varela (1971) have suggested that sensory cilia MT act as engines driven backwards. When an external force moves the microtubules of the mechanoreceptor cilium, they suggest that the MT release ions like calcium which can regulate cellular activities. Sensory transduction, guidance and alignment are "intelligent" MT activities which are vital to biological growth, embryological development, secretion, synapse formation and many other important biological functions. Temporal and spatial control of MT deployment are achieved by two mechanisms of pattern formation: directed assembly from MT organizing centers (MTOC) and self assembly of multitubular arrays by means of intertubule linkers (Figure 5.6). MT are the scaffolding, conveyor belts, and computers of living cells.

### 5.2.2 Microtubule Assembly and the Generation of Form

As changeable cellular superstructures, MT can assemble into rigid tubular rods when and where they are needed, and then disassemble into subunits which may be transported away by other MT. Like many viruses, MT self assemble from their subunits into orderly polymers. The large increase in order, or negative entropy associated with MT assembly would appear to

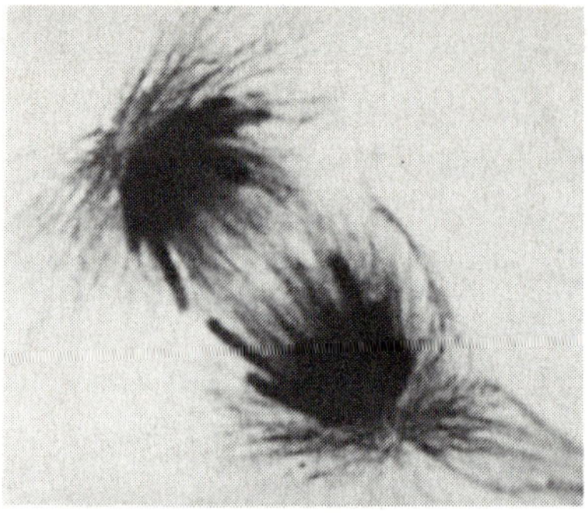

Figure 5.7: Microtubule mitotic spindles (wispy filaments separating chromosomes (heavy rods) establishing daughter cell polarity during cell division. Centrioles and MTOC are at nidus of each array. With permission from Marc DeBrabander (1982).

flow upstream against the second law of thermodynamics which, in general, states that order tends towards disorder. The increase in order observed during MT assembly can be related to the dispersal (disordering) of tightly bound structured water from the subunits as they polymerize, consistent with the concept that MT subunits associate by hydrophobic interactions (Chapter 6).

MT vary markedly in stability and function. Structurally similar MT may be highly stable, or fleetingly transient. This variability can result from differences in tubulin chemistry, from effects of secondary proteins attached at specific points on MT surface lattices (microtubule associated proteins: MAPs), local cytoplasmic influences, membrane interactions and many other factors. Highly stable MT are found in cilia and flagella in which they are assembled into complex patterns and carry out their functions without disassembly. The other extreme are MT within mitotic spindles which separate chromosomes and establish daughter cell shape in cell division ("mitosis"— Figure 5.7). These "labile" MT not only assemble before mitosis and disassemble afterwards, but apparently undergo localized disassembly/reassembly during mitosis ("dynamic instability"). In some animal cells and during plant cell mitosis, assembly of MT subunits into MT appears to take place in the cytoplasm without any relation to a particular organelle. In other cells, structures including centrioles, basal bodies, centrosomes, and kinetochores facilitate assembly of microtubules at a specific site and orientation. These structures are called microtubule organizing centers (MTOC), and they consist of a pair of centrioles and a dense granular material. In general, negatively charged ends of MT attach to the MTOC, and positively charged ends grow distal to the MTOC, establishing cell polarity, shape and orientation.

The MT assembly process may be tracked by parameters that crudely reflect the polymer level such as turbidity or viscosity. Generally, there is first a lag phase when no microtubules form, then a phase of exponential growth and finally a stable plateau. At low concentrations, no microtubules are formed. Above a certain critical concentration (Cc) of tubulin, microtubules increase in relation to the total tubulin concentration. However, MT assembly is more complex than being simply at equilibrium with a pool of unassembled subunits.

Spontaneous assembly of tubulin dimers depends on physiological conditions which alter the critical concentration (Cc) of subunits required. Spontaneous MT assembly thus depends on many cofactors: calcium ion concentration, temperature, presence of microtubule inhibitors or stabilizers (Figure 5.8), microtubule associated proteins (Figure 5.12), the presence of microtubule organizing centers (MTOC—Figure 5.9) and the availability of GTP. Like ATP (adenosine triphosphate-required for actin assembly), GTP (guanosine triphosphate) is a source of biochemical energy. As will be described in Chapter 6, ATP and GTP can each donate phosphate bond energy by being "hydrolyzed" to their "diphosphates" ADP and GDP, respectively. In the case of assembly of both MT and actin filaments, the presence of GTP or ATP *without being hydrolyzed* is an important requirement for assembly. Nonhydrolyzable analogs of GTP are equally effective in promoting assembly of MT; hydrolysis and energy input take place *after* incorporation of a dimer into a tubule. MT formed with nonhydrolyzable analogs are more stable than those with GTP, so hydrolysis energy may be related to disassembly. In actin filaments, hydrolysis also occurs *after* the formation of the polymer. Therefore a paradox exists in that GTP *binding* is required for microtubule assembly, but GTP *hydrolysis* occurs later and is not required for assembly. An identical situation occurs with actin and ATP. Depletion of ATP (for actin) and GTP (for MT) strongly inhibits disassembly but does not hamper assembly per se. However, in energy depleted cells, *random assembly* prevails instead of organized assembly. The energy from phosphate bond hydrolysis, the main energy currency in all biological systems, is unaccounted for within the cytoskeleton, the dynamic organizer of cell function. Perhaps the energy is used to generate communicative lattice vibrations, coherent excitations, or "solitons" in MT and the cytoskeleton in general (Chapters 6 and 8).

MT assembly also requires the presence of magnesium ion and a low

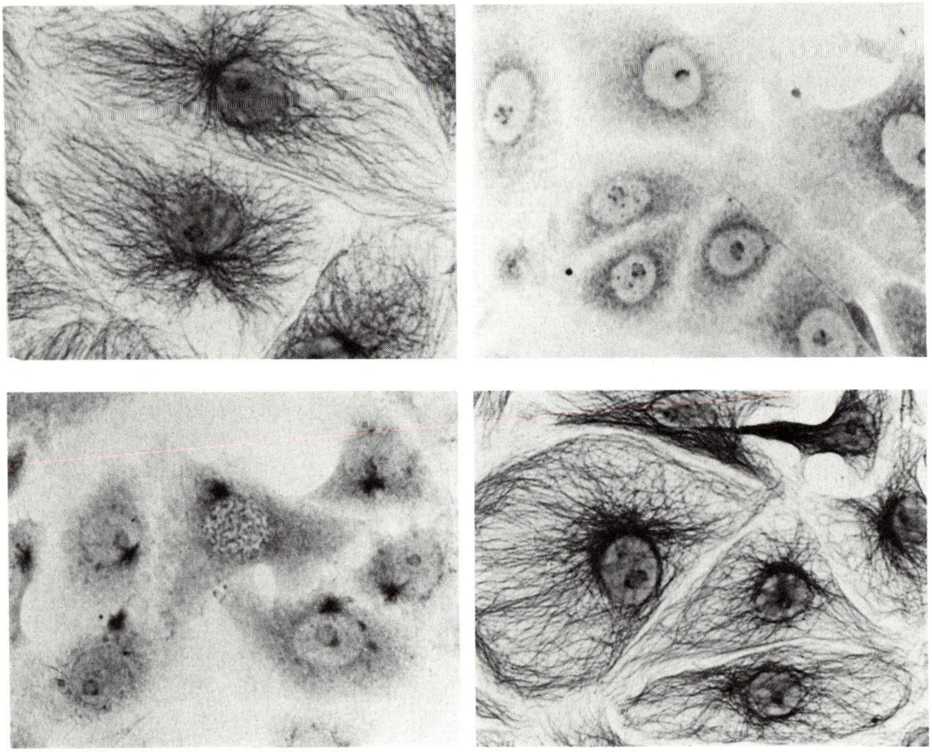

Figure 5.8: Effects of MT disassembly drug nocodazole in PtK2 cells. MT visualized with peroxidase-antiperoxidase (PAP) method. Upper left: organized MT radiate from dense MTOC region near nucleus. Upper right: cells have been treated with nocodazole ($2 \times 10^{-5}$ M, 4 hours) and MT depolymerized. Lower left: cells washed after nocodazole, 5 minutes later MT growing from MTOCs. Lower right: cells 20 minutes after wash have reorganized MT system. With permission from DeBrabander, Geuens, Nuydens, Willebrords and DeMey (1981).

concentration of calcium ion. Calcium is an important messenger system within many forms of cytoplasm. Waves of calcium can be caused by membrane and cytoskeletal activities, and can induce conformational changes in proteins and dissolve cytoplasmic gel, converting it to a more aqueous state:

"sol." Tubulin dimers loosely bind 16 calcium ions per dimer (Hayashi and Matsumara 1975); an excess of calcium, however, causes MT disassembly. In the presence of abundant zinc ions, tubulin polymerizes into flat expansive sheets rather than cylinders.

The dimer subunits alpha/beta tubulin are all arranged in the same direction. The beta subunit protrudes at the "plus" end and contains the exchangeable GTP binding site. "Cc" for assembly is lower at the plus end than it is at the minus end. At steady state in the presence of hydrolyzable GTP, a net incorporation of dimer subunits is seen at the plus end, and a net loss of dimer subunits is seen at the minus end. Subunits thus move from the plus end to the minus end in MT—a phenomenon called "treadmilling." In MT anchored at one end to organizing centers, treadmilling is a net movement of subunits within the MT lattice, and may correspond to the slow 1 millimeter per day component of axoplasmic transport. MT thus appear to be continually growing, perhaps twisting, at a rate of about 10 nanometers per second. Labile MT can polymerize their way through the cytoplasm, adding GTP-tubulin at the beta plus end, and dumping GDP tubulin at the alpha minus end. These types of MT can thus behave like mobile tractors, cytoskeletal caterpillars (Figure 5.10).

Disassembly or separation of tubulin dimer subunits from MT is induced when terminal dimers bind GDP. When GTP is hydrolyzed, a phosphate group is lost and guanosine triphosphate (GTP) becomes guanosine diphosphate (GDP). GTP bound dimers at open ends of MT (GTP "caps") are stable: they stay assembled. However, when open ends of MT contain subunits binding GDP, they tend to release and shorten the MT cylinder. MT whose distal ends have exposed GDP tubulin, and are not anchored by structures like MTOC, shrink and depolymerize. Generally, free MT polymers (those not anchored by MTOC) are "chimeric" with stretches of unhydrolyzed GTP at the ends (GTP caps) and GDP in the interior. Growing polymers have "protective" GTP subunit caps at their end; shrinking polymers lose their GTP caps and expose GDP subunits at the end, causing disassembly.

Microtubules thus appear to exist in two populations: the majority growing at an appreciable rate and the minority shrinking very rapidly and providing new subunits for growth. This concept has been called "dynamic instability" with the rapid shrinkage of MT referred to as "microtubule catastrophes" (Kirschner and Mitchison, 1986). The faster the growth rate, the larger the GTP cap and the lower the probability of the cap disappearing and the microtubule depolymerizing. When the cap disappears and GDP subunits are exposed, the polymer enters a rapid depolymerizing ("catastrophic") phase. MT contain thirteen parallel protofilaments and it is unclear how many exposed GTP or GDP containing subunits at MT terminals are required for stability or instability. An essential factor is the rate of GTP hydrolysis which results in GDP tubulin and induces disassembly. The utilization of GTP hydrolysis energy by MT and the cytoskeleton is a significant portion of biological energy consumption, yet remains unappreciated. Theories of protein conformational state regulation such as solitons and coherent excitations which could account for the useful consumption of hydrolysis energy will be described in Chapter 6. Solitons, coherent lattice vibrations, and cooperative resonance present possibilities for collective, and possibly intelligent effects within cells.

Japanese investigators Horio and Hotani (1986) have directly observed growing and shrinking MT using dark field microscopic video. They observed that both ends of a microtubule can exist in either the growing or the shortening phase and alternate quite frequently between the two phases in an apparently random manner. Further, growing and shortening ends can coexist on a single microtubule: treadmilling. One end may continue to grow simultaneously with shortening at the other end. The two ends of any given microtubule have remarkably different characteristics. One "active" end (presumably the beta, plus end) grows faster, alternates in phase between growing and shrinking more frequently, and fluctuates in length to a greater extent than the inactive end. Microtubule associated proteins ("MAPs") suppress the phase conversion and stabilize microtubules in the growing phase.

There are many apparent mechanisms for stabilizing polymerized microtubules: they may be capped or bound to various structures. Binding at the proximal end to MTOCs stabilizes MT and prevents their depolymerizing at the opposite end. MT need to be stabilized at merely one end to predominate within cells because, even if they depolymerize, another will reform in the same place and orientation (DeBrabander, 1985). Thus MTOC provide cells with a means of selective retention of a subclass of specifically oriented MT.

MTOC oriented in a specific direction can determine cell polarity, including extension of cells in embryological growth, formation of probing appendages called filopodia in locomotory cells and growth of nerve processes. Signals at the cell periphery apparently cause an asymmetry of the microtubule cytoskeleton and overall cell polarity. How can a peripheral clue lead to reorganization deep in the cell? One possible explanation is that a signal is relayed through the cytoskeleton to the MTOC, leading to a change in its orientation and directed nucleation of microtubules. Another is that a signal at the periphery affects the MT distribution directly. Since the entire cytoskeletal array is dynamic, it might only be necessary to transiently stabilize a particular subset of microtubules for the cellular cytoskeletal array to rapidly transform. Kirschner and Mitchison (1986) have proposed that the dynamic microtubule array probes many regions at random. By stabilizing certain MT configurations as they arise, they believe the cell can arrive at a structure that is not precisely defined by genetic information but one that adapts to fulfill a particular functional role.

Dynamic structural rearrangement of the MT cytoskeleton appears to require intelligence. Like the brain as a whole, cytoskeletal intelligence has features of connectionism, parallelism, distributedness, and hierarchy. The apparent cytoskeletal commanders are MTOC, of which the critical structures are the organelles which may have hijacked evolution: centrioles.

### 5.2.3 Microtubule Organizing Centers (MTOC) and Centrioles

MTOC and their chief components, centrioles, are the specific apparatus within living cells which trigger and guide reorganization of cytoplasm such as occurs during growth, generation of form and function ("differentiation") and cell movement. The enigmatic MTOC determine where, when, and how these functions occur (Figure 5.11).

MTOC (or "centrosomes") contain centrioles and "pericentriolar substance" which facilitates tubulin assembly by somehow lowering Cc. Centrioles are the common structure in all of these cellular control centers. Centrioles are composed of two similar cylinders; their diameters are 0.2 microns or 200 nanometers. Each cylinder possesses a 9 fold radial symmetry and is constructed essentially of 9 triplets of microtubules fused longitudinally. "Satellites," electron dense proteins, appear to orbit the centrioles. A cartwheel filamentous structure ("pinwheel") connects all the microtubules within each centriole at one end. One centriole begets another by replication perpendicular to the cylindrical surface of the centriole. The first step in cell division is maturation of the "mother" cell, followed by separation of pairs of centrioles and migration to establish architecture of "daughter" cells. Centriole mechanisms of perpendicular replication, orientation and guidance are unknown.

Between cell division cycles ("interphase"), the MT system is relatively quiescent and "radial." Most cell MT are anchored at one end to the MTOC,

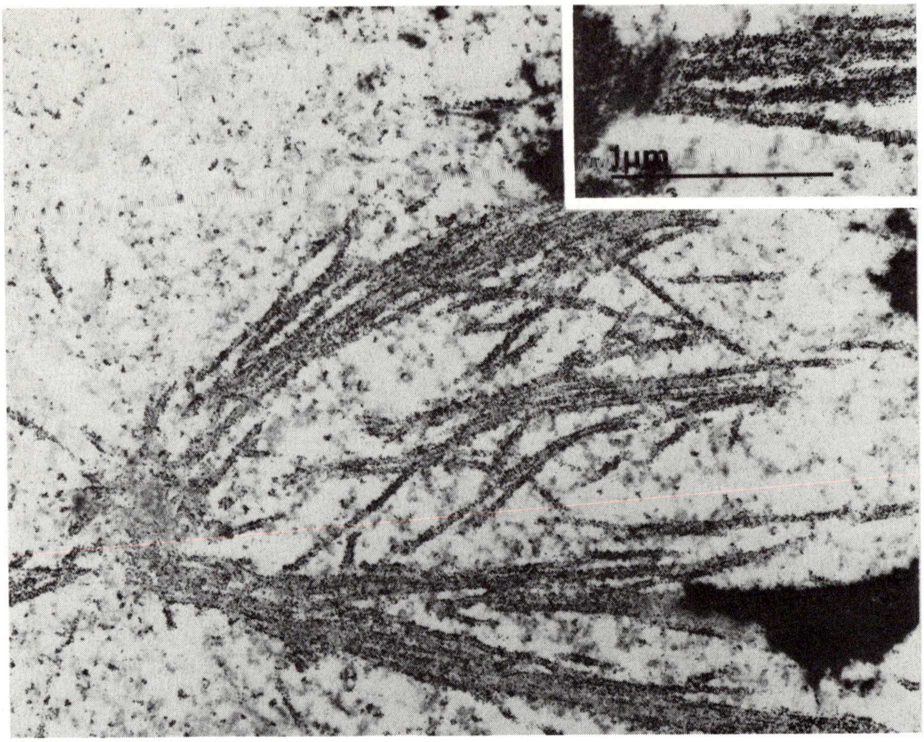


Figure 5.9: Microtubules in mitotic PtK2 cell labeled with tyrosine tubulin immunogold. The spindle pole region (MTOC) at left is a focus for spindle MT which radiate toward chromosomes (dark material). Insert upper right: kinetochore MT attaching chromosome. With permission from Geuens, Gundersen, Nuydens, Cornellisen, Bulinski, DeBrabander (1986).

although they don't appear to contact any structures. Rather, MT minus ends are somehow stabilized by the "pericentriolar material." This favors MT assembly and protects against disassembly by binding and capping the minus end of MT. Most tubulin is polymerized during interphase, however free MT not radiating from the MTOC occur near the periphery of the cell. Except for free MT, the network is rather stable with low turnover rates and minimal treadmilling and dynamic instability.

During "prophase", the cytoskeleton prepares for the separation of duplicated chromosomes so the cell can make a copy of itself.

DeBrabander (1985):

> It appears that nature selected a near fail safe microtubule based

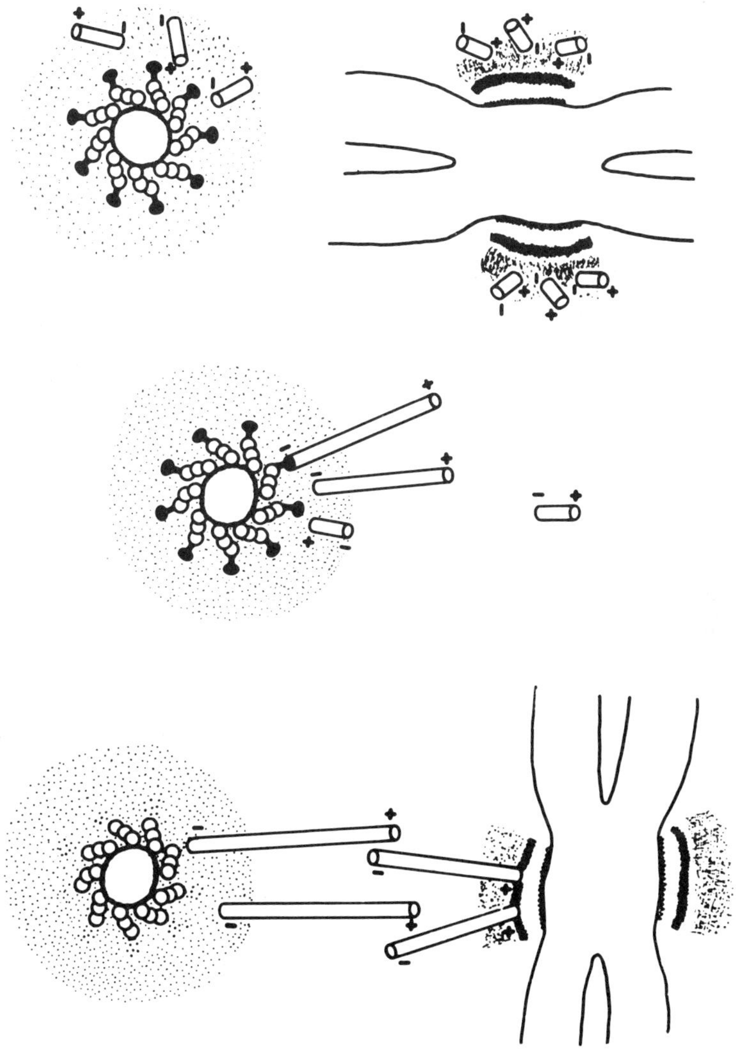

Figure 5.10: DeBrabander's model of MTOC action. Upper left: A centriole is surrounded by a dense material that lowers Cc, the critical concentration of tubulin required for microtubule assembly. At the beginning of mitosis, new MT are preferentially nucleated near the MTOC. Upper right: Kinetochore plates associated with chromosomes also have the Cc lowering material but anchor plus ends; new MT form. Middle: MT minus ends are stabilized by the pericentriolar dense material and plus ends grow outward. Free MT are unstabilized and remain short due to consumption of tubulin subunits by MTOC stabilized MT. Bottom: stabilized MT slide along one another to pull chromosomes apart. With permission from Marc DeBrabander (1982).

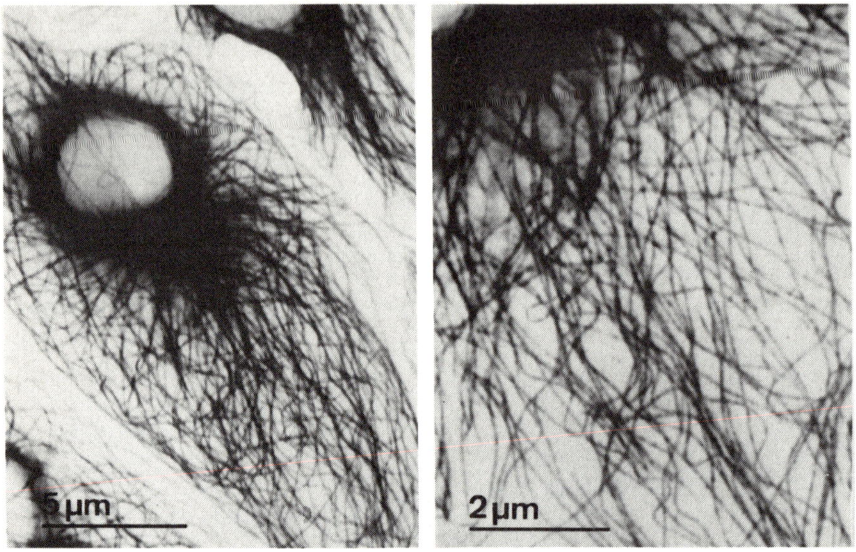


Figure 5.11: Microtubules in PtK2 cells illustrated by tubulin antibody at two different magnifications. Nucleus (left) is devoid of MT, which densely emanate from nearby MTOC. With permission from Geuens, Gundersen, Nuydens, Cornellisen, Bulinski, and DeBrabander (1986).

mechanism to allow the eukaryotic cell to handle more and more complex genetic libraries.

As the chromosomes condense in prophase, the MT system changes its display and turnover rate. Interphase MT are gradually shortened while new MT start to grow from the centrosome. At this point, the system is extremely sensitive to polymerization inhibitors. Drugs which are used against cancer are often mitotic inhibitors which prevent the rapid polymerization of prophase MT from MTOC. Kinetochores are mobile MTOC which attach chromosomes and bind MT at their *plus* end (Figure 5.10). MT assemble between centrosomes (which bind their minus end) and kinetochores (which bind their plus end) and separate the genetic material towards the daughter cell poles in the mitotic cycle. The delicate array of the two centrioles, connected MT spindles and "star-like astral projections" (MT which "overshoot" the centrioles) have suggested to many observers some type of electo-

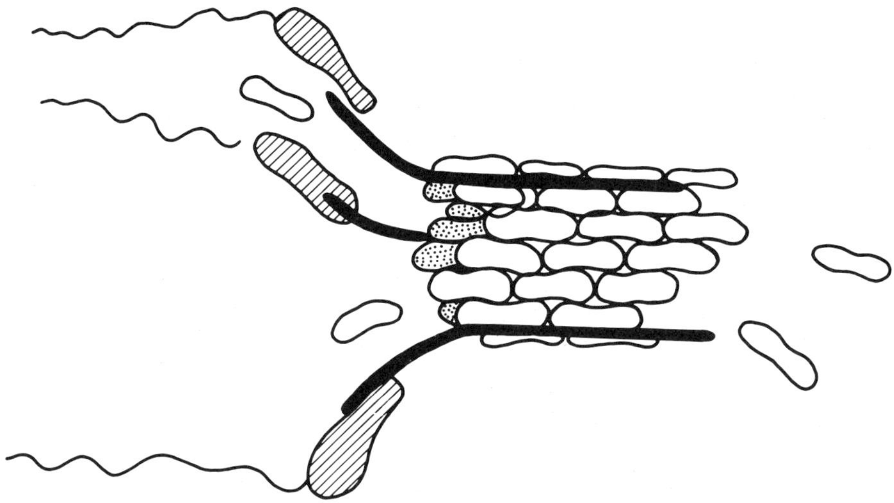

Figure 5.12: Schematic representation of a microtubule associated protein (MAP dark lines) facilitating assembly of a microtubule (MT). The longitudinal binding retards dissociation of subunits and stabilizes the MT structure. With permission from DeBrabander, DeMey, Geuens, Nuydens, Aerts, Willebrords and Moerman (1985).

magnetic field because of the resemblance to a magnetic field pattern. The contractile ring formed perpendicular to the axis of mitosis by the microtrabecular lattice establishes a cleavage furrow which separates the daughter cells. The chromosome material decondenses while a new nuclear membrane is zipped onto it, and cellular elements assume again an organized aspect. Later, the cells flatten and gradually reassume their normal interphase shape.

MTOC/centrioles are involved in orientation, shape, timing of division and growth. With MT and other cytoskeletal structures, MTOC/centrioles determine what cells do, when and how they do it, and what kind of cells they are! Are MTOC the command centers of the cytoskeleton? Chapter 3 described the significant position of centrioles and MT in evolution, and Chapter 8 will describe theories which explain centrioles' enigmatic capabilities. These include consideration (Bornens, 1979) that centrioles rotationally oscillate with a *gyroscopic inertia* which senses and establishes cell orientation to neighbor cells, tissue, environment, chemical gradient fields, gravity, and perhaps other factors. Centrioles seem to be a relatively immobile anchor around which other cytoskeletal elements move and are organized. Bornens suggests that centrioles communicate with actin filament networks by propagating conformational effects, and thus regulate cellular activities. Another centriole theory has been lodged by Northwestern University's Guenter Albrecht-Buehler (1977,1985), who proposes that centrioles detect signals (such as infra-red, ultrasound, microwave) propagating linearly throughout the cytoplasm. Albrecht-Buehler has considered the design principles for a nanoscale directional signal detector and finds that centrioles are appropriate. Centrioles, or their future hybrids, may have interesting technological applications (Chapter 10).

### 5.2.4 Microtubule Associated Proteins (MAPs)

The full range of MT functions is achieved by the actions of various proteins which bind in precise fashions at specific tubulin dimers in MT lattices. These ***microtubule associated proteins*** (MAPs) include electromechanical enzymes which generate force and movement, communicative crossbridges to other cytoskeletal filaments and organelles, MAPs which enhance MT assembly (Figure 5.12) and a class of MAPs within neurons whose functions are not understood.

In many cases, the attachment patterns of MAPs to MT lattice walls have a precise geometrical configuration which appears related to the function of the MAP-MT complex (Figures 5.13 thru 5.16). For example Allen (1979, 1985) and colleagues showed that smooth passage of vesicles in axoplasmic transport requires contractile MAPs spaced between 15 and 25 nanometers apart along MT lengths. Many other MAP binding patterns are spirals which wind around MT cylinders in super-helices, and several theories suggest how these sites can be determined. These include linear sequencing of varying tubulin isozymes along protofilaments (Kim, 1986), crystal symmetry describing each MAP terrain (Figure 5.17, Koruga, 1986) or by the lattice steps necessary to get from one MAP site to its nearest neighbor MAP. For example, electron micrographs of MAP-MT binding by Burns (1978) clearly show MAP binding which may be described as "over 3, up 1." That is, if the tubulin subunit to which the MAP is bound is the starting point, moving 3 monomers along the leftward row helix, and then up 1 monomer along the protofilament will indicate the next attachment site. A number of geometrical super-helical patterns have been observed which match other "chess-like" movement rules. These are shown in Figures 5.12–5.15. In other instances such as brain MAPs, attachment patterns are irregular and heterogeneous,

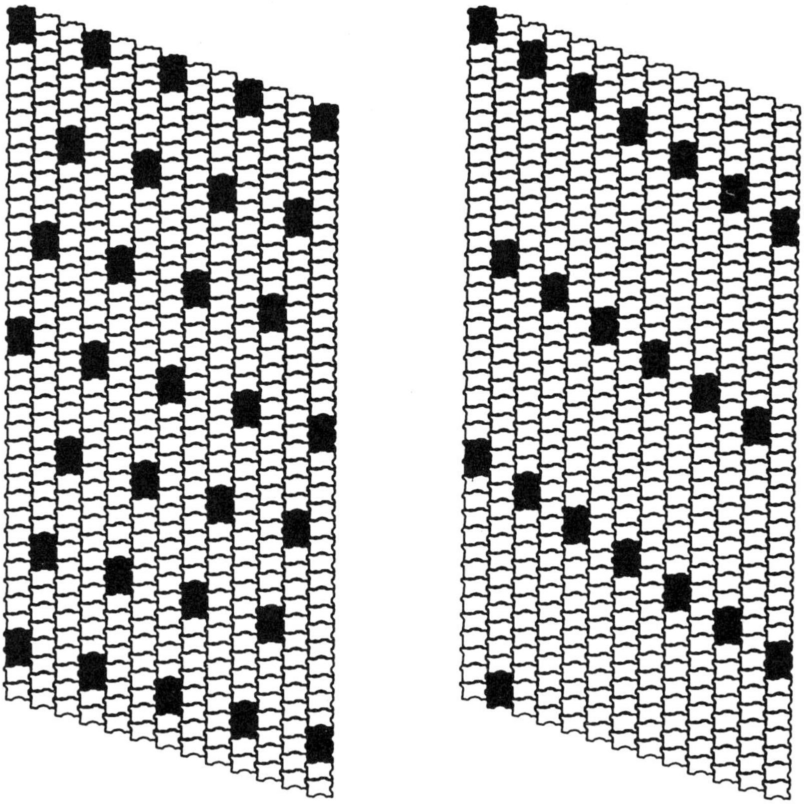

Figure 5.13: Patterns of MAP attachment to microtubules observed by electron microscopy. (Burns, 1978; Kim et al., 1986; and others).

and thus are capable of representing information.

Rod shaped bridges which occur between MT are of at least two sorts. Motion producing protein "arms" which consume ATP hydrolysis energy to generate force are analogous to myosin bridges of skeletal muscle. Moving arms attached to MT are made of proteins called kinesin and dynein. Dynein arms on MT which contract in organized sequences to produce collective movements were first described and characterized in cilia and flagella by Ian Gibbons (1968). The arms reside at periodic intervals along the outer MT within cilia and flagella. They possess "ATPase activity," that is they contract conformationally due to hydrolysis of ATP. Dynein arms contract in waves and mediate sliding between the adjacent outer tubules to drive

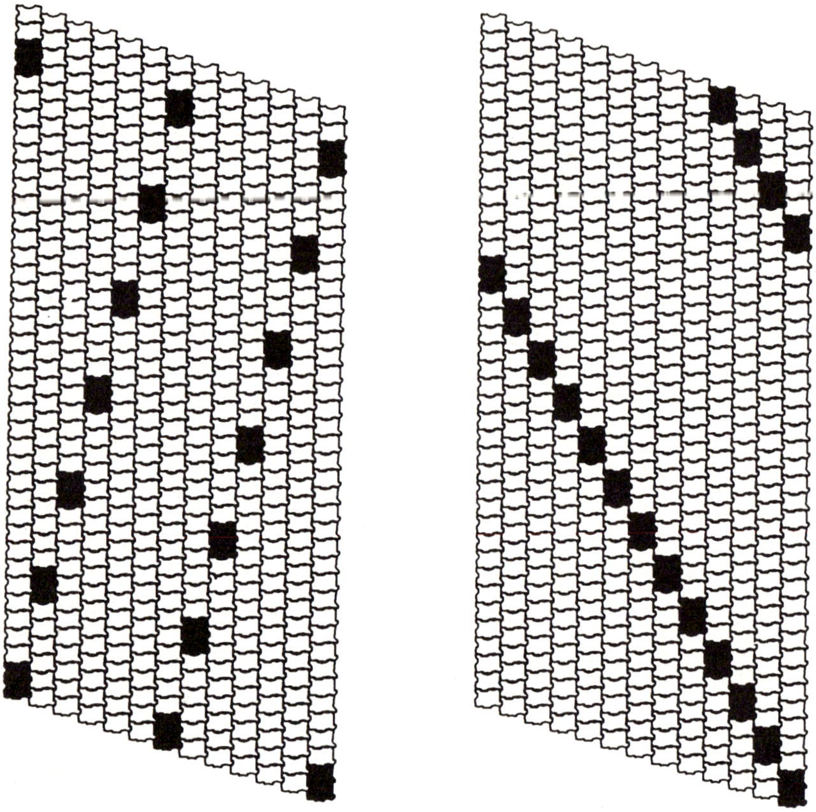

Figure 5.14: MAP attachment patterns.

ciliary and flagellar motility. Dynein arms attached to MT in neuronal axons act collectively to pass material in "bucket brigade" axoplasmic transport. These and other MT "bending sidearms" will be discussed in Section 5.5.2.

Other MAPs are bridges between parallel MT, filaments, organelles and the "microtrabecular lattice" and appear to integrate microtubules with the rest of the cytoplasm (Figure 5.6). MAPs from mammalian brain neurons have been characterized into three main groups. These are high molecular weight (greater than 100 kilodalton) MAP 1 and MAP 2, and a number of closely related 56–62 kilodalton proteins designated as "tau." Studies in rats show an increase in number and an altered distribution of the various forms of tubulin, high molecular weight MAPs, and tau which correlate with rat brain development and learning. Both MAP 2 and tau are heat stable, stim-

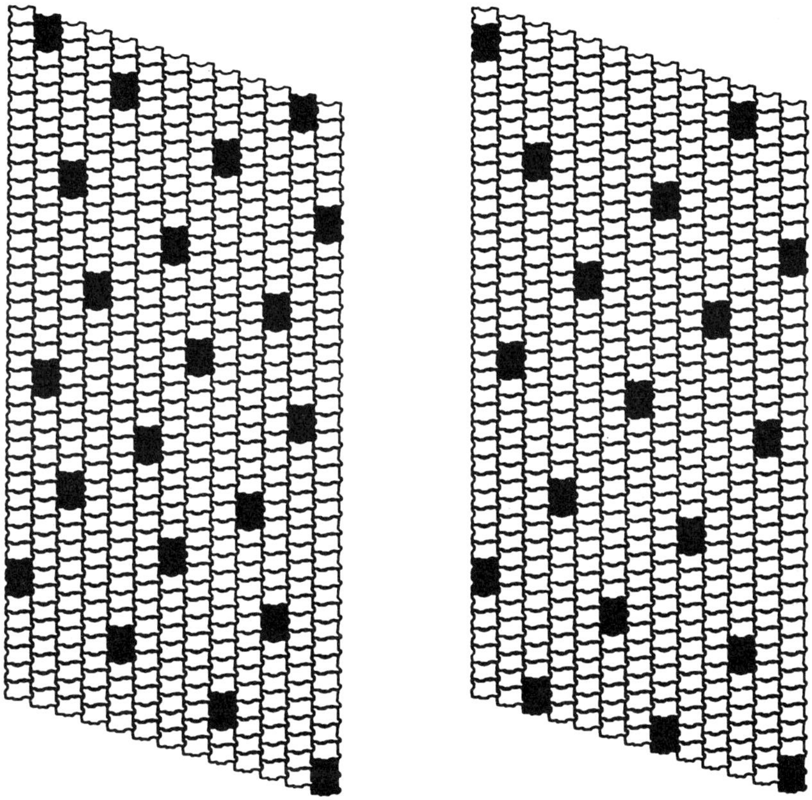

Figure 5.15: MAP attachment patterns.

ulate MT polymerization (lower Cc) and share binding domains on intact MT; they are heterogeneous protein families whose functions are unknown. MAP 2 is concentrated in cell bodies and dendrites of neurons, found in a ratio of MAP 2 to tubulin of one to twelve, occurs in only small amounts in axons and is absent in glia. Conversely, brain tau is largely confined to axons. Both MAP 2 and tau are phosphorylated by cyclic AMP dependent protein kinases, proteins which amplify effects of calcium ions. Intracellular levels of calcium ions, in turn, are regulated by diverse extracellular signals including neurotransmitters, hormones and electrical factors to regulate cell shape, motility, secretory processes and other functions. Thus these MAPs can mediate diverse inputs into the cytoskeleton. Each MAP attached to an MT may function like a "synapse" in a parallel, laterally interconnected

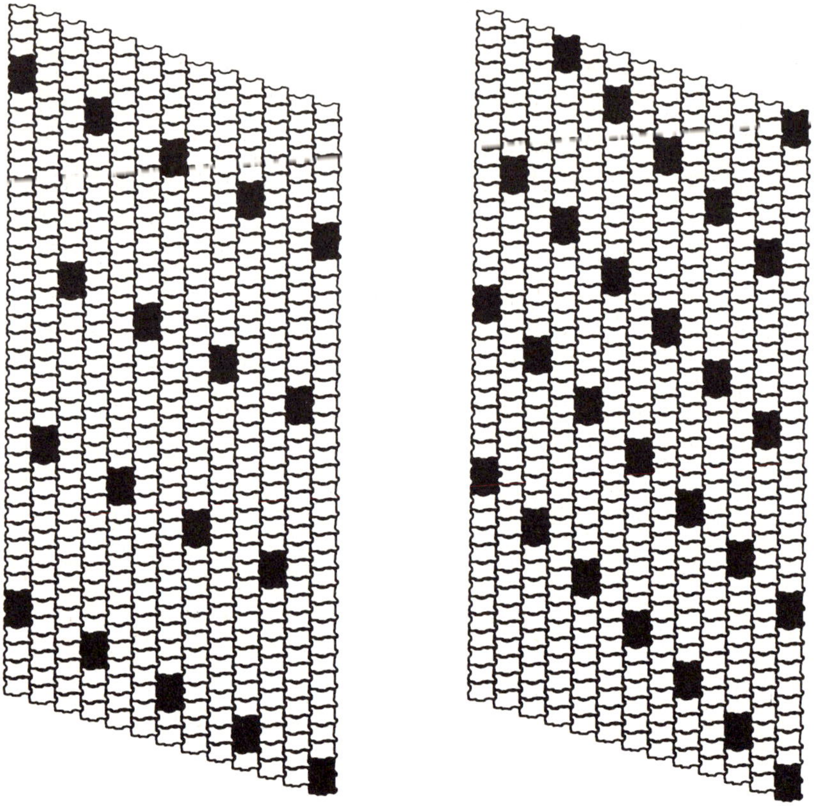

Figure 5.16: MAP attachment patterns.

network.

Tau heterogeneity varies during brain development and these changes result in the predominance of different tau polypeptides in mature cerebral cortex and cerebellum. Binder, Frankfurter, and Rebhun (1986) propose that MAP 2 binding sites result from genetically determined forms of tubulin which are predominant in neural cell bodies and dendrites, while the tau binding domains are specified by axonal tubulins. They suspect that different tau species appearing during development may herald the cytoskeletal differentiation of unique axonal subpopulations whose MT are committed to slightly different tasks. Variability and responsibilities of MAPs may be extensive. They are functional appendages, structural and communicative links ("arbiters"), and essential integrators of cytoskeletal and cellular func-

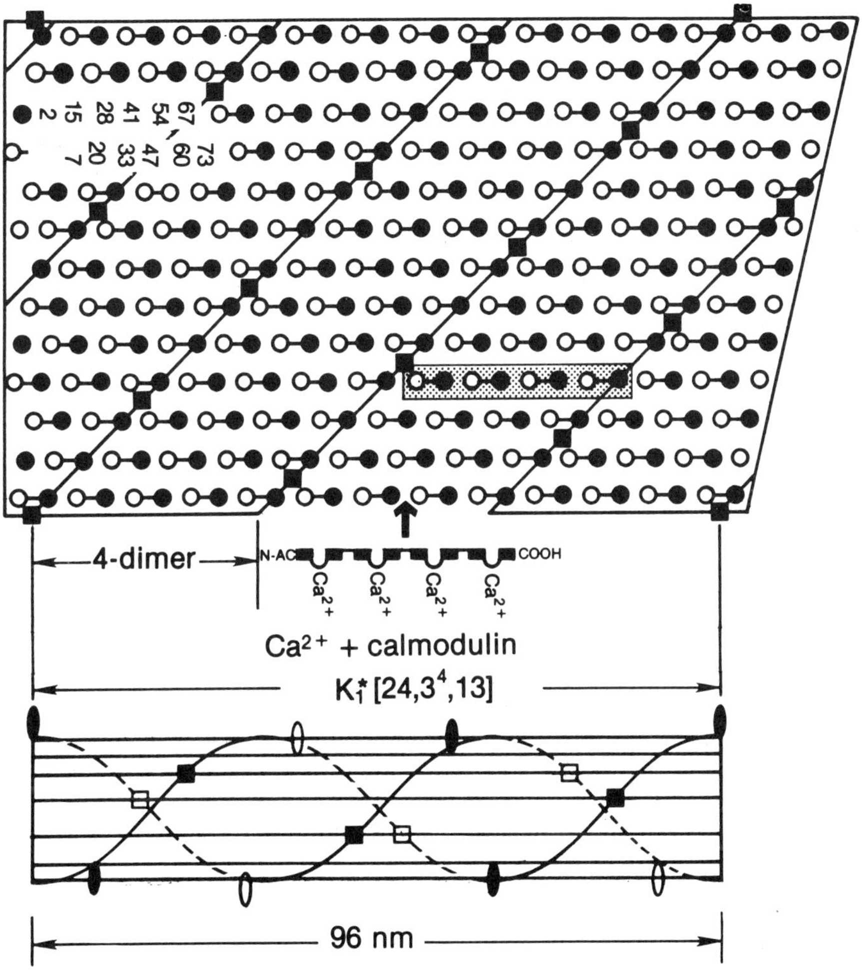


Figure 5.17: MAP attachment patterns as part of Koruga (1986) microtubule information code (Chapter 8). With permission from Djuro Koruga.

tion.

## 5.3 Intermediate Filaments

The major filamentous components of the cytoskeleton are MT, actin filaments, intermediate filaments (IF), and a class of delicate interconnecting fibrils called the microtrabecular lattice (MTL) which will be described later in this chapter. Here the "unknown" members of the cytoskeleton, *in-*

*termediate filaments* are reviewed (Lazarides, 1980). Intermediate filaments represent the most nebulous and chemically variable subgroup among the cytoskeleton. Five classes of IF have been distinguished which are built from polypeptides containing mostly alpha helix rod domains (Chapter 6). Under the electron microscope, IF appear as relatively featureless, 8–12 nanometer wide unbranched filaments. Treatment with certain fixative agents can cause IF to unravel into several 2–3 nanometer protofilaments, or 4–5 nanometer protofibrils whose number may vary from filament to filament, and even along the same filament. IF may form from parallel coiling of the alpha helix domains and under some circumstances, can form a polygonal meshwork with a 52 nanometer repeat of 8–10 nanometer wide filaments. These can also aggregate into crystal-like arrays with 24 nanometer spaced transverse bands. As such, IF may be involved in structures otherwise described as microtrabecular lattice, or cytomatrix.

Lazarides (1980) has shown that different types of IF associate with specific types of cells. For example, subunit structure defines five major classes of IF: 1) keratin ("tonofilaments"), which are found in epithelial cells, 2) desmin filaments predominantly found in smooth, skeletal and cardiac muscle cells, 3) vimentin filaments, found in mesenchymal cells, 4) neurofilaments, found in neurons (Figure 5.18), and 5) glial filaments, found in glial cells. Often two or more of these classes co-exist in the same cell.

Neurofilaments appear to function as a three dimensional structural lattice providing tensile strength to axons. Extruded axoplasm is a highly structured gel rich in neurofilaments; exposure of the gel to calcium ion results in degradation of neurofilaments and conversion of the gel to a more watery sol state, a phenomenon generally attributed to dissolution of actin filaments.

All IF including neurofilaments appear to be phosphorylated although the function that this could serve is not understood. IF are a distinct class separate from MT and actin filaments, and Lazarides argues their biochemical and morphological properties indicate they are involved in mechanically integrating the various components of the cytoplasmic space. IF remain poorly understood; their dense presence in neurons is mysterious, perhaps they participate in some way in the cognitive functions of the nervous system. In Chapter 8, Barnett's theory of neurofilaments as "string transform" memory banks coupled to MT will be described.

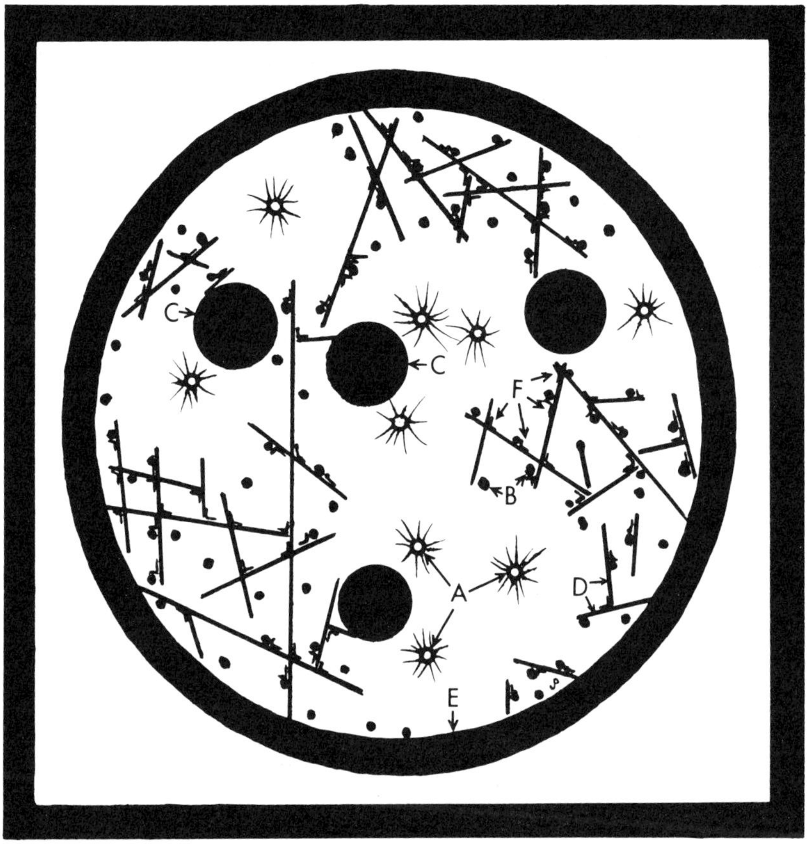


Figure 5.18: Cross section of small nerve axon A. Microtubules with radial MAPs. B. Neurofilaments, outnumbering microtubules about 10:1. C. Vesicle being transported by microtubule axoplasmic transport. D. Crosslinked actin/neurofilament gel (microtrabecular lattice). E. Myelin layers. F. Filamin bracketing actin. By Paul Jablonka.

## 5.4 The Cytoplasmic Ground Substance

MT and IF are not the finest texture of cytoplasmic organization. Smaller, more delicate structures branch and interconnect in "gel" state networks which comprise the substance of living material.

Three distinct cytoskeletal component systems have been well characterized: MT, IF and actin filaments. Actin is the most versatile component. In conjunction with other proteins it can polymerize in string-like filaments, form dynamic branching nanoscale meshworks or geodesic polyhedrons. Even more evanescent than labile MT, assembly of actin and associated proteins create transient configurations of cytoplasm for specific purposes. The roots of intelligence may well be grounded in dynamic cytoplasmic expressions such as contractile rings which divide the cytoplasm in cell division, probing lamellipodia, and dendritic spines and synapses in neurons.

The nature and structure of the "ground floor" of organization, the cytoplasmic ground substance, has been explored from a number of orientations resulting in overlapping descriptions. These include the microtrabecular lattice, cytomatrix, and cytoplasmic solid state.

### 5.4.1 The Microtrabecular Lattice (MTL)

New techniques in electron microscopy developed by Keith Porter and colleagues (1981) at the University of Colorado at Boulder have led to observation of an irregular three dimensional lattice of slender strands throughout the cytoplasm, interconnecting nearly everything in the cell. The interlinked filaments appear to suspend the various cell systems, organelles, and larger cytoskeletal elements such as microtubules and filaments with a matrix material continuous with the individual lattice filaments. The lattice is suggestive of the trabecular structure of spongy bone and it was named the microtrabecular lattice (MTL, Figure 5.19).

Porter and colleagues took advantage of the properties of the high voltage electron microscope at Boulder. This massive device is capable of accelerating electrons across a potential drop of a million volts, ten times that of standard electron microscopes. The extra voltage gives the electrons sufficient energy to penetrate thick specimens or even intact cells up to several thousand nanometers thick. Previously, cells had to be sliced into sections thinner than 200 nanometers, and artifacts due to cell destruction were more prevalent. The high voltage electron microscope provides more information about depth, giving a three dimensional view of cell organization. Another innovation, the critical point drying method (Ellisman, 1981) avoids the distorting effect of surface tension which causes cells to collapse when they are dried in air. High voltage electron microscopy and critical point drying have now demonstrated the MTL in all eukaryotic cells which have been examined.

The MTL is a three dimensional lattice of slender strands called microtrabeculi. They range in diameter from 4 to 10 nanometers with lengths of 10 to 100 nanometers. The MTL is a finely organized meshwork that divides the ground substance into two phases, a protein rich polymerized phase comprising the MTL, and a water rich fluid phase that fills the intratrabecular spaces. At high magnification, microtrabeculi of the ground substance are seen to crosslink many elements in the cytoplasm. For example, they connect microtubules with the smooth endoplasmic reticulum. The MTL not only crosslinks, but appears to be dynamically active in moving elements through the cytoplasm. The dynamic MTL moves material in a saltatory manner at velocities equivalent to axoplasmic transport. A co-pioneer with Porter in the observation of cytoplasmic ground structure, Mark Ellisman (1981) of the University of California at San Diego has observed that MT and filaments are analogous to rigid skeletal elements such as bone, and that the microtrabecular lattice may be appropriately equated with work producing components such as skeletal musculature. Thus Ellisman suggests a more appropriate description of the MTL might be "cytomusculature."

In neurons, wispy MTL strands radiate from MT and neurofilaments at right angles into the cytoplasm, often interconnecting MT with other MT and neurofilaments with neurofilaments. At the synapse, the MTL appears intimately involved in both neurotransmitter release and post synaptic receptor mechanisms. Presynaptic terminals demonstrate notable MTL crosslinking among synaptic vesicles and among vesicles and synaptic membranes. In post synaptic dendrite MTL, linkages are visible among filamentous subsynaptic densities and the other elements of dendritic cytoplasm. Where axonal cytoplasm enters synaptic regions, transitions occur in the form of the MTL. In axons, cross linkages are clearly exhibited between MT, filaments, and other components. However, in presynaptic terminals and dendrites, the MTL network is more similar in structure to isolated actin-myosin gels observed in non neuronal cells. Raymond Lasek (1981) of Case-Western Reserve University has shown that the cytoplasm of nerve axon growth cones, which differs from axonal cytoplasm, is rapidly converted from the axonal variety to the growth cone variety in axons severed from their cell bodies. Thus "local" factors appear able to transform axoplasm to synaptic terminal cytoplasm, or axoplasm to growth cone cytoplasm. The MTL is the fine structure and texture of cytoplasm.

MTL activities appear to be dependent on calcium ion concentration. Ellisman's work shows that high calcium exposure causes the MTL to shorten and deform, leaving free ends or "nubbins." In normal axoplasm, different forms of MTL corresponding to both high and low calcium ion concentration

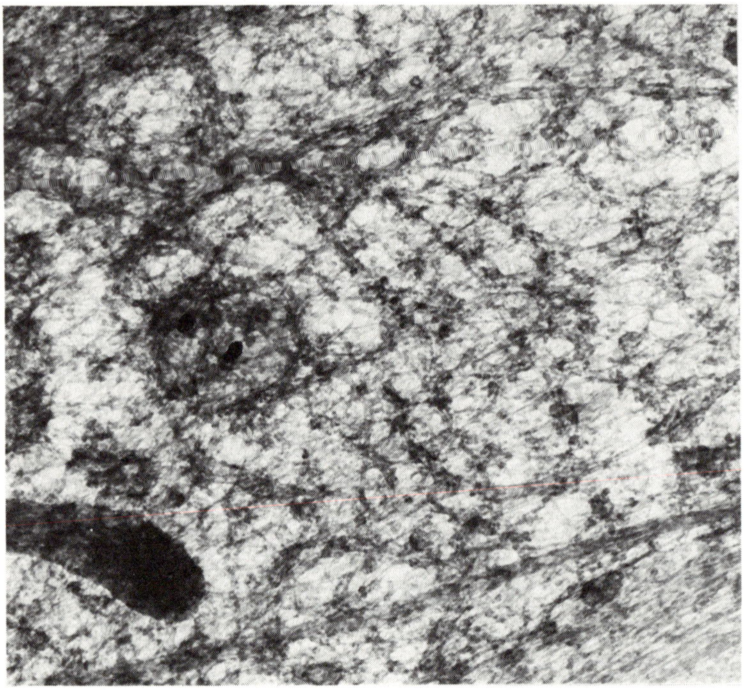

Figure 5.19: Microtrabecular lattice (MTL) in PtK2 cell cytoplasm. Microtubules can be seen (one rising left to right on bottom) in the background of microtrabecular lattice (MTL). Two 40 nanometer gold particles are trapped in a vesicle within the MTL. With permission from DeBrabander (1985).

appear to be found concurrently. These may correspond with "sol" and "gel" states determined by other techniques. Ellisman suggests that calcium may be the coupling agent for the dynamic aspects of the MTL in terms of location, extent and form of crossbridging of cellular constituents. Calcium regulated by cytoskeletal structures, or membranes, and the MTL itself may be locally dynamic and patterned. Standing waves, dissipative patterns, or holograms "hardwired" in MTL could be representations of calcium coupled states in cytoplasm.

Ellisman raises the question of the importance of the cytoskeleton and MTL in the nervous system. One consideration is modulation of membrane related events such as receptor activity and neurotransmitter extrusion. Ellisman proposes that the MTL is triggered by calcium ions to release neurotransmitter vesicles. In addition, the MTL participates in other cytoskeletal functions also involving MT and filaments. These are axoplasmic transport, turnover and maintenance of membrane proteins and receptors, and the availability of neurotransmitters and enzymes at synapses. Ellisman has predicted some of the functions and/or behavior of the cytoskeleton and MTL within the nervous system. These are: 1) the MTL is a substrate to maintain cell shape changes including synaptic formation, and to regulate excitable properties of neurons. One example would be changes in distribution and volume of dendritic spines and synapses with implications for synaptic connections and perhaps learning and memory. Thus the MTL and other cytoskeletal elements can modify synaptic function by participating in trophic maintenance and turnover of membrane proteins to modulate membrane excitability and neuronal signaling. 2) The MTL, according to Ellisman, can also buffer small molecules and ions, maintaining these in specialized locations and patterns for metabolic and electrical functions. Thus the MTL may compartmentalize the cell, forming regions whose environments may vary. Buffering and control of calcium ion flux may be particularly important, and directly relate to cognitive functions. 3) The MTL can differentiate specific zones of cytoplasm and membrane, for example controlling the types of receptors or channels at synaptic zones by specific cytoskeletal linkages. The MTL and cytoskeleton also control the distribution of organelles, for example keeping the endoplasmic reticulum from entering axons, restricting the Golgi apparatus to perinuclear zones and keeping the synaptic bouton the right composition of axoplasm. 4) The MTL can mediate embryological development or morphogenesis through linkages with specific hormone receptors or tissue factors resulting in variations in developmental patterns. 5) The MTL can transduce chemical or mechanical work for intracellular transport and processes such as axoplasmic transport and translocation of synaptic vesicles to release sites on the presynaptic membrane. To Ellisman, the MTL regulates the rest of the cytoskeleton and the cell at large. In his view, it is the dynamic ground substance capable of intelligent behavior. One could argue, however, that MT regulate the MTL. What is most significant is the question of how they communicate.

The most labile and transient of the cytoskeletal levels of organization, the MTL is the current microfrontier of living material organization. The MTL is a network within a network of cytoskeletal proteins which, in the case of the nervous system, is a network within a network of neurons. In Chapter 8, a model of information processing will be discussed in which the MTL represents standing wave patterns of calcium coupled sol-gel states resulting from dynamic excitations of the cytoskeleton. Coherent excitations in the cytoskeleton could result in "holographic" standing waves which may be the bottom level of an information hierarchy: "infoplasm." An analog picture in which the MTL is both paint and canvas could be the texture of consciousness.

### 5.4.2 The Cytomatrix

Others have viewed the fine structure of cytoplasm and described it in different terms. Among these are Peter Satir (1984) of Albert Einstein University whose work has focused on the protein makeup and functional organization of the MTL, or "cytomatrix." Satir sees a functional integration of the cytomatrix, noting that protein-protein interactions have geometrical as well as biochemical and biophysical consequences. Certain interactions occur only at the ends of cytoskeletal polymers, others require or produce parallel arrangements of such elements by interacting only with their lateral surfaces, and still others may require orthogonal arrangements. In concert with centrioles, MTOC, and MT, cytoplasmic structures appear and function where and when needed.

The primary building block of the cytomatrix/MTL appears to be actin, a ubiquitous protein whose versatility describes the very nature of cytoplasm. Interaction with a variety of other proteins ("actin-binding proteins") unleash actin's full capabilities. When crosslinked by proteins such as fimbrin, actin can form rigid bundles which provide structural support. When associated with myosin (a "mechanoenzyme"), useful muscle-like contraction occurs. Also important are proteins such as talin, spectrin, vinculin, ankyrin and fodrin which connect the cytomatrix with membrane proteins, and calmodulin which mediates effects of calcium ion fluxes on the cytomatrix.

Proteins such as alpha actinin, troponin, and filamin link actin in networks which cause a gelatinous consistency to cytoplasm—a "gel" (Figure 5.20). In the presence of calcium ions and actin fragmenting proteins such as villin and gelsolin, the gel network liquefies to a solution—"sol." Other actin regulatory proteins include actin capping proteins, which stabilize polymerized actin and promote a "gel" condition, and profilin which binds actin subunits, prevents polymerization and maintains "sol" conditions. The layer of cytoplasm immediately below cell membranes is in a continuous gel state ("cortical gel"), but elsewhere dynamic transitions can occur. In the presence of myosin, actin gels can undergo contraction and streaming which are important in cytoplasmic movement such as amoeboid locomotion. The

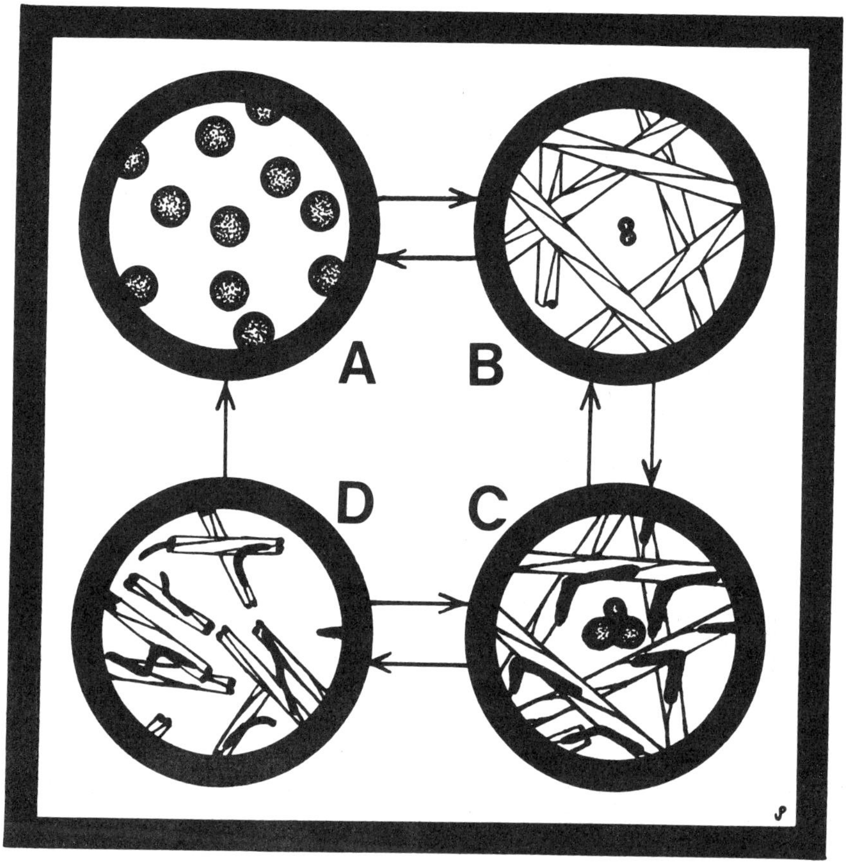


Figure 5.20: Components of the cytomatrix. Assembly of actin and related proteins form cytoplasmic "gel" states. A) monomeric actin, B) assembled actin filaments which are helical pairs of monomeric strands; center-cross section of actin filament, C) the presence of filamin (dark strands) causes actin cross-linking and dense "gel state," D) calcium and/or actin fragmenting proteins causes liquification to a "sol state." By Paul Jablonka.

myosin appears to pull actin filaments against each other. A rise in calcium ion causes a sharp drop in viscosity of actin networks, through actin fragmenting proteins such as gelsolin. The same rise in calcium activates myosin to pull actin filaments against one another, resulting in gel contraction and vigorous streaming in adjacent "sol" regions. George Oster (1984) of the University of California at Berkeley has described a mechanism for amoeboid movement in which waves of calcium ion trigger transient sol/gel regions. The net flow of material is towards the calcium flux, which serves to pull the cell onward.

Cytomatrix structures give rise to amoeboid and other types of cytoplasmic movement as well as polyhedral structures. Calcium mediated sol-gel transition and gel contraction can be important in many cell functions, including perhaps the transient representation of short term information and memory within neurons.

### 5.4.3 Cytoplasmic Solid State

Molecules such as proteins and RNA are heterogeneously distributed throughout cytoplasm and passively flow at rates below that expected for normal diffusion. Actively transported molecules such as those carried by axoplasmic transport travel at velocities far in excess of diffusion. Initially, the slow diffusion of biomolecules which are not actively transported was explained on the basis of cytomatrix barriers and channels. However, Gershon, Porter and Trus (1985) studied molecular diffusion through cytoplasm and came to a different conclusion. They found that the cytoskeleton/cytomatrix/MTL comprised from 16 to 21 percent of cytoplasmic volume, and that the cytoskeleton/cytomatrix/MTL *surface area* was from 69,000 to 91,000 square microns per cell (69 to 91 billion square nanometers). Their data has been interpreted to suggest that the cell "solid state" (cytoskeleton/cytomatrix/MTL) is *not* a barrier to diffusion since the aqueous phase occupies 4/5 of cell volume. They conclude that proteins and other molecules are dynamically bound to the solid state, accounting for the slow diffusion (Figure 5.21).

Biologist James Clegg (1981) has discussed how the binding of "soluble" enzymes to the solid state could account for the efficiency of various enzymatic processes. A sequence of enzymes in a complex biochemical pathway is much more efficient if physically arranged so that a cascade of reaction products occurs. If the product of one enzyme is the precursor for its neighbor enzyme, transfer time is minimized and the reactions are usefully facilitated. This also allows the possibility of dynamic regulation of these enzymes by the solid state structures. Clegg has also accrued evidence regarding the role of water within cytoplasm, particularly surrounding solid state structures. Using neutron diffraction and other techniques, Clegg has found that water adjacent to the cytomatrix is "ordered," that is aligned with polar bonds on the protein surfaces. Thus a layer of ordered water extends at least 3 nanometers from the billions of square nanometers of solid state surface area within each cell. This ordered water may be coupled by dipole oscillation to dynamics of the solid state, inhibit the thermal dissipation

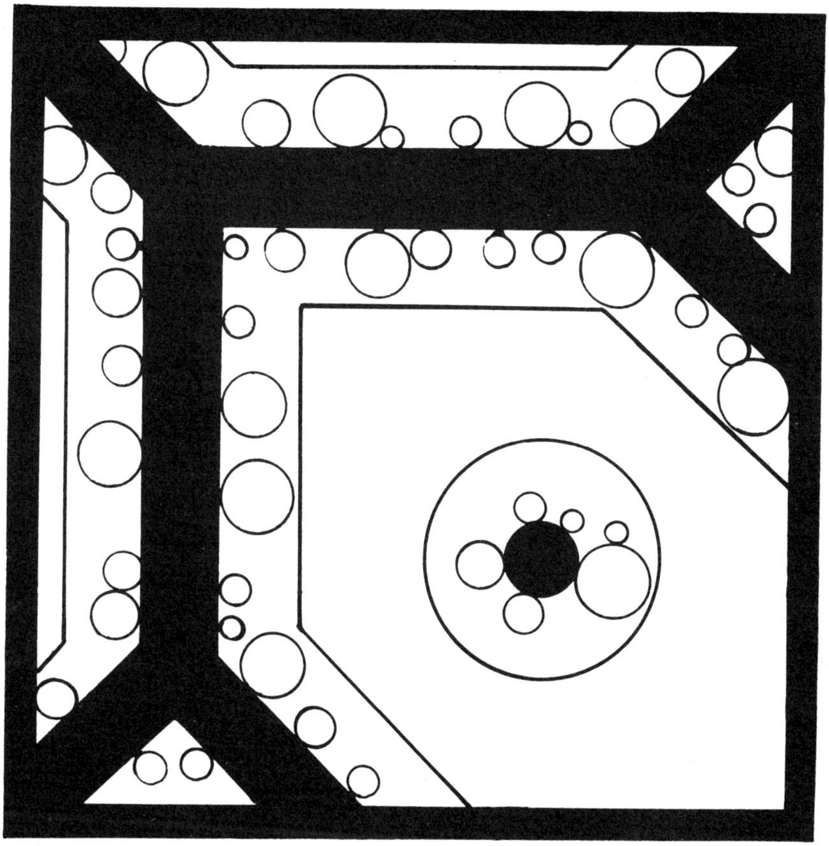

Figure 5.21: "Soluble" intracellular enzymes (circles) governed by the microtrabecular lattice (MTL: solid dark) and bordering region of ordered and vicinal water (within thin line). The dark circle is part of a cross sectional view. Clegg (1981) has proposed that dynamic activity within the MTL can regulate the enzymatic activity. By Paul Jablonka.

of protein oscillation energy from within the solid state, and shield calcium and other ions from random solid state interactions. Clegg also suggests that additional layers of slightly less ordered water ("vicinal water") extending further from the solid state would limit true "aqueous" water within cells to narrow cellular "sewage channels."

Is the MTL/cytomatrix the *final* microfrontier of biological organization? Current imaging techniques would not reveal smaller levels of organization, just as it took advances in electron microscopy to discover first the MT cytoskeleton and then later the MTL. Perhaps smaller, more intricate networks ranging down to the level of single peptide filaments might be present. With layers of ordered water and charged ions, such a solid state "infoplasm" could occupy nearly the entire volume of biological material.

## 5.5 Cytoskeletal Motility

Albrecht-Buehler (1985):

> Cell movement ... appears to be determined by some kind of chemical computer, the nature of which is beyond our present understanding.

Cell motility was discovered by van Leeuwenhoek who observed flagella propelled sperm cells swimming in the 17th century. A variety of cytoplasmic movements include "amoeboid" creeping over a surface, internal streaming, axoplasmic transport, muscle contraction, ciliary and flagellar bending, and cell shape changes. These maneuverings are all engineered by a relatively small group of proteins whose net collective effects can be quite spectacular. For example, lymphocytes and macrophages are white blood cells which combat infection by migrating from the blood stream into an open wound to engulf invading bacteria or foreign material. Cells of developing embryos perform precisely choreographed movements that give rise to different tissues. Internal movement such as the streaming of cytoplasm, secretion of cell product vesicles (i.e. neurotransmitters), engulfment of matter, and the separation of paired chromosomes in cell division are routine functions whose complexity, organization, and precision generally boggle biologists.

Also, well controlled muscle contraction which depends on the conformational bending of myosin head molecules can result in strenuously running a 4 minute mile, or delicately painting a Mona Lisa. Four categories of cytoplasmic movement and force generation will be considered with attention to their regulatory mechanisms. These are: cytoplasmic probing, bending sidearms, ciliary and collective movements, and geodesic tensegrity structures.

### 5.5.1 Cytoplasmic Probing

In the early 20th century, W. H. Lewis, an embryologist at Carnegie Institute of Washington, took time lapse motion pictures of cells grown in culture (Lazarides and Revel, 1979). Lewis treated the fibroblasts (active cells which repair wounds) with protein cleaving enzymes to break up adhesive contacts

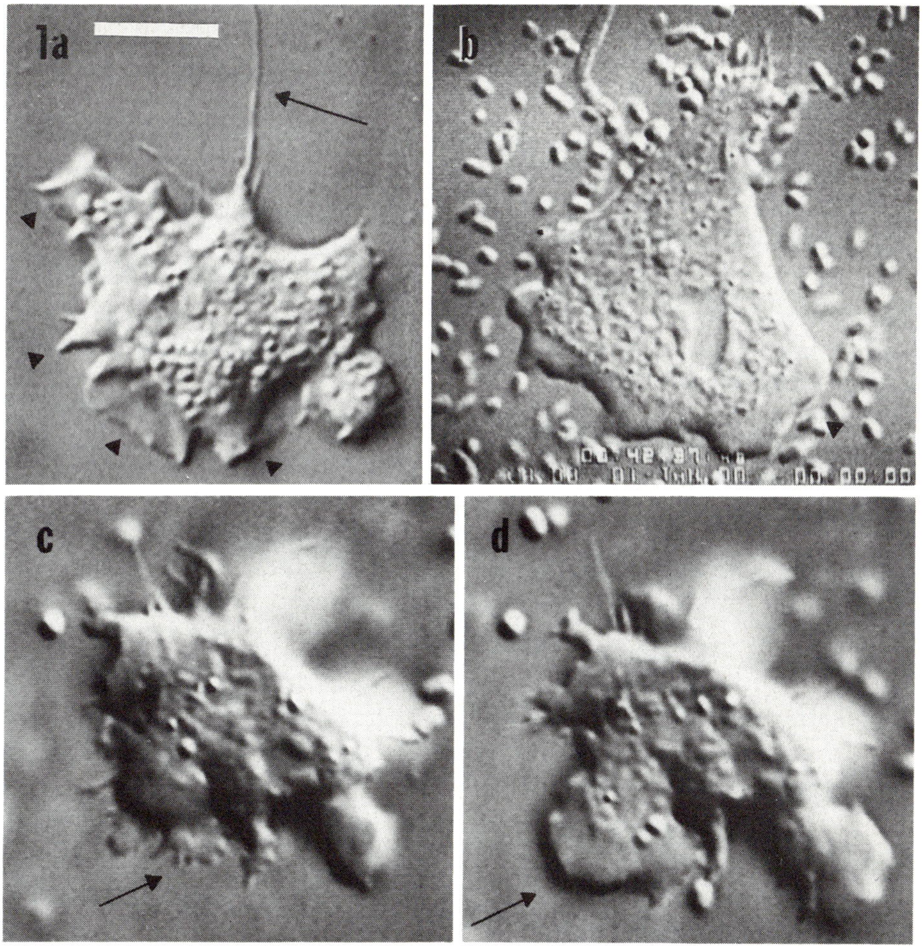


Figure 5.22: Amoeboid movement in a time-lapse series by U. P. Roos at the University of Zurich. 1a) Probing lamellipodia (arrow-heads) in direction of cell movement. Retraction fiber (arrow) shows trailing end of cell. b) Cytoplasm has flowed to fill out lamellipodia. c) Finger-like filopodia (arrow) probe direction. d) Cytoplasm advances, moving cell. With permission from U. P. Roos (1984).

and networks of extracellular material holding the cells together. He separated the resulting clumps and grew individual cells in a nutrient medium. After a few hours, Lewis' fibroblasts had acquired a polygonal shape. Along the shortest side of the polygon were lamellipodia, delicate sheet-like extensions of the cytoplasm. He exclaimed: "the ruffled lamellipodia slowly bend back and forth like the ruffles of a dress in a slight breeze" (Lazarides and Revel, 1979). Later, Ingram and Abercrombie at University College London studied the membrane coated ruffled lamellipodium, fibroblasts' main locomotion organ. The lamellipodium forms strings and contacts with the substrate on which the cell rests, stretching portions of the rest of the cell among several temporary adhesions. When the cell is spread out to its fullest extent it maintains one or two primary ruffles, each of which tend to lead the cell in a particular direction. The cell then migrates with changes in direction which typically come at intervals of several hours. If a ruffle does not stick and the cell move in another direction, the ruffle folds back on the upper surface of the cell, collapses onto it and disappears into the cell (Figure 5.22, 5.23 and 5.24).

Using the electron microscope, Lazarides and Revel (1979) described the lamellipodium as a somewhat irregular sheet of cytoplasm about one tenth of a micrometer (100 nanometers) thick. It is decorated along the edge by small protrusions resembling stubby fingers which attach themselves to the surface of the dish, and can quickly extend several times their initial length into rod-like filopodia. These are highly dynamic structures that can change from a fluid state to a rigid one in less than a second by virtue of the sudden assembly of actin filaments. Growing from the cell surface, they provide spatial guides between which the cell membrane and cytoplasm flows to form the ruffle. The filopodia either attach themselves to the substrate and become rigid, or melt back into the cell. The cell adheres to the surface below it at only a few distinct sites, rather like a hand contacting a table top only with finger tips. Permanent contacts of the ruffle with substrate result in adhesion plaques; as the cell moves the plaques are continuously formed and broken. In addition to their role in cell adhesion and motility, the ruffles and particularly the filopodia serve a sensory function. In the developing embryo when the filopodium of one migrating nerve cell makes contact with the surface of another, the cells stop moving toward each other. Similarly, when the filopodium of a cell spreading in tissue culture touches the surface of a cell that is already flattened, the membrane of the first cell will flow in a direction opposite to the point of contact. This type of sensitive regulation of growth processes is defective in malignant cancer cell.

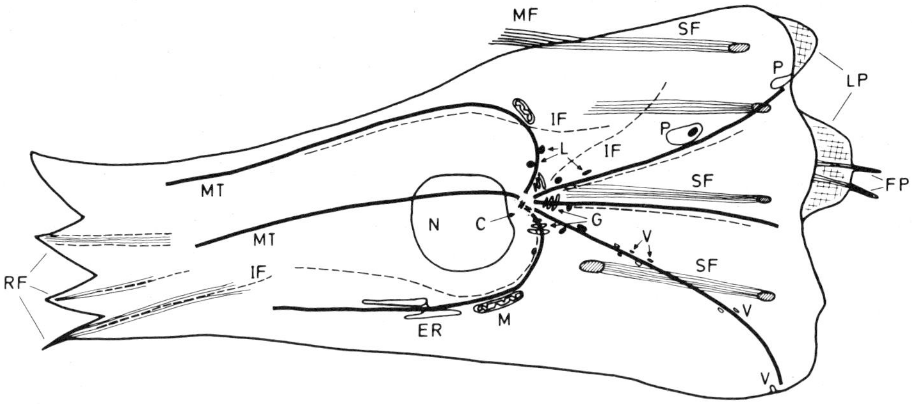


Figure 5.23: Cytoplasmic organization of a cultured cell moving to the right. Microtubules (MT) radiate from the centrosome (C), located near the nucleus (N), Golgi elements (G), and lysosomes (L). Mitochondria (M) and endoplasmic reticulum (ER) are aligned along MT. Phagocytic and pinocytic vacuoles (P) and secretory vesicles (V) move along MT between the centrosome and periphery. Intermediate filaments (IF) roughly parallel MT. Stress fibers (SF) are the cytoplasmic anchors for extracellular matrix fibers (MF) which contact the substrate. At the leading edge of the moving cell, lamellipodia and filopodia, comprised of actin bundles and meshwork, probe the future. At the trailing end, retraction fibers (RF) contain actin bundles and intermediate filaments. With permission from Marc DeBrabander (1982) and Janssen Pharmaceutica Research Laboratories.

Northwestern University biologist Guenter Albrecht-Buehler (1977, 1980) has studied isolated cytoplasm and beholds intelligent behavior. Examples include the movements of single cell organisms alternating between tumbling and smooth motions (Adler, 1969), swimming of motile bacteria in chemotactic environments (Macnab and Koshland, 1972), backward swimming of paramecium after collision with an object, and the amoeboid directed locomotion of cultured 3T3 mouse cells. Albrecht-Buehler classifies three major types of amoeboid motion: 1) force generation (contraction of an actin-myosin complex, assembly-disassembly of filament polymers, dynein-MT sliding, filament rotation), 2) transmission of motile forces to the framework of a cell (cytoskeleton in general, anchorage of fibrous elements to the cell surface in particular) and 3) generation of traction (adhesion plaques,

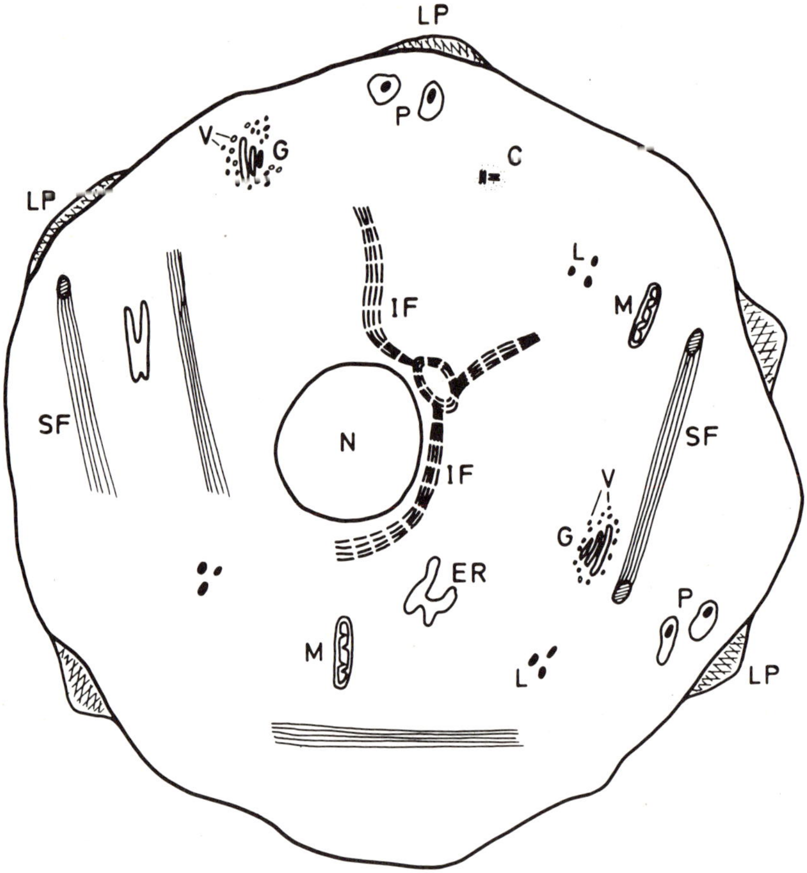


Figure 5.24: Cell treated with microtubule assembly inhibiting drug. The cell has lost its polarity. Lamellipodia form around the periphery of the "blob"ular cell. The centrioles (C) are separated from the nucleus. Golgi elements (G), lysosomes (L), mitochondria (M) and endoplasmic reticulum (ER) are disorderly scattered throughout the cytoplasm. With permission from Marc DeBrabander (1982) and Janssen Pharmaceutica Research Laboratories.

extracellular matrix, basement membranes).

Albrecht-Buehler further divides the *control* of cell migration into three subtopics: 1) generation of a front-rear axis (polarity involves MTOC and MT), 2) cell body coordination—how the forty billion protein molecules comprising each cell act cooperatively to move as a unit, and 3) logic of migration—the rules of changing direction. Migrating cells make decisions and globally assess their environment. Albrecht-Buehler asserts that the control of amoeboid cell migration provides evidence of cytoplasmic intelligence because it requires coordination of many cell domains and navigation which involves assessment of the environment.

But the cell is not the indivisible unit of intelligent behavior. Albrecht-Buehler contends there are hierarchical strata of information processing within cells. He has shown that small fragments of cytoplasm comprising about two percent of cell volume can be torn out of living fibroblasts (Albrecht-Buehler, 1980). These "microplasts" are devoid of any genetic input, yet can move filopodia, ruffle lamellipodia and produce blebs. Some fragments are so small they appear as isolated active filopodia, ruffles, or blebs and suggest that cells contain cytoplasmic domains capable of autonomous amoeboid movement. Albrecht-Buehler asserts that these cytoplasmic activities within cells are controlled by a superior stratum of organization to permit coherent locomotion. This level of control is not located in the genes since cells with their nuclei removed still exhibit migration behavior. There is intelligence in the cytoplasm. The cell has a driver's license.

### 5.5.2 Bending Sidearms

Several important cytoplasmic movements occur due to bending of contractile proteins attached along rigid cytoskeletal filaments. In muscle, arrays of parallel, interdigitating "thin" filaments (actin) interact with "thick" filaments (myosin). Tiny sidearms or crossbridges ("myosin heads") attached to the myosin thick filaments extend across a gap of about 13 nanometers and "cyclically row like banks of tiny oars" to move the filaments relative to each other. The energy from hydrolysis of ATP drives the conformational changes which causes the myosin molecules to curl. The precise utilization of ATP hydrolysis energy to cause contractility and other conformational changes remains poorly understood, but will be discussed in Chapter 6.

Each myosin filament carries about 500 heads, each of which cycles about 5 times per second. There is some evidence of cooperativity in that once a myosin head has detached, it is carried along by the action of other myosin heads along the thick filament. The myosin head rowing is initiated and coordinated by waves of calcium ion released from a reservoir (the sarcoplasmic reticulum) triggered by membrane electrical activity. Rises in calcium ion causes an actin-bound regulatory protein called troponin to shift its position and allow actin-myosin ratcheting.

Many MT related activities generate force, locomotion and movement of vesicles and other material; axoplasmic transport is one well studied example (Lasek, 1981; Ochs, 1982). Parallel MT within axons are polarized with their fast growing "plus-ends" distal from the cell body facing the synapse. Force generating sidearms occur about every 16–18 nanometers along MT lengths. These contractile crossbridges generate directional movement of material along MT by undergoing a sequence of conformational changes involving attachment of crossbridges to vesicles, ATP dependent force generation by the crossbridges, and detachment of crossbridges from vesicles. Detachment occurs only at "plus" ends near synapses. The process is similar to rowing of myosin heads to slide actin and myosin filaments past each other and shorten muscle fibers. MT based dynein activities, however, are far more variable, flexible, and interactive than the repetitive nanoscale events in muscle. For example, in MT dependent axoplasmic transport different material is simultaneously transported in the opposite direction, from the synapse to the cell center. This "retrograde" axoplasmic transport is thought to provide feedback to the protein synthesis machinery as to what enzymes or material are required, and/or to allow "recycling" of some molecules (Figure 5.25).

Robert Allen (1985) was among the first to suspect that MT and MAPs served as intracellular conveyor belts. He and his colleagues studied isolated MT and MT fragments which, with available biochemical energy in the form of ATP, "glide" along glass cover slips at velocities of 150 to 450 nanometers per second. The velocity is independent of MT fragment length, occurs essentially in a straight path, and is in the direction of retrograde axoplasmic transport. The straight paths of gliding MT segments suggest that the force generating enzymes are deployed in linear, rather than helical paths along the MT, and that the stroke that causes gliding is parallel to the MT with a spacing interval of about 17 nanometers. Reducing the available ATP concentration slows the gliding speed significantly, but does not affect the number or behavior of gliding MT. Gliding MT almost never interact when they cross paths, and when the forward progress of a gliding MT is blocked by an obstacle, it "fishtails" slowly from side to side through a series of serpentine maneuvers. Allen and colleagues concluded that the forces observed in their slithering free MT as well as in axoplasmic transport and ciliary bending are due to "force generating enzymes" directly attached to MT. Dynein, which functions to cause binding in cilia and flagella, is one MT force generating enzyme and kinesin is another motor for organelle transport along microtubules. Latex beads coated with kinesin translocate along microtubules similar to organelles, although at a slower velocity. Purified kinesin can increase the frequency of axoplasmic organelle movement along purified MT.

Allen and colleagues proposed the "backstroke hypothesis" which states that the force generating enzyme (dynein or kinesin) makes an elliptical stroke which imparts some force in both directions. Allen's dynein backstroke model is capable of carrying vesicles in opposite directions simultaneously through sufficiently separated pathways so that they seldom collide. Further, the motion generated can be continuous, not interrupted by cycles of attachment and detachment. The mechanical cycle of each sidearm includes a radial stroke that moves vesicles in the anterograde direction toward the microtubule plus end. This part of the cycle also causes isolated MT to glide "retrograde." The return stroke is tangential to the MT surface and transports the larger organelles in a retrograde pathway, and propels gliding MT toward their "anterograde" plus end.

The mechanisms by which the force generating protein arms may use ATP energy to contract will be discussed in Chapter 6. Even less well understood is the signaling and communication which orchestrates contractile activities of rows of arms spatially arrayed on MT lattices. Collective communication among MT lattice subunits (solitons, coherent excitations, lattice vibrations) could explain this orchestration.

The backstroke model is currently favored more than another model: microstreams. Shimizu and Haken (1983) had proposed a dynamic cooperativity of cytoskeletal elements which generated hydrodynamic microstreams conveying cellular materials. They specifically focused on actin-myosin interactions to generate these microstreams. New techniques such as Nanovideo Microscopy developed by Marc DeBrabander and colleagues (1986) at Janssen Pharmaceutica Research Laboratories in Belgium show direct tracking of immunolabeled particles along MT, questioning the significance of microstreams. Particles are seen to travel in opposite directions along the same MT, passing each other like two railroad trains on adjacent tracks. Microstreaming does not appear dominant in axoplasmic transport, but could be important in other phenomena. The primary site of coordinated transport and its complex orchestration appears to rest solely in the province of MT.

### 5.5.3 Ciliary and Collective Movement

The structure of cilia and flagella is a core of parallel microtubules in a cylindrical "9+2" arrangement of doublet or triplet microtubules. The arrangement is similar to centrioles which are also cylinders of 9 MT triplets,

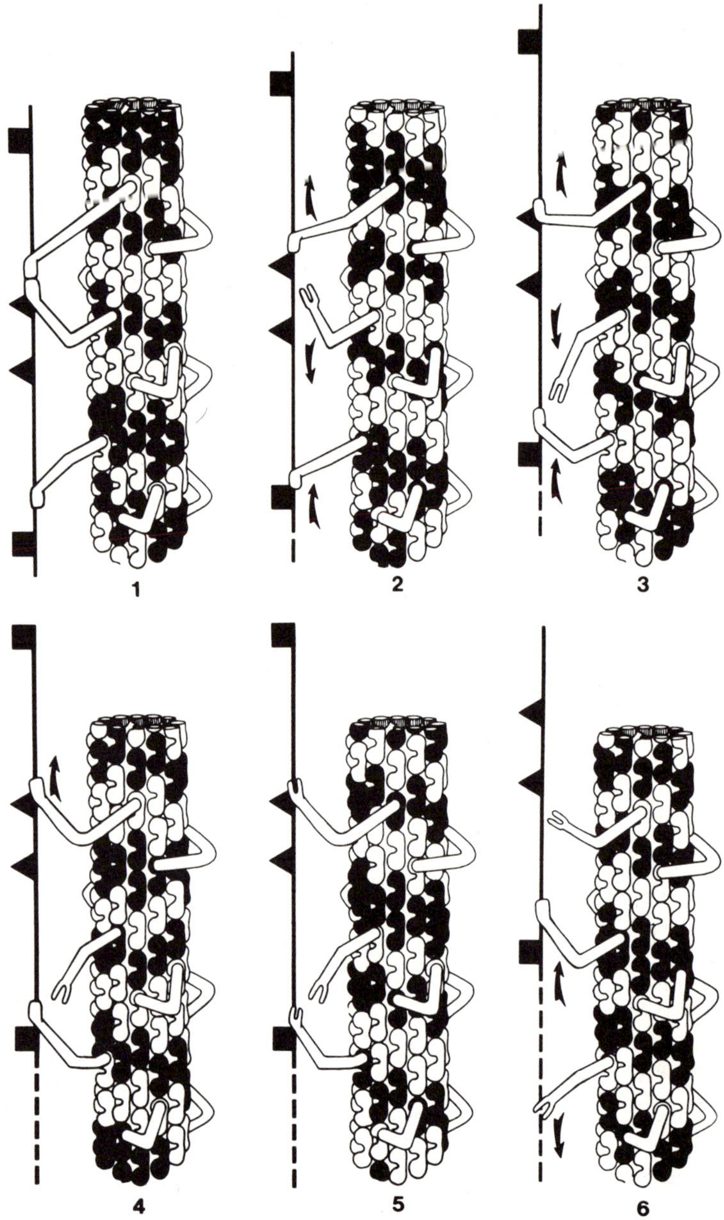


Figure 5.25: Axoplasmic transport occurs by coordinated activities of sidearm, contractile proteins ("dynein"), which cooperatively pass material in a "bucket brigade." The orchestration mechanism is unknown, but shown here as the consequence of signaling by "soliton" waves of tubulin conformational states. By Fred Anderson.

but without the additional central pair. Motor cilia and flagella have actin wound around their central pair of MT. The side arm proteins which connect the parallel microtubules are called links, spokes, or sidearms and are comprised of dynein, the contractile protein which utilizes ATP energy to produce force. Cilia and flagella are anchored inside the cell to basal bodies which are also composed of parallel microtubules and form a short cylinder with the same outer diameter and nine fold symmetry as cilia, flagella and centrioles. It has been clearly shown that ciliary mechanisms can function without influence from the cell to which they are attached. Flagella or cilia severed from cells by laser beams continue to propagate normal bending movements, if the surrounding medium contains ATP and either magnesium or calcium ions. Cilia function to move fluid over the surface of a cell or to propel single cells through a fluid. Single cell organisms use cilia for the collection of food particles and for locomotion. In the human lung and respiratory tract, epithelial lining contains about a billion cilia per square centimeter which act to sweep layers of mucous, trapped particles, and dead cells towards the mouth where they may be coughed up or swallowed. The cilia bend in coordinated unidirectional waves in which each cilium moves as a tiny whip. There is a forward stroke in which the cilium is extended and exerts maximal force on the surrounding liquid medium, followed by a recovery phase when it returns to its initial position. The movement is a rolling motion which minimizes viscous drag and requires about 0.1 to 0.2 seconds. Cycles of adjacent cilia are not quite in synchrony, and the small delay produces wave-like patterns of the entire ciliary complex. Simple flagella of sperm and single cell organisms are much like large cilia in their internal structure but are usually longer and propagate in quasi-sinusoidal waves rather than faster whip-like movements. The mechanism of flagellar beating in eukaryotes (but not of bacteria) is very similar to that of cilia.

Muscle contraction, axoplasmic transport, and ciliary and flagella motion all occur by the contractile bending of sidearm proteins attached along cytoskeletal filaments. In the case of muscle contraction, the stable cytoskeletal filaments are myosin with appendages called myosin heads crawling along parallel actin filaments. In the case of axoplasmic transport and ciliary and flagellar bending, microtubules are the stable filaments from which dynein contractile sidearms crawl or row along other MT (in the case of cilia and flagella), or pass along material or vesicles (in the case of axoplasmic transport). The energy for all of these mechanisms is supplied by the hydrolysis of ATP, but its precise utilization, transfer of conformational states and the temporal orchestration required to control these sidearm appendages are unknown. Figure 5.25 shows one possible cooperative control mechanism in which a wave of tubulin conformational states (soliton) travels along the MT cylindrical surface lattice to trigger the dynein activities.

### 5.5.4 Geodesic Tensegrity Gels

Actin-myosin contraction also occurs in non-muscle cells. Soon after actin and myosin were identified in muscle, Loewy found that extracts of slime mold cytoplasm responded to ATP with a decrease in viscosity. He hypothesized that ATP provided chemical energy which was converted into mechanical work required for slime mold locomotion. After H. E. Huxley and colleagues at University College London showed that muscle contraction was achieved by ATP dependent sliding of actin and myosin filaments past one another, Hatano and Osawa at Nagoya purified actin from slime mold. Subsequent investigations turned up myriads of cytomatrix proteins including myosin which were related to contractility and its regulation (Lazarides and Revel, 1979).

Using fluorescent antibodies to "light up" specific proteins, Lazarides and Revel (1979) have observed actin filaments organized into strikingly regular networks looking remarkably like *geodesic domes*. They describe icosahedral geodesic networks that encompass the area above and around cell nuclei and other cell regions after the cells are treated with fluorescent antibodies illuminating actin, alpha actinin and tropomyosin. Alpha actinin is localized primarily at the vertices of the geodesic network and tropomyosin is localized along the actin fibers connecting the vertices. Longer fibers attach to network vertices and extend into filopodia, lamellipodia, and membranes. The geodesic network and vertices act as organization centers involved in maintaining cell structure (Figure 5.26).

Some actin networks are transient dynamic structures which serve fleeting, but vital functions during the cell cycle. During the final stages of cell division, after duplicated chromosomes have been pulled apart by the mitotic spindle, a ring of constriction encircles the equator of the mother cell perpendicular to the spindle axis. Work by this "cleavage furrow" constricts the cytoplasm until it is divided into two daughter cells. The constricting tension has been estimated from observations in sand dollar and sea urchin eggs to be about $3 \times 10^5$ dynes per square centimeter, a value comparable to the force generated in muscle (Lazarides and Revel, 1979). The cleavage furrow contractile ring is a temporary structure that exists for only about 10 minutes; it rapidly assembles and disassembles. Although the width and

Figure 5.26: Geodesic actin cytomatrix gel surrounding cell nucleus (from Lazarides and Revel (1979). Vertices of the icosahedral nuclear dome attach filaments which extend to the cell membrane. By Paul Jablonka.

thickness of the contractile ring remain constant during constriction, the volume decreases. The ring must disassemble even as it contracts, meaning that any sliding interaction of the actin and myosin filaments must be followed by disaggregation of some of the filaments. This disassembly must take place uniformly through the ring during its entire brief lifetime. Therefore, a simple sliding filament model is insufficient to explain the cell cleavage function of the contractile ring.

Steve Heidemann and colleagues (Joshi, Chu, Buxbaum and Heidemann, 1985) at Michigan State University have examined compressive and tensile properties of cytoplasm. They find that semi-rigid microtubules are under compressive forces generated by interwoven contractile actin filaments. A balance between parallel compressive and tensile forces leads to a self-supporting property characterized by Buckminster Fuller (1975) as "tensegrity." Tensegrity can provide cell support even if the rigid parallel element are not in direct contact. Tensegrity in the cytoskeleton might explain the self-supporting structural properties of cytoplasm in which the rigid parallel elements are not in direct contact.

Robert Jarosch (1986) has proposed that contractile actin winds and unwinds microtubules by a "torque drive," causing rotational oscillations and perhaps tuning and detuning of the microtubule system. Dynamic compression/tension may also be important in the regulation of membrane receptors whose mobility are limited by anchoring MT. Conversely, contractile actin filaments can redistribute the receptors unless they are restrained.

Dynamic tensegrity (Chapter 8) may be an important mechanism in many biological functions. In the cytoplasm, complex structures assemble, perform, and vanish into soluble subunits.

## 5.6 The Cytoskeleton and Development

The cytoskeleton performs critical functions in reproduction and development. These include meiosis (division of duplicated chromosomes in sperm and egg cells), sperm motility, mitosis, cell proliferation, cell migration and changes in cell shape which accompany *differentiation,* the expression of cell form and function. All cells in a given organism have the same genetic capabilities, but take on the roles of specific tissues by the mysterious processes of trophism and differentiation. Cells within developing organisms or embryos can be moved and will grow and assume the characteristics of the new tissue in which they are placed.

The cytoskeleton is crucial to all steps in reproduction, growth, trophism and differentiation. If chromosomes are maldistributed in meiosis or mitosis, nonviable or abnormal offspring can occur. Such maldistribution may be related to cytoskeletal dysfunction, an early theory for the cause of cancer (Boveri, 1929). Variability in tubulin isozymes and MAPs could explain tissue and cell specificity based on differences in the molecular composition and activities of the cytoskeleton.

The origins of form, growth patterns, and differentiation comprise the biological science of embryology, first addressed by the Greek Aristotle. Two possible explanations for the development of form and patterns from what appeared as nothingness were considered by Aristotle. The first notion was that embryos derive from formless, tiny masses through a process of continuous unfolding he termed "epigenesis." The second notion was that individual patterns exist in miniature within the parent, and development consists in steady growth of its dimensions, the process of "preformation." Aristole (Book VI of Historia Animilium) described his experiments in which he interrupted the incubation of a hen's egg in various stages. After three days,

> the heart appears like a speck of blood in the white of egg, ... [it] beats and moves as though endowed with life and from it two vein ducts with blood trend in a convoluted course.

Only seven days later are the "chick and all its parts distinctly visible." Aristotle favored epigenesis, however the concept of preformation dominated science for centuries until 1759 when Friedrich Wolff, a German physician living in St. Petersburg published "Theoria Generationis." With better optics than Aristotle, Wolff noted that the chick started as "globules" or cells that seemed to have no functional relation to one another. With time, distinct groups of *cells* emerged to form the structures that become the hen's principle organs. Further evidence against the preformation theory came in 1900, when Hans Adolf Driesch demonstrated that a sea urchin embryo, when cut in half, develops into two complete embryos (Burnside, 1974).

Aristotle's epigenesis theory of differentiation states that homogeneous spheres begin to divide, yielding more spheres which themselves divide. The "cells" begin to exhibit differences in structure and function and a "collective coherence." Each cell knows where to go, when to go there, which cells to congregate with, and how to work cooperatively. Excluding the possibility that a transcendental intelligence personally oversees the development of each organism, the conclusion is that the original cell contains enough information to orchestrate the entire process. Aristotle would probably be even more amazed to discover that during early embryological development each cell is totipotential: it can take on the role of any tissue in the body. We now know that the chromosomal DNA in each cell contains all genetic information. We are beginning to understand how specific genes are turned on and off. Cells appear to take on the role of their environment due to morphological "trophic" influence conveyed by cytoskeletal axoplasmic transport within their "innervating" nerve cells.

Much of the mystery of biological form and differentation thus focuses on the growth, pathfinding and trophic capabilities of embryological neurons. The form, architecture, and connections which nerve processes make not only establish tissue and body shape, but synaptic connections and brain "hardware."

How do neuronal processes know where and when to send their processes and establish connections? The mystery has been likened to the "Indian rope trick" fable in which a "fakir" tosses a rope into the air where it mysteriously stays rigid. He then climbs up the rope, pulls the lower portion up, and disappears! The embryological Indian rope trick depends on neuronal shape and orientation: the number of young axons and dendrites ("neurites") which arise from the cell body, their direction, degree of branching, how far they grow, and where and how they form synapses. All these factors are manifested by the cytoskeleton. The centriole/MTOC establishes cell polarity and orientation and presumably "launches" the extending processes. The tips of lengthening neurites are specialized areas of cytoplasm called "growth cones" which not only grow outward, but participate in decisions about direction, branching, and termination of extension. Growth cones can sense changes in local environment and, acting in concert with other cytoskeletal elements within the neurite, shift direction, branch, or form synapses. Growth cone activities are remarkably similar to movement in amoeba and other simple organisms. They continually probe their surroundings by sending out and then retracting delicate ruffles known as lamellipodia, and finger-like projections called filopodia or microspikes. These dynamic micro-appendages are comprised of meshworks and parallel arrays of actin filaments. MT and neurofilaments from the neurites splay into the growth cones, but generally stop short of the actin-rich areas (Bunge, 1986).

The roles of MT and actin in axonal extension have been studied by Yamada, Spooner and Wessells (1970). Addition of cytochalasin B, which is known to alter actin filaments, blocks neurite outgrowth if added before extension has begun. Treatment of extending axons with cytochalasin B results in cessation of the "ruffling" activity of the growth cone with little immediate effect on the neurite shaft. In contrast, antimicrotubule agents such as colchicine leave growth cone activity unaffected but block neurite extension. Continued exposure results in retraction of the cell process with subsequent formation of one or more growth cones from the cell body. These observations together with ultrastructural evaluations of treated cells, show that MT are necessary for the growth of neurites, while actin filaments are essential to growth cone protrusion, a phenomenon similar to amoeboid motion. A complex interplay of dynamic activities of actin and MT are required for neurite sprouting and growth cone extension. A composite view of growth cone activities is that, under direction of the MT cytoskeleton, actin assembly generates protruding lamellipodia and filopodia. Upon contact with an appropriate external substrate, filopodia adhere and actin bundles form from the filopodium tip into the growth cone. Myosin colocalizes with actin and the actin-myosin interaction produces a "muscle-like" tension which provides an anchorage for the growth cone to the filopodium adherence site. Lamellipodia, sheet-like ruffles, dart out among the finger-like filopodia, particularly near points where decisions about branching are required.

Growth cone activities related to embryological differentiation are at the very ground floor of intelligence. Evidence suggests that expression of tubulin, MAPs, and other cytoskeletal elements are also important for determination of cell form and function. Barra and colleagues (1974) have shown changes in alpha and beta tubulin during brain development. The relative amount of alpha and beta tubulin in rat brain peaks at about the 14th day after birth and then declines to adult levels as a result of reduction in the rate of synthesis. The pattern of alpha and beta tubulin genetic variants (isozymes) undergoes marked changes during brain development with an increase in the variety of tubulin. For example, 3–4 different types of beta tubulin are observed in embryonic mouse brain compared to 13 types in adult mice. Modification of alpha tubulin isotypes begins in the embryonic brain whereas the beta tubulin modification begins to occur after birth and coincides with a period of extensive outgrowth of processes and synaptogenesis in the developing brain.

MAPs are also implicated in development and expression (Barra et al., 1974). In the case of rat brain, two tau polypeptides are present in the first few days of birth. By 35 days the adult pattern has emerged in which four major tau polypeptides are apparent. The tau proteins of maturity are more adept at promoting MT assembly than are the tau proteins of immaturity. MAP 1 can be resolved into three groups of MAP 1A, 1B, and 1C. In chicken brain development, MAP IA is initially present at low levels and increases during late embryonic development and post hatching. Similarly, MT from 10 day old rat brains are depleted of MAP 1A in comparison to adult brains. MAP 2 is restricted to dendrites and cell bodies in adult brain and is particularly concentrated in dendritic tips, Purkinje cell dendrites and all neuronal cell bodies, but absent from axons. Localization of MAP 2 to the dendritic cytoskeleton begins at the earliest times of appearance of dendritic outgrowths. In the adult brain, MAP 2A and MAP 2B show up at different times and brain regions and are localized in neurofilament rich axons.

All these changes are due to localized cytoskeletal mechanisms in addition to genetic expression. The symphony of alterations in tubulin and MAP isozymes has the score written in DNA, but its performance requires collective cooperative interactions among the conductor and orchestra. For example, "tyrosination" of beta tubulin occurs due to signaling and conditions present in the cytoplasm. Modifications are involved in the control of the outgrowth of axonal and dendritic processes, transport of constituents to the tips of the growing processes, cell migration and division, and establishment and maintenance of synapses and the adult form of the neuron. The temporal relationships between changes in neuronal MT proteins and differentiation is essential to the final product: the brain/mind.

## 5.7 The Cytoskeleton and Medicine

Defects related to microtubules are specifically linked to several human diseases. One example is "immotile cilia syndrome" (Afzelius, 1979) which is caused by altered dynein and results in an inability to expel secretions from the lungs, leading to recurrent bacterial infections. Another is developmental disability in infants which is caused by abnormal MT function induced by defective MAPs (Purpura, 1982). The cytoskeleton participates in the effects of various diseases (malignancy, Alzheimer's disease, viral infections), drugs, toxins and the body's response to disease.

The pathway to understanding MT led through the disease gout, a painful swelling of joints caused by the body's response to accumulation of urate crystals. Lymphocytes and macrophages, the body's immune cells, leave the bloodstream and migrate by amoeboid locomotion towards the urate crystals which often lodge in a joint of the big toe. Gout may be precipitated by purine containing foods which are metabolized to urate. The urate crystals are not particularly harmful except for the painful inflammatory immune response they trigger. By luck, a drug called colchicine was found to be helpful in relieving the pain and tenderness. Later it was discovered that colchicine worked by depolymerizing microtubules and preventing the locomotion and engulfment behavior of the lymphocytes and macrophages.

In addition to cell migration, the cytoskeleton is clearly involved in the establishment of normal cell architecture, function, and control of cell division and growth. In malignancy, control of cell reproduction is lost and growth proceeds without regard to the needs of the organism. Cancerous cells exhibit a tendency to break away from their anchorage and set up growth elsewhere in the body. Malignant cytoskeleton is disorganized with formation of oscillating aggregates of contractile material that aids dislodgement of the cell from its anchorage, instability in chromosome number and loss of growth regulation. All these effects *could be* primarily cytoskeletal in origin, and Puck (1977) has proposed that malignancy is a disease of the cytoskeleton. Clearly, the cytoskeleton is involved in the expression and processes of malignant cells. In cancer, cell division goes out of control: multipolar or asymmetric mitotic spindles are commonly observed. Boveri observed in 1929 that such aberrant distribution of genetic material could result in any combination or permutation of genes, most of which would be nonviable. However, some permutations may be sufficiently viable and have the abhorrent traits of malignancy. Many other factors leading to genetic alterations can account for the same results, so the precise etiology cannot be pinpointed. Indeed, cancer is probably caused by a number of etiologies including viruses which infiltrate and usurp the genetic apparatus and cytoskeleton.

Much of our current knowledge of the cytoskeleton has been learned from experimental perturbations by toxins, poisons, or drugs. Claude Bernard said in 1875:

> The poison becomes an instrument which dissociates and analyzes the special properties of different living cells; by establishing their mechanisms and causing cell death or changes in cell function we can learn indirectly much about the relation between molecular structure in the physiological process of life.

Peripheral nerve MT and axoplasmic transport are vulnerable to toxin and drug effects. Vinca alkaloids (vincristine, vinblastine) are commonly employed in battling cancer because they disrupt MT mitotic spindles as they polymerize. Because the cancerous cells are dividing so much more rapidly than normal cells, the mitotic spindle poisons are effective against malignancy. However, they can cause side effects including peripheral nerve damage by injuring MT dependent axoplasmic transport. This results in peripheral nerve damage in many patients who receive the anti-microtubule drugs. Sometimes severe neurological dysfunction limits the use of the vinca drugs. Fortunately, neither vinca alkaloids nor colchicine cross the blood brain barrier so central nervous system problems are limited.

No agents are known to selectively disrupt intermediate filaments (although 2,5 hexane dione *may* act in this way). In intact cells, vanadate combined with drugs that disassemble microtubules cause vimentin type intermediate filaments to collapse around the nucleus. A toxin known as beta, beta prime iminodiproprionitrile (IDPN) disrupts microtubule/neurofilament organization in axons, which results in colocalization of certain MAP II subgroups with intermediate filaments. Wisniewski and colleagues (1966) showed that injection of aluminum salts into the brain or cerebrospinal fluid of experimental animals induced marked accumulation of 10 nanometer filaments in cell bodies, dendrites and initial axon segments of large neurons. When silver stained, these accumulations superficially resemble the neurofibrillary tangles that are encountered in a number of human disease states, notably Alzheimer's disease, Parkinsonism dementia and senility. The filamentous accumulations are not precisely identical to those of the human diseases but raise interesting questions about the role of cytoskeletal proteins in these afflictions. Alzheimer's disease and related forms of senile dementia are characterized by these neurofibrillary tangles which result in various symptoms including a shortage of acetylcholine at synapses of "cholinergic" neurons. This shortage may be due to a breakdown in the cytoskeletal transport mechanisms which deliver acetylcholine precursors to the synapses. The cognitive dysfunction which characterizes the dementias is somehow related to disruption of the cytoskeleton. Other toxins whose mechanisms have been ascribed to cytoskeletal effects include hexanedione and hexacarbons, carbon disulfide, acrylamide, methylmercury, and mineral fiber asbestos.

When cells are injured by lack of oxygen, acidosis, toxins, or other causes, there is an irreversible point beyond which cell death is inevitable. Recent evidence points to a pathological elevation of cytoplasmic calcium ion as the irreversible point (Farber, 1981). The excess calcium may originate outside the cell, diffusing across damaged membranes, or may be released from cytoplasmic reservoirs including the cytoskeleton. Regardless of its source, the elevated calcium causes depolymerization of microtubules and other cytoskeletal components. The net result is a loss of cytoplasmic organization, enzyme function, and structural integrity leading to cell death. Dimethysulfoxide (DMSO) is a solvent which stabilizes polymerized microtubules, and which has been shown to preserve integrity of cells and tissues against damaging effects of radiation, low temperature, and other insults. Other compounds (taxol, poly-lysine, etc.) also fortify or preserve MT and the cytoskeleton and could be important in future therapies.

The effects of a number of other drugs may be mediated via the cytoskeleton; among these are anesthetics which cause a reversible cessation of consciousness. In the 19th century, Claude Bernard noted that an anesthetic (chloroform) inhibited "protoplasmic streaming" in slime molds, a function of the cytoskeleton. Allison and Nunn (1968) showed that sufficient concentrations of the general anesthetic halothane reversibly depolymerized MT. Sleep producing barbiturates, local anesthetics, and major tranquilizers such as thorazine also bind to brain MT. Chapter 7 will describe how anesthesia is related to inhibition of cooperative activities related to information processing in the cytoskeleton and connected membrane proteins. The future emergence of nanotechnology (Chapter 10) may permit medical intervention to be more specifically tuned to functions and dysfunctions of the cytoskeleton.

## 5.8 Intelligence in the Cytoskeleton

Cytoskeletal gel networks have complex repertoires. Motile events within non-repetitive muscle cells are dynamic, often transient and variable and not strictly analogous to muscle contraction. Cells contain from two to five different types of actin and these may combine up to ten different ways depending on the presence of binding proteins and other factors. Certain actins polymerize in the presence of calcium or magnesium ions whereas other actins polymerize only in their absence. Non-muscle cells possess control mechanisms that dynamically switch between monomeric actin, actin in a filamentous bundle form, and actin in a geodesic gel meshwork form. Some cytoplasmic motility is the result of actin myosin crawling, whereas examples depend on explosive polymerization of actin, rapid disaggregation of actin filaments, or interconversion of filament bundles into a meshwork (Satir, 1984). Microtubules, filaments, and centrioles are also involved in these same aspects of cell movement and are particularly important to orientation and directional guidance.

How can these diverse modes be coordinated? Guenter Albrecht-Buehler (1985) cites two basic requirements for cytoplasmic intelligence. These are compartmentalization, which separates components engaged in various functions to prevent chaos, and the information content. According to Shannon (1948) information is not concerned with the meaning of the message but only with its formal structure. As Marshal MacLuhan said: "the medium is the message"; the cytoplasm is both!

Information can have spatial and temporal content: spatially a message can be in the form of a letter or magnetic tape, and/or can have a temporal structure such as a radio signal or movie. It is the very coupling of spatial and temporal components in a dynamic sense that provides the medium of information. Shannon suggested that intended signals be "unpredictable." For example, a meaningful text is a sequence of words and letters that the reader cannot anticipate. In contrast, text consisting of an uninterrupted string of a single letter doesn't carry much information. Similarly in temporal messages such as radio signals, the strictly periodic and therefore predictable carrier wave of a transmitter carries no information. Only after the periodic oscillations of the carrier wave have been modulated with unpredictable changes of frequency (FM), or amplitude (AM) can speech or music be heard. At the opposite end of the spectrum, random stochastic noise is equally devoid of information.

Albrecht-Buehler observes that cytoplasm is neither totally regular, like a periodic crystal, nor totally random like a boiling liquid, but is an organized piece of matter. Complex, intricate activities occur "in the midst of drowning thermal noise all around and within." The cytoplasm not only keeps its cool against the thermal background, it routinely couples spatial and temporal components to manifest information. One example of cytoplsamic information which is independent of DNA/genetic control is the pattern of ciliary orientation in paramecium (Figure 5.27). Extrinsically altered, "non-genetic" patterns are maintained through one hundred mitotic generations (Aufderheide, Frankel and Williams, 1977).

Albrecht-Buehler describes three possible and progressively enlightened approximations to account for the intelligent actions of cytoplasm. 1) "The secret of cytoplasmic complexity is sought in the very randomness of thermal chaos combined with the observed specificity of biochemical reactions." This view is doubtful since the cytoskeleton and other structures take on elaborate, nonrandom forms. 2) Cytoplasmic intelligence stems from "supramolecular topology, architecture and dynamics rather than freely swarming inhibitors and promoters with their competing binding constants." The complex spatial arrangements of protein subunits and other molecules in macromolecular assemblies strongly suggests cooperativity between biochemical events over large intracellular distances. This leads to consideration of cytoplasm as a "giant multienzyme complex" based on cooperative actions of actin, IF, and MT. This implies an automatic, robot-like machine function of the cytoplasm, presumably set in motion and directed by the genetic apparatus. This view is embraced by many biologists who deify DNA as the prime mover in all biological activities and neglect the "real time" cytoplasmic activities of organisms. 3) Albrecht-Buehler suggests an inherent intelligence within cytoplasm. Intelligence implies the ability to collect and process data and make decisions on the basis of these data. Also important are intrinsic criteria that distinguish between desirable and nondesirable outcomes. Intelligence implies the ability to assess global situations, not merely reacting to local stimuli whenever and wherever they occur, and it implies communication of data with other intelligent objects and appropriate adjustment of actions. Albrecht-Buehler suspects that computers were discovered, rather than invented, and that cytoplasm is a "chemistry based gel or even liquid data processing system."

Cytoplasmic intelligence may depend on collective dynamics of cytoskeletal subunits. Parallel arrays of MT and neurofilaments provide a framework around which microtrabecular lattice structures could form with varying durations of existence as correlates of learning, information, memory, and consciousness. Mechanical contractility of actin-myosin and other proteins within the cytomusculature/cytomatrix could impart mechanical vibrations and cooperative resonances, solitons, or interference wave patterns. Geodesic tensegrity nets of MT, actin and their ordered water may be pulsating in the nanoscale at this very moment within all living cells. Polymerization patterns of actin and other proteins may also be regulated by calcium induced sol-gel state phase differences or harmonic coupling with other microtrabecular and cytoskeletal structures. Coherent nanosecond excitations and propagating solitons are additional mechanisms which have been proposed to occur within the cytoskeleton. In the brain, reinforcement from higher levels of parallel processing (neuron level, neural net, brain) could fortify specific substructures such as wave patterns of calcium coupled sol-gel states in neural cytoplasm.

The "genetic code" was decrypted by Marshal Nirenberg and colleagues (1961, 1964) who were able to equate DNA base pair patterns with specific amino acid sequences in synthesized proteins. The "real time" information codes in the cytoskeleton may be understood when cytoskeletal nanoscale events just beyond our current capabilities become approachable with the advent of nanotechnology in the next decades.

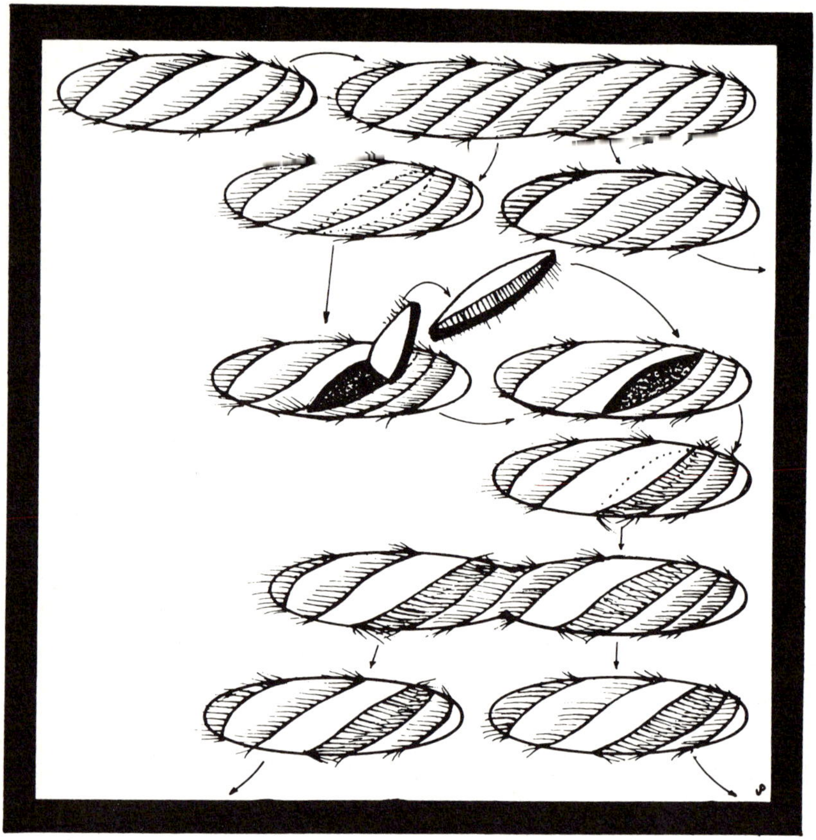

Figure 5.27: Paramecium mitosis and cytoskeletal "heredity." Top: a single paramecium elongates and divides into two daughter cells. The surface of paramecium are covered with hair-like cilia, each of which is a collective assembly of microtubule doublets or triplets. Cilia bend cooperatively to propel the paramecium, and to propel liquid environment past the organisms. All the cilia are oriented in the same direction, to the left. Second row: Experiments done by Aufderheide, Frankel and Williams (1977) are illustrated. A segment of cell membrane and underlying cytoskeleton which anchors cilia is transplanted in reverse orientation to a sister paramecium. The abnormal cilia orientation persists in the next generation, and for 100 generations to follow! By Paul Jablonka.

# Chapter 6

# Protein Conformational Dynamics

Proteins are the structural and organizational elements of "the living state." Their essential functions are intrinsically linked to their structure and dynamic switching among different conformational states. For example, ion channel proteins embedded in a membrane may be either in an "open" conformation, through which specific ions diffuse along a gradient, or they may be "closed." This type of conformational switch does not require biochemical energy such as ATP hydrolysis, but utilizes energy stored in transmembrane voltage gradients and occurs in response to an appropriate trigger. Proteins such as enzymes, receptors, cytoskeletal filaments, muscle myosin and hemoglobin undergo important conformational changes in response to a variety of stimuli. Dynamic patterns of conformational states among cytoskeletal subunits may represent information, exert control over routine biological functions, and provide the "grain of the engram."

## 6.1 Protein Structure

Proteins have several hierarchical levels of structural organization which determine their dynamic and functional capabilities. Protein primary structure is determined by a sequence of 22 possible amino acids held together by peptide bonds. An amino acid consists of carbon atoms bound to nitrogen atoms (peptide bond) and their attendant side groups. The 22 amino acids found in proteins differ in their side groups although each contains a carbon-oxygen double bond known as "amide 1." Strings of amino acids

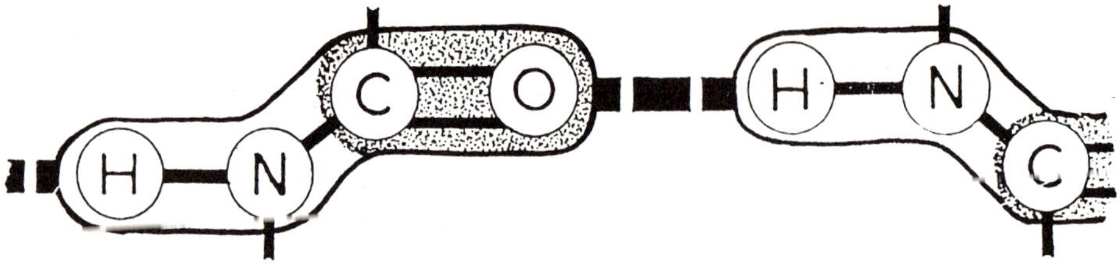


Figure 6.1: Hydrogen bond between hydrogen and oxygen in amide one resonance bond. With permission from Bolterauer, Henkel and Opper (1986).

called "peptides" perform many physiological functions including acting as neurotransmitters and circulating hormones. The amino acids impart specific properties according to the sequence of their incorporation into the peptide string. Amino acids differ in their size depending on their side groups; their molecular weights can range from 75 daltons (a dalton is the mass of a hydrogen atom, a twelfth of a carbon atom) for glycine, to about 200 daltons for tryptophan which contains a double aromatic ring. Functional proteins also range in size: 55 kilodaltons (one kilodalton = one thousand daltons) for tubulin monomers to 630 kilodaltons for thyroglobulin, to several million kilodaltons for protein assemblies which comprise large viruses and infectious agents of certain diseases like psittacosis and lymphogranuloma venereum (Harper,1969). Thus proteins are comprised of several hundred to millions of amino acids. The precise sequence of amino acids is determined by DNA and they are assembled on ribosomes. A polypeptide "backbone" of 200 amino acids would have $22^{200}$ different possible *primary* structures. The mass required to include one example of each of this number of protein structures would far exceed that of the universe!

Amino acids which comprise polypeptide chains and primary protein structure can form linkages with other amino acids within the polypeptide by hydrogen bonds and disulfide bonds. These linkages cause bending of polypeptide chains into coiled or folded structures which determine protein *secondary* structure. The most stable and commonly observed secondary structure is a right handed coil called an "alpha helix" in which hydrogen bonds (Figure 6.1) form between an oxygen and a hydrogen separated by 3 or 4 amino acids on the polypeptide chain. Alpha helices can interact among themselves in a protein, stacking in parallel or antiparallel arrangements or forming left handed "coiled-coils." Alpha helices (Figure 6.2) are common occurrences in a wide variety of proteins; tubulin is about 40 percent alpha helix in most circumstances (Dustin, 1978). The alpha helix has been proposed as an important site of energy and information transfer in proteins. Davydov has proposed that the energy of ATP hydrolysis utilized in actin-myosin muscle contraction and many other biological events is conveyed along alpha helix pathways via propagating "solitons" which occur due to anharmonic coupling in the carbon-oxygen double bonds (Section 6.7).

Hydrogen bonds forming among amino acid-side groups on polypeptide chains in *parallel* result in planar secondary protein configurations called beta-pleated sheets. First proposed by Linus Pauling in 1951, beta pleated sheets may themselves align in parallel, antiparallel or mixed formations.

Alpha helices, beta pleated sheets and other secondary structures interact to define protein *tertiary* structure, upon which most protein functions depend. Hydrophobic forces, charge distribution, disulfide bridges and secondary structure result in folding into globular regions which define the shape of single protein units. Alpha helix and beta pleated sheets pack into arrangements which help determine the tertiary structural domains which often have specific functions like binding a molecule, or associating with other domains to create larger structures. Assembly of groups of tertiary structure determines *quaternary* structure, like tubulin subunits assembling into microtubules. Generally, hydrophobic interactions like Van der Waals forces are important in quaternary protein assemblies.

Levels of protein assemblies include, at the simplest level, monomeric enzymes. Weighing from 20 to 90 kilodaltons, these enzymes have a single active site at which specific molecules undergo chemical reactions facilitated ("catalyzed") by the enzyme. Some enzymes may require metal ions, organic molecules or specific cofactors to function. Reduced enzymatic activity can result from occupancy of the active site by molecules resistant to enzymatic action, a phenomenon called "competitive inhibition." Active site binding corresponds with specific conformational states of enzymes.

A more complex system is the oligomeric enzyme, an example of which is the acetylcholine receptor: a four subunit oligomer which is activated by binding of acetylcholine. Being composed of several subunits leads to collective properties which arise from the organization and interactions of the components. Many or perhaps most enzymes are oligomeric with molecular weights ranging from 35 to several thousand kilodaltons. Oligomeric subunits often have two binding sites; one is the active site for its enzyme action and the other is a regulatory site which controls the active site by a change in subunit shape or conformation. Regulatory, or effector molecules can change not only the catalytic site activity on that subunit (allosterism) but also on adjacent subunits (cooperativity). Oligomer subunits may be alike (homopolymers) or different (heteropolymers) and coenzymes may be necessary. A single coenzyme molecule can "turn on" the binding capabilities of multiple subunits within an oligomeric enzyme (Roth and Pihlaja, 1977).

Examples of oligomeric enzymes include glutamate dehydrogenase, composed of six 55 kilodalton subunits forming a complex with approximate dimensions of 100 by 100 by 80 nanometers. An example of a more complex oligomer is aspartate transcarbamylase. It contains six regulatory units of about 34 kilodaltons, each composed of 3 dimers. An additional six 100 kilodalton catalytic subunits are each composed of two trimers. Six zinc atoms are necessary in each of the regulatory subunits. Another cooperative oligomer is hemoglobin, composed of two "alpha" and two "beta" subunits; many other oligomeric enzyme patterns have also been observed. Oligomeric enzyme subunits have cooperative interactions which means that the function of one subunit facilitates the conformational actions of other subunits. In hemoglobin, conditions sufficient for binding of one oxygen molecule to one subunit causes the remaining subunits to bind oxygen more easily.

More complex protein assemblies include cytoskeletal polymers and virus coats. The cooperative interactions of *their* subunits can lead to information processing and thus to intelligent behavior.

## 6.2 Protein Conformation

At the tertiary structural level, most protein molecules are able to shift reversibly between several different but related stable conformations and are known as *allosteric* proteins. Such proteins may be able to form several alternative sets of hydrogen bonds, disulfide bridges and Van der Waals forces of about equal energy among their constituent amino acid side chains. Each alternative set of intraprotein bonds and charge distribution requires a change in the spatial relationships between components of the polypeptide chains and thus motion occurs among the states. Only certain distinct conformations are energetically favorable and any intermediate conformations are unstable and unlikely. Charge redistribution (i.e. dipole oscillation) can also couple to conformational switching. Thus step-like, nonlinear jumping occurs between specific conformations so that only a discrete number of alternative conformations exist for a given protein, and random switching

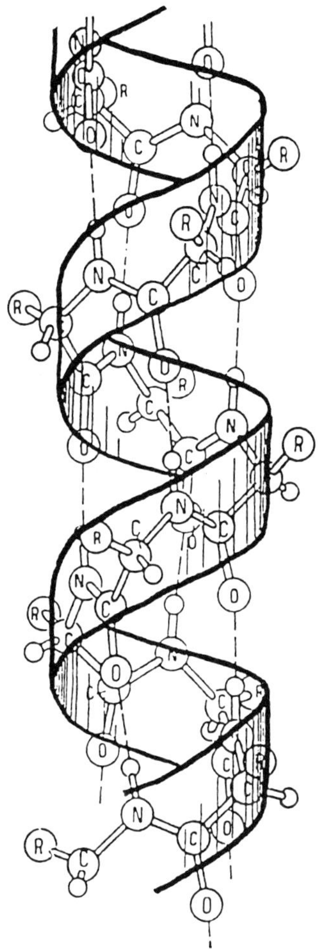

Figure 6.2: Alpha helix with three "spines" formed by hydrogen bonds (dashed lines) between hydrogen and oxygen atoms. With permission from Bolterauer, Henkel and Opper (1986).

from conformation to conformation is limited.

Each distinct conformation of an allosteric protein has a different surface and a different ability to interact with other binding molecules or "ligands." Only one of several conformations may have a high affinity for a particular ligand and the presence or absence of that ligand can determine the conformation that the protein adopts. Two distinct ligands may bind specifically to different surfaces of the same protein, so that the concentration of one ligand may change the affinity of the protein for the other. Such allosteric mechanisms facilitate fine tuning and regulation of many biological processes in which proteins transduce, or integrate various modes of signaling and information. Allosteric proteins are especially effective signaling devices when they exist as *aggregates of identical subunits*. The conformation of one subunit can influence that of neighboring subunits, producing a collective effect similar to amplification or switching in a computer. The collective behavior of hemoglobin in its binding of oxygen is one example; state transformations in microtubules may be another.

The conformational state of a specific protein is regulated by a variety of factors and may have profound importance for functional activity. Hemoglobin is an iron containing protein in red blood cells which carries oxygen from the lungs to the various body tissues like brain, heart, or muscle. Hemoglobin which binds an oxygen molecule is in a different conformation ("oxyhemoglobin") than hemoglobin without oxygen ("deoxyhemoglobin"). The conformational state of hemoglobin found inside red blood cells determines the protein's capacity to bind or release oxygen. Hemoglobin conformation is determined by factors which include the availability of oxygen, temperature, pH, presence of certain other molecules (i.e. 2,3 DPG), and the amount of nearby oxyhemoglobin. The latter describes a "collective" effect; when a critical amount of hemoglobin binds oxygen, the oxyhemoglobin conformation is easier to attain for the remaining protein molecules. This results in a "sigmoid" shape to the curve which describes oxygen/hemoglobin binding. The nature of this collective phenomenon is an indication of the behavior of protein assemblies.

Different frequencies of conformational changes coexist cooperatively in the same protein (Table 6.1). Some totally reversible conformations like oxyhemoglobin persist for tens of seconds; others can be very short-lived (i.e. femtoseconds: $10^{-15}$ seconds), or very long (minutes to hours). The oxyhemoglobin conformation persists until the oxygen molecule is delivered to the tissue whose mitochondria use it to produce ATP and GTP. To reach their binding sites within hemoglobin's central core, oxygen molecules diffuse through transient packing defects in the protein's structure. Nanosecond scale, conformational "breathing" permits the oxygen molecules to slip in and out, dependent on surrounding pH, temperature, hormones and other factors. There thus appear to be at least two levels of conformational states in hemoglobin. The "functional" conformation (binding vs no binding of oxygen) are relatively long lasting (long enough to carry oxygen from the

| | |
|---|---|
| Amplitude of motions | 0.001 nanometer to 10 nanometers |
| Energy | 0.1 kilocalories to 100 kilocalories |
| Time Range | $10^{-15}$ sec (femtosecond)<br>to $10^{3}$ sec (many minutes) |
| Time for collective, functional steps | $10^{-9}$ sec (nanoseconds) |
| Types of motion | local atom fluctuations |
| | side chain oscillations |
| | displacements of loops, arms, helices, domains and subunits |
| | collective elastic body modes,<br>coupled atom fluctuations,<br>solitons and other nonlinear motions,<br>coherent excitations |

Table 6.1: Protein Conformational Dynamics—Motions of Globular Proteins at Physiological Temperature (Modified from Karplus and McCammon, 1984).

lungs to tissue for delivery). In addition there are more rapid conformational vibrations such as the nanosecond "breathing" which facilitates diffusion of oxygen molecules through hemoglobin to reach their binding sites.

The vast majority of processes of biological interest are in the time scale greater than one nanosecond, which also lies in the collective mode realm for protein dynamics (Karplus and McCammon, 1984). Protein conformational changes in the nanosecond time frame are able to be coupled to a stimulus and result in a functional conformational change. Three fundamental features appear to control these functional states: hydrophobic interactions, charge redistribution and hydrogen bonding. Appropriate stimuli including neurotransmitters, voltage alterations, hormones, ions like calcium, and enzyme substrates can induce conformational changes within specific proteins. Conformational transduction is the sensitive link in the mechanism of anesthetic gas molecules (Chapter 7) and indicates that cognitive functions relating to consciousness depend in some way on protein conformational regulation. In many cases, proteins "integrate" a number of factors to result in a new conformation (Figure 6.3). Portions of proteins may be dynamically

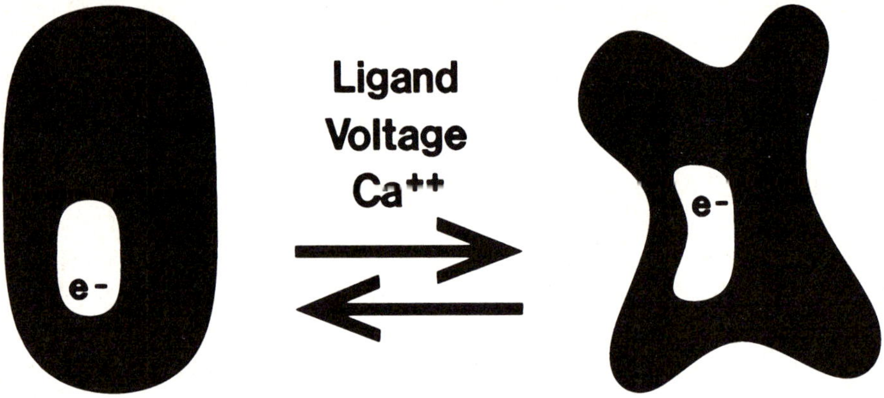


Figure 6.3: Protein switching between two different conformational states induced by binding of ligand, calcium ion, or voltage change. One mechanism proposed by Fröhlich is that dipole oscillations ($e^-$) within hydrophobic regions of proteins may be a trigger, or switch for the entire protein.

active with functional importance. For example, electron transfer processes such as that which occur in the mitochondrial production of ATP may rely on vibrational coupling and fluctuations which alter the distances between electron donors and acceptors. Relative motions of distinct structural domains within proteins are important in their activities related to muscle contraction, enzyme activities, antibody functions, and assembly and activities of supra-molecular structures such as viruses and cytoskeletal proteins. Dynamic conformational changes of proteins *are* the dynamics of living organisms.

## 6.3 Proteins and Energy

All movements within cells depend on forces generated by proteins. A typical allosteric protein can adopt two alternative conformations—a low energy form and a high energy form that differ by about the energy available from forming a few hydrogen bonds on a protein surface. The low energy conformation will be favored by about 1,000 to 1, and the protein will tend to be in this "inactive" conformation unless influenced by other factors. It can be "pulled" into the active high energy conformation by binding to a ligand which binds only to the high energy state. Also, an input of chemical energy can be used to "push" the protein into the high energy conformation. A common mechanism involves the transfer and covalent linkage of a phosphate group from biochemical energy sources like ATP or GTP to one of several amino acid side chains (serine, threonine or tyrosine) in the protein. This "phosphorylation" can create a distribution of charged amino acid side chains which are more favorable to one conformation than another, leaving the nonphosphorylated conformation unavailable. Ten percent of all mammalian cell proteins undergo phosphorylation as a regulatory or programming mechanism (Alberts et al., 1983).

The laws of thermodynamics demand that free energy be depleted when work is performed. Therefore, protein molecules cannot show net movement (as do microtubules and other cytoskeletal proteins) without some added source of energy. Phosphorylation is one way by making one step irreversible. Another is to drive allosteric changes by the hydrolysis of an ATP molecule, converting it to ADP (or hydrolyzing GTP to GDP). The energy released in the hydrolysis reaction is imparted to the protein, pushing it to a higher energy state. The precise mechanism by which this energy is utilized and transferred by proteins is unknown.

Besides generating mechanical force, allosteric proteins can use the energy of ATP phosphorylation and hydrolysis to do other forms of work, such as pumping specific ions into or out of the cell. An allosteric protein known as "sodium potassium ATPase" which is found in the membrane of all animal cells including nerve cells, pumps three sodium ions out of the cell and two potassium ions in during each cycle of conformational change driven by ATP mediated phosphorylation. This ATP driven pump creates ion gradients across the cell membrane. The energy stored in ion gradients is harnessed by conformational changes in other membrane proteins: ion channels. When these are triggered to open in nerve membranes, a conformational change initiated by a voltage change, ligand binding, or allosteric effect from membrane or cytoskeletal protein permits ions to passively diffuse. The passage of charged ions depolarizes the membrane as part of a local gated potential or traveling action potential.

The ion flow may also drive other membrane bound protein pumps that transport glucose or amino acids into the cell. Energy available in proton/hydrogen ion gradients across the inner mitochondrial membranes is used to synthesize most of the ATP used in the animal world. Actin and myosin and other proteins create contractile force in muscle by cooperative conformational changes induced by ATP hydrolysis. Within bending cilia, contractile protein bridges which span between microtubules hydrolyze ATP to drive their mechanical activities. ATP and other energy producing molecules are the fuel of biological protein engines. Less well understood, however, is their control, guidance and orchestration.

## 6.4 Protein Cooperativity—Historical View

An interesting idea regarding the control of protein conformational state and utilization of energy was proposed by Szent-Gyorgyi (1948). He suggested that proteins behave as semiconductors and that electrons "hop" between specific intraprotein regions. Experimental data, however, showed that the intraprotein energy band gap was too great to support such a concept. Utilization of dynamic biological and protein conformational energy remained enigmatic through the 1950's and 1960's.

A focal point in the history of attempts to understand protein conformational dynamics was a 1973 meeting of the New York Academy of Science (which also hosted a landmark meeting concerning microtubules in the same year). The twin mysteries of how ATP energy was utilized to produce mechanical protein events, and how energy and information were transferred in proteins and organelles were addressed by a gallery of scientists. The prevalent theme was "a crisis in bioenergetics," in that cooperative processes of obvious importance in biological systems were unexplained. One idea which generated controversy was that electromagnetic resonance energy was transferred between periodically arrayed excitation sites. At that meeting, C. W. F. McClare (1974) proposed that ATP induced excited states ("excitons") were coupled by infrared dipoles in protein biostructures to provide a communicative medium. McClare concluded:

> there is yet another level of organization in biological systems: a tuned resonance between energy levels in different molecules that enables bioenergetic machines to operate rapidly and yet efficiently.

This concept was supported by several allies including Wei (1974) who suggested dipole coupling among membrane proteins to explain nerve functions. McClare's general model of resonant dipole coupling was greeted by skepticism, however. Highly respected biochemist Gregorio Weber (1974) raised two objections. He noted that at least some of the emitted energy following molecular excitation is emitted in $10^{-12}$ seconds, too short to be utilized by biomolecules. The second objection referred to the transfer of quantum vibrational energy through the surrounding medium. The process of dipole-dipole coupling involves a similarity of the energies emitted and received, and requires that the interacting oscillators be surrounded by a medium transparent to the wavelength of the transferred energy quantum. Water surrounds every biomolecule and provides a medium which appeared to be opaque to the infrared energy McClare described. Weber implied any emitted energy would be dissipated as heat to the water environment. McClare countered, unconvincingly, that systems could have evolved to be able to slow down the relaxation and energy emission, and that such systems may be somehow separated from the aqueous phase. A consensus was that energy needed to be stored for a nanosecond or longer to be useful.

The mode of energy and information transfer was described in other comparable ways by different scientists. Several considered propagating packages of conformational or acoustic energy: "phonons." A phonon can be defined as a propagated lattice vibration, an intranuclear vibration of a carbon-nitrogen bond that propagates in a protein lattice. The transmission of phonons occurs at the speed of sound, so they are not resonance transfer phenomena but a propagated vibration in a periodic lattice. In 1967 Straub postulated that protein lattices contain phonons which must be isolated from their aqueous environment. At the New York Academy meeting in 1973 he suggested that hydrophobic interactions like Van der Waals forces could contain phonons in the lattice and shield them from the aqueous bath. The concept of "conformon," a quantum of protein conformational energy coupled to discrete protein states was independently described by Green (1970) and Ji (1974) in America, and Volkenstein (1972) in the Soviet Union. Later, A. S. Davydov of the Soviet Union proposed that mechanical conformational movements in proteins were nonlinearly coupled to electronic bond distortion or charge movement, resulting in propagating waves called "solitons." These models all faced a common problem of shielding from the aqueous environment. One possible resolution of this problem is that the water surrounding proteins may be "ordered" and resonantly coupled rather than thermally dissipating protein excitation energy. Another is that hydrophobic interactions exclude water from areas where important bioenergetic interactions occur.

## 6.5 Living Water and Hydrophobic Interactions

Biomolecules have evolved and flourished in aqueous environments, and basic interactions among biomolecules and their pervasive hosts, water molecules, are extremely important. The properties of intracellular water are controversial. Many authors believe that more than 90 percent of intracellular water is in the "bulk" phase—water as it exists in the oceans (Cooke and Kuntz, 1974; Schwan and Foster, 1977; Fung and McGaughy, 1979). This traditional view is challenged by others who feel that *none* of the water in living cells is bulk (Troshin, 1966, Cope 1976, Negendank and Karreman, 1979). A middle position is assumed by those who feel that about half of "living" water is bulk and the other half "ordered" (Hinke, 1970; Clegg, 1976; Clegg, 1979; Horowitz and Paine, 1979). This group emphasizes the importance of "ordered" water to cellular structure and function.

Many techniques have been used to study this issue, but the results still require a great deal of interpretation. Nuclear magnetic resonance (NMR), neutron diffraction, heat capacity measurement, and diffusion studies are all inconclusive. Water appears to exist in *both* ordered and aqueous forms within cells. The critical issue is the relation between intracellular surfaces and water. Surfaces of all kinds are known to perturb adjacent water, but within cells it is unknown precisely how far from the surfaces ordering may extend. We know the surface area of the microtrabecular lattice and other cytoskeleton components is extensive (billions of square nanometers per cell) and that about one fifth of cell interiors consist of these components. Biologist James Clegg (1981) has extensively reviewed these issues. He concludes that intracellular water exists in three phases. 1) "Bound water" is involved in primary hydration, being within one or two layers from a biomolecular surface. 2) "Vicinal water" is ordered, but not directly bound to structures except other water molecules. This altered water is thought to extend 8 to 9 layers of water molecules from surfaces, a distance of about 3 nanometers. Garlid (1976, 1979) has shown that vicinal water has distinct solvent properties which differ from bulk water. Thus "borders" exist between water phases which partition solute molecules. 3) "Bulk water" extends beyond 3 nanometers from cytoskeletal surfaces (Figure 6.4).

Drost-Hansen (1973) described cooperative processes and phase transitions among vicinal water molecules. Clegg points out the potential implications of vicinal water on the function of enzymes which had previously been considered "soluble." Rather than floating freely in an aqueous soup, a host of intracellular enzymes appear instead to be bound to the MTL surface within the vicinal water phase. Significant advantages appear evident to such an arrangement: a sequence of enzymes which perform a sequence of reactions on a substrate would be much more efficient if bound on a surface in the appropriate order. Requirements for diffusion of the substrate, the most time consuming step in enzymatic processes, would be minimal. Clegg presents extensive examples of associations of cytoplasmic enzymes which appear to be attached to and regulated by, the MTL. These vicinal water multi-enzyme complexes may indeed be part of a cytoskeletal information processing system. Clegg conjectures that dynamic conformational activities within the cytoskeleton/MTL can selectively excite enzymes to their active states.

The polymerization of cytoskeletal polymers and other biomolecules appears to flow upstream against the tide of order proceeding to disorder which is decreed by the second law of thermodynamics. This apparent second law felony is explained by the activities of the water molecules involved (Gutfreund, 1972). Even in bulk aqueous solution, water molecules are somewhat ordered, in that each water molecule can form up to 4 hydrogen bonds with other water molecules. Motion of the water molecules (unless frozen) and reversible breaking and reforming of these hydrogen bonds maintain the familiar liquid nature of bulk water. Outer surfaces of biomolecules form more stable hydrogen bonding with water, "ordering" the water surrounding them. This results in a decrease in entropy (increased order) and increase in free energy: factors which would strongly inhibit the solubility of biomolecules if not for the effects of hydrophobic interactions. Hydrophobic groups (for example amino acids whose side groups are non-polar, that is they have no charge-like polar groups to form hydrogen bonds in water) tend to combine, or coalesce for two main reasons: Van der Waals forces and exclusion of water. Combination of hydrophobic groups "liberate" ordered water into free water, resulting in increased entropy and decreased free energy, factors which tend to drive reactions. The magnitude of the favorable free energy change for the combination of hydrophobic groups depends on their size and how well they fit together "sterically." A snug fit between groups will exclude more water from hydrophobic regions than will loose fits. Consequently, *specific* biological reactions can rely on hydrophobic interactions. Formation of tertiary and quaternary protein structure (including the assembly of microtubules and other cytoskeletal polymers) are largely regulated by hydrophobic interactions, and by the effect of hydrophobic regions on the energies of other bonding. A well studied example of the assembly of protein subunits into a complex structure being accompanied by an *increase* in

Figure 6.4: Microtrabecular lattice (MTL-solid dark) with layers of ordered and vicinal water. "Soluble" enzymes (circles) are held within ordered or "vicinal" water and may be functionally regulated by the MTL. Modified from Clegg (1984) by Paul Jablonka.

entropy (decrease in order) is the crystallization of the tobacco mosaic virus. When the virus assembles from its subunits, an increase in entropy occurs due to exclusion of water from the virus surface. Similar events promote the assembly of microtubules and other cytoskeletal elements.

The attractive forces which bind hydrophobic groups are distinctly different from other types of chemical bonds such as covalent bonds and ionic bonds. These forces are called Van der Waals forces after the Dutch chemist who described them in 1873. At that time, it had been experimentally observed that gas molecules failed to follow behavior predicted by the "ideal gas laws" regarding pressure, temperature and volume relationships. Van der Waals attributed this deviation to the volume occupied by the gas molecules and by attractive forces among the gas molecules. These same attractive forces are vital to the assembly of organic crystals, including protein assemblies. They consist of dipole-dipole attraction, "induction effect," and London dispersion forces. These hydrophobic Van der Waals forces are subtly vital to the assembly and function of important biomolecules.

Dipole-dipole attractions occur among molecules with permanent dipole moments. Only specific orientations are favored: alignments in which attractive, low energy arrangements occur as opposed to repulsive, high energy orientations. A net attraction between two polar molecules can result if their dipoles are properly configured. The "induction" effect occurs when a permanent dipole in one molecule can polarize electrons in a nearby molecule. The second molecule's electrons are distorted so that their interaction with the dipole of the first molecule is attractive. The magnitude of the induced dipole attraction force was shown by Debye in 1920 to depend on the molecules' dipole moments and their polarizability. Defined as the dipole moment induced by a standard field, polarizability also depends on the molecules' orientation relative to that field. Subunits of protein assemblies like the tobacco mosaic virus have been shown to have high degrees of polarizability. London dispersion forces explain why all molecules, even those without intrinsic dipoles, attract each other. The effect was recognized by F. London in 1930 and depends on quantum mechanical motion of electrons. Electrons in atoms without permanent dipole moments (and "shared" electrons in molecules) have, on the average, a zero dipole, however "instantaneous dipoles" can be recognized. Instantaneous dipoles can induce dipoles in neighboring polarizable atoms or molecules. The strength of London forces is proportional to the square of the polarizability and inversely to the sixth power of the separation. Thus London forces can be significant only when two or more atoms or molecules are very close together (Barrow, 1966). Lindsay (1987) has observed that water and ions ordered on surfaces of biological macromolecules may have "correlated fluctuations" analogous to London forces among electrons. Although individually tenuous, these and other forces are the collective "glue" of dynamic living systems.

## 6.6 Electret, Piezo, and Pyroelectric Effects

Assemblies whose microscopic subunits and macroscopic whole both possess permanent electric dipoles are known as electrets. They exhibit properties known as piezoelectricity and pyroelectricity which may be useful in biological activities. In these crystals, dipolar elementary subunits are arranged in such a way that all the positive dipole ends point in one direction and all the negative dipole ends are oriented in the opposite direction. In microtubules, the positive ends of tubulin dimer subunits point away from microtubule organizing centers (MTOC) toward the cell periphery, and the negative ends point toward MTOC. Electrets can store charge and polarization and have now been identified in a variety of nonbiological materials such as ionic crystals, molecular solids, polymers, glasses, ice, liquid crystals and ceramics. Biological tissues demonstrating electret properties include bone, blood vessel wall materials, keratin, cellulose, collagen, gelatin, artificial polypeptides, keratin, DNA, cellulose and microtubules (Athenstaedt, 1974). An electret effect accounts for specific properties such as anti-blood clotting in biomaterials and the non-stickiness of teflon. Sources of polarization or charge storage in macromolecules are dipoles, ionic space charges, or ordered surface water.

Electret materials are piezoelectric (Gubkin and Sovokin 1960). Piezoelectric materials change their shape or conformation in response to electrical stimuli, and change their electrical state in response to mechanical stimuli. Koppenol (1980) has shown that the dipole moment orientation of an enzyme changes in concert with its functional activity. Electrets are also "pyroelectric," in that any change in temperature alters the electrical and conformational characteristics of the molecule. The permanent electric dipole moment of pyroelectric bodies results from the parallel alignment of elementary fixed dipole moments. Any change in temperature modifies the length of the pyroelectric body and alters its elementary dipole moments (pyroelectric effect). In the same way, any mechanical change in length or any deformation of a pyroelectric body produces a modification of its dipole moment (piezoelectric effect). Thus every pyroelectric body is at the same time piezoelectric. The two requisites for such behavior are an electric property which causes a permanent dipole moment in the molecules and a morphological property that favors a parallel alignment. Microtubules and other cytoskeletal structures appear to be appropriately designed electret, pyroelectric, piezolectric devices.

The electret state within bone has been well studied and is able to store large amounts of polarization of the order of $10^{-8}$ coulombs per square centimeter (Mascarenhas, 1974, 1975). The limit of charge separation (equivalent to maximal information density) has been calculated (Gutman, 1986) to be about $10^{17}$ electronic charges per cubic centimeter, while there may be about $10^{21}$ total molecules per cubic centimeter. One cubic centimeter of parallel microtubules densely arrayed 100 nanometers apart contains about $10^{17}$ tubulin subunits and therefore may contain $10^{17}$ dipoles equivalent to the maximal density of charge! Electret and related properties can impart interesting and potentially useful properties to biomolecules including cytoskeletal polymers. Among these are the potential capacity to support the propagation of conformational waves such as solitons.

## 6.7 Solitons/Davydov

Important biological events involve spatial transfer of energy along protein molecules. One well known example is the contractile curling of myosin heads in muscle contraction, fueled by the hydrolysis of ATP molecules. These quanta of biological energy are equivalent to 0.43 electron volts, only 20 times greater than background energy and insufficient to excite molecular electronic states. Consequently, under usual conditions biological systems do not emit photons. This implied to Davydov (1977) during the 1970's that ATP hydrolysis energy is transferred by vibrational excitations of certain atomic groups within proteins. Davydov focused on the alpha helix regions of proteins and identified the "amide 1" (carbon-oxygen double bond) stretch vibration of the peptide group as the most likely "basket" in which energy may be carried. His selection was based on the "quasi-periodic," or crystal-like arrangement of amide 1 bonds, their low vibrational energy (0.21 electron volts—half the energy of ATP hydrolysis) and their marked dipole moment (0.3 Debye). According to linear analysis, energy transported by this means should spread out from the effects of dispersion and rapidly become disorganized and lost as a source of biological action. However Davydov analyzed *nonlinear* aspects of the amide 1 stretch and concluded that amide 1 vibrations are retroactively coupled to longitudinal soundwaves of the alpha helix, and that the coupled excitation propagates as a localized and dynamically self sufficient entity called a solitary wave, or "soliton." Davydov reasoned that amide 1 vibrations generate longitudinal sound waves which in turn provide a potential well that prevent vibrational dispersion, thus the soliton "holds itself together."

Solitary waves were first described by nineteenth century naval engineer John Scott-Russell in 1844. While conducting a series of force-speed experiments for boats on a Scottish canal, an accident happened. A rope broke and a canal boat suddenly stopped, but the bow wave which the barge caused kept on going. Russell galloped alongside the canal and followed the wave for several miles. He described the exceptional stability and automatic self-organization of this type of wave. Mathematical expressions of solitary waves can be given as particular solutions of some nonlinear equations describing propagation of excitations in continuous media which have both dispersive and nonlinear properties. These solitary wave solutions have "particle-like" characteristics such as conservation of form and velocity. Such traits led Zabusky and Kruskal (1965) to describe them as "solitons." As an answer to the problem of spatial transfer of energy (and information) in biological systems, Davydov applied the soliton concept to biological systems in general, and amide 1 vibrations within alpha helices in particular. For solitons to exist for useful periods of time and distance, certain conditions must be met. The nonlinear coupling between amide 1 bond vibrations and soundwaves must be sufficiently strong and the amide 1 vibrations be energetic enough for the retroactive interaction to take hold. Below this coupling threshold, a soliton cannot form and the dynamic behavior will be essentially linear; above the threshold the soliton is a possible mechanism for virtually lossless energy transduction.

Computer simulation and calculation of solitons have led to assumptions about the parameters necessary for nolinearity and soliton propagation. A critical parameter is the "anharmonicity," or nonlinearity of the coupling of intrapeptide excitations with displacement from equilibrium positions (Figure 6.5). Anharmonicity is the nonlinear quality which determines the self capturing of the two components of the soliton. The degree to which the electronic disturbance nonlinearly couples to the mechanical conformational change of the protein structure is the crux of the soliton question. If the two are coupled as a step-like functioning switch (nonlinear) rather than a direct linear correlation, they can provide a "grain" to represent discrete entities capable of representing and transferring information. An index for soliton viability, the anharmonicity parameter, is known as $\chi$. For $\chi$ greater than $0.3 \times 10^{-11}$ newtons, solitons *in computer simulation do* propagate through the spines of an alpha helix at a velocity of about $1.3 \times 10^3$ meters per second. The distance of 170 nanometers corresponding to the length of a myosin head in striated muscle would then be traversed by a soliton in about 0.13 nanoseconds. Computer simulations by Eilbeck and Scott

(1979) demonstrate, for above threshold values for the coupling parameter $\chi$, soliton-like excitations propagating along alpha helix spines in the form of a local impulse with a size of a few peptide groups. Experimental data supporting such biological solitons is slowly emerging. Although results from biomolecular light scattering (Webb, 1980) seemed to provide confirmation (Lomdahl et al., 1982), follow up work (Layne, et al., 1985) indicated another source of the experimental observations. Optical experiments on crystalline acetanilide (a hydrogen bonded, polypeptide crystal) however, do provide an unambiguous demonstration of a localized state similar to that discussed by Davydov (Careri, et al., 1984; Eilbeck, et al., 1984; Scott, et al., 1985; Scott, 1985). Definitive resolution of the existence of solitons in biological materials may await the imminent advent of nanotechnology (Chapter 10).

Davydov has considered other types of solitons such as those in solid three-dimensional crystals which have phase transitions. He contends that sufficient anharmonicity in these lattice structures will produce soliton type excitations representing themselves as local displacements of the equilibrium positions moving along the molecular chains. These are called acoustic solitons. Other, "topological" solitons are described as symmetry "kinks" which travel through an ordered medium.

Toda (1979) invoked the concept of solitons to describe local displacements from equilibrium positions of molecules in one dimensional lattices. He studied molecular chains and assumed that displacement of individual molecules within the chain interacted with neighboring molecules. Mathematical evaluation of Toda lattices by Davydov show localized excitations described by a bell shaped function characterizing a reduction in the distance between molecules in the excitation region of the lattice. These are called "supersound acoustic solitons," or "lattice solitons" and have been modeled in proteins by Bolterauer, Henkel and Opper (1986).

Davydov's work further suggests that excess electrons can be captured by supersound acoustic solitons and conveyed along with them, giving rise to "electrosolitons." Electron transfer between donor-acceptor pairs of proteins are found in photosynthesis, cell respiration (ATP generation) and the activity of certain enzymes. These arrangements of structures are often called electron transport chains. Davydov observes that electron transfer has traditionally been assumed to be accomplished by quantum mechanical tunneling as first proposed by Britton Chance and colleagues (Devault, Parkas, Chance, 1967). In these systems, electrons are generally transferred between centers spaced about 3–7 nanometers apart. Davydov argues that this electron transfer can be better explained by an electrosoliton.

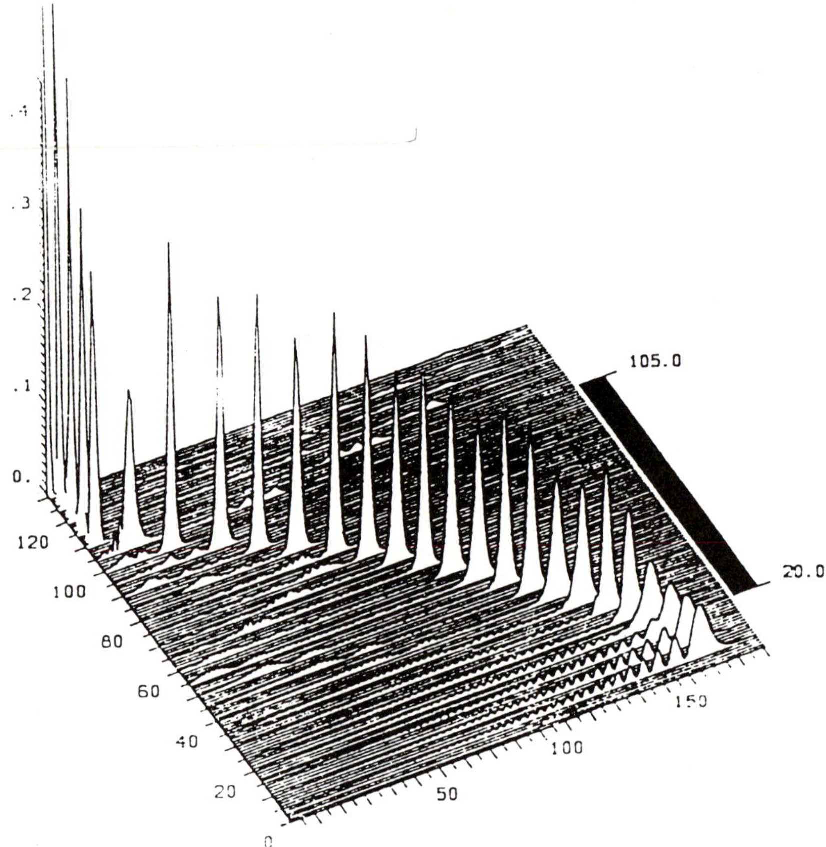


Figure 6.5: Computer simulation showing energy (vertical axis) localization along distance (southwest to northeast axis) as function of anharmonicity (southeast to northwest axis). Over a specific range, energy becomes localized and travels as solitary wave, or soliton. With permission from Bolterauer, Henkel and Opper (1986).

Concrete evidence exists for solitons as giant waves in or underneath the ocean, as optical solitons in laser fiber optics and in other systems. Current technologies are incapable of proving or disproving biological solitons. If Davydov is correct about myosin heads, then solitons are responsible for the molecular level filamentous contractions that drive every move we make. Propagating solitons in the cytoskeleton could be the dynamic medium of biological information processing. If so, solitons would be to consciousness what electricity is to computers.

## 6.8 Coherent Excitations/Fröhlich

Protein conformational states can register dynamic biological information and control the real time functions of cytoplasm. The mechanisms of conformational regulation are not clearly understood, primarily because technology has not (quite) yet reached the nanoscale. Proteins are clearly vibrant, dynamic structures in physiological conditions. A variety of recent techniques (nuclear magnetic resonance, X-ray diffraction, fluorescence depolarization, infrared spectroscopy, Raman and Brillouin laser scattering) have shown that proteins and their component parts undergo conformational motions over a range of time scales from femtoseconds ($10^{-15}$ sec) to many minutes.

The most significant conformational vibrations are suggested by Harvard's Karplus and McCammon (1984) to be in the middle of this range: nanoseconds. Such fluctuations are appropriate for conformational motion of globular proteins (4 to 10 nanometer diameter), consistent with enzymatic reaction rates and "coupled modes" like solitons. As an example, Karplus and McCammon describe a rotation of a tyrosine ring deep inside a globular protein called bovine pancreatic trypsin inhibitor. The side chain of the amino acid tyrosine includes a six carbon "aromatic" hexagonal ring with an electron resonance cloud. The rotation of the ring has been studied experimentally by nuclear magnetic resonance and Karplus and McCammon have done a computer simulation based on that data which shows the protein changing conformational state as the tyrosine ring rotates 90 degrees. The switch occurs in the nanosecond time scale and is collectively coupled to movement in the polypeptide backbone chain.

Collective nanosecond conformational states have been elegantly woven in a theory of *coherent* protein excitations by Professor Herbert Fröhlich who presently divides his time between Liverpool University and the Max Planck Institute in Stuttgart. Recognized as a major contributor to the modern theory of superconductivity, Fröhlich turned to the study of biology in the late 1960's and came to several profound conclusions. One is that changes in protein conformation in the nanosecond time scale are triggered by a charge redistribution such as a dipole oscillation within hydrophobic regions of proteins (Fröhlich, 1975). Another Fröhlich (1970) concept is that a set of proteins connected in a common voltage gradient field such as within a membrane or polymer electret such as the cytoskeleton would oscillate *co-*

*herently* at nanosecond periodicity if energy such as biochemical ATP were supplied. Fröhlich's model of coherency can explain long range cooperative effects by which proteins and nucleic acids in biological systems can communicate. A major component of Fröhlich's theory suggests that random supply of energy to a system of nonlinearly coupled dipoles can lead to coherent excitation of a single vibrational mode, provided the energy exceeds a critical threshold. Frequencies of the order of $10^9$ to $10^{11}$ Hz are suggested by Fröhlich, who maintains that the single mode appears because all others are in thermal equilibrium. Far reaching biological consequences may be expected from such coherent excitations and long range cooperativity.

Conformation of proteins and their dipole moments in aqueous, physiological environment are dominated by interaction of their charge groups with surrounding water and ions. Some biomolecules may possess excited states with very high dipole moments. These levels, according to Fröhlich, tend to become stabilized (become "metastable states") through internal and external deformations and through displacement of "counter" ions like calcium. Metastable states, which correlate with functional conformations, are thus collective effects involving the molecule and its surroundings. A molecule might be lifted into a metastable state through the action of electric fields, binding of ligands or neurotransmitters, or effects of neighbor proteins. Thus rapid, nanosecond oscillations may become "locked" in specific modes which correspond to useful conformations of a protein. For example an ion channel, receptor, enzyme, or tubulin subunit may stay in one conformational state for relatively long periods, on the order of milliseconds. Fröhlich characterizes these conformations as "metastable" states.

Fröhlich observes that the high electric field of the order of $10^7$ volts per meter maintained in many biological membranes (100 millivolts / 10 nanometers = $10^7$ volts/meter) requires an extraordinary dielectric property of the membrane components including lipids and proteins. Similar requirements would exist for cytoskeletal proteins in an electret. Ordinary material would suffer dielectric breakdown in such fields unless specially prepared. Fröhlich contends the biological evolution of this dielectric strength on a molecular scale must have strong significance. Biological organisms are relatively impervious to effects of electromagnetic radiation, yet can be exquisitely sensitive to it in some circumstances. Some biological functions border on the limit imposed by quantum mechanics. Our eyes are sensitive to single photons and certain fish are sensitive to extremely weak electric fields. Such performances, according to Fröhlich, require the use of *collective* properties of assemblies of biomolecules and certain types of collective behavior such as coherent vibrations should be expected.

If coherent oscillations representing dipole vibrations within molecular systems do coherently oscillate in the range of $10^9$ to $10^{11}$ Hz, it should be possible to excite these modes by electromagnetic radiation. In order to couple and excite the biological vibrations, radiation should be matched to the biological frequency and the wavelength should be large compared to the dimension of the oscillating object. A significant amount of evidence supports this notion. Irradiation of a great variety of biological objects with coherent millimeter waves in the frequency region of $0.5 \times 10^{11}$ Hz can exert great influences on biological activities provided the power supply lies above a critical threshold (Grundler and Keilman, 1983). According to Fröhlich, the biological effects are not temperature effects. They show very sharp frequency resonances which indicates that localized absorption in very small spatial regions contributes to the biological actions.

The sharp resonance of this sensitive window has a frequency width of about $2 \times 10^8$ Hz. The layer of ordered water and ions subjacent to membranes and cytoskeletal structures (the "Debye layer") absorbs in the region of $10^8$ Hz. This suggests that the Debye layer is closely involved with the dynamic functional activities of the biostructures which they surround. Green and Triffet (1985) have modeled propagating waves and the potential for information transfer in the dynamics of the Debye layer immediately beneath membranes and cytoskeletal proteins. They have hypothesized a holographic information medium due to the coherent vibrations in space and time of these biomolecules. The medium they consider is the ordered water and layers of calcium counter ions surrounding the high dipole moments in membranes and biomolecules. Thus they have developed a theory of ionic bioplasma in connection with nonlinear properties which relates to the existence of highly polar metastable states. The small scale and ordering would minimize friction in these activities. Fröhlich observes: "clearly the absence of other frictional processes would present most interesting problems." He suggests the possibility of propagating waves due to the lack of frictional processes ("superconductivity") in the biomolecule itself as well as the layer of ordered water or Debye layer (Kuntz and Kauzmann, 1974). Until recently, superconductivity has been considered to occur only in certain ordered materials at temperatures near absolute zero. Recent discoveries, however, have shown that superconductivity can occur in materials at higher temperatures due apparently to coherent ordering and coupling among localized and collective lattice vibrations (Maddox, 1987; Robinson, 1987).

Expanding on Fröhlich's work, Wu and Austin (1978) conclude that oscillating dipoles within a narrow band of resonance frequencies with large enough coupling constants may be expected to cause strong long range (about 1 micron) attractive forces among dipoles. In a dense microtubule array, 1 cubic micron (one billion cubic nanometers) would encompass about 160,000 tubulin subunits—an array sufficiently large for collective effects.

Evidence for such "long-range" effects have been observed in the behavior of red blood cells. Discoids of eight micron diameter (8 thousand nanometers), red blood cells tend to array themselves in stacks called "Rouleaux formation." Rowlands (1983) has studied Rouleaux formation and found that attractive forces begin when the red cells are about four microns (4 thousand nanometers) apart, a distance several orders of magnitude greater than the range of attractive chemical forces. Rowlands views this behavior as consistent with Fröhlich's coherent excitations and long range cooperativity. Rowlands also projects the significance of Fröhlich's theory to communication in the nervous system.

Rowlands (1983) notes that

> A communication band extending from $10^{10}$ to $10^{11}$ Hz ... could pack over a million FM radio stations ... or 150,000 television broadcasts ... the action potential may be just a crude fast transmitter of urgent messages ... . Fröhlich vibrations might be transmitted along the membrane of the nerve fibers, but they would be interrupted by action potentials. It is more likely that *microtubules* in the axon are used.

## 6.9 Massless Bosons, Cytoskeletal Self-Focusing

The complexity of biological systems has attracted the interest of nonbiologists who possess mathematical tools useful for "many body problems." In addition to Fröhlich, these include a group of scientists from the University of Milan (Del Giudice, Doglia, Milani and Vitiello, 1986). Viewing living matter as a sea of electric dipoles, they have taken advantage of mathematical and computer tools used to keep track of the countless particles involved in nuclear reactions. Using a mathematical approach called quantum field theory, the Milan group considers electret states and the consequent ordering of water around biomolecules as sets of dipoles whose states, order and symmetry have collective properties. In a typical *non*biological system, component dipoles are random and disordered, resulting in an overall symmetry.

Such a system would look the same when viewed from any angle ("rotationally invariant"). In living systems, order is induced by reduction of tridimensional symmetry to a rotational alignment along filamentous electrets such as cytoskeletal structures. According to the Milan group quantum field theory and the "Goldstone theorem" require that the symmetry breaking ("Bose condensation") results in long range interactions among system components (dipoles) conveyed by massless particle/waves ("Goldstone bosons"). The Milan group argues that the energy required to generate massless bosons is invested in the electret states of biomolecules and correlated fluctuations of their surrounding water and ions.

Celaschi and Mascarenhas (1977) showed that electret activation energy of biomolecules (0.2–0.4 electron volts) is equivalent to the hydrolysis of one ATP or GTP molecule and what Davydov predicted for initiation of solitons. Consequently solitons, massless bosons, and Frölich's coherent polarization waves may be synonymous.

Pursuing their quantum field approach, Del Giudice and his colleagues came to an astounding concept of self-focusing of electromagnetic energy within cytoskeletal filaments. Electromagnetic energy exceeding a threshold and penetrating into cytoplasm would be confined *inside filaments* whose diameters depended on the original symmetry breaking ("Bose condensation") of ordered dipoles. Any electric disturbance produced by thermally fluctuating dipoles or by any other source would be confined inside filamentous regions. Ordering is preserved outside the filaments and is disrupted only inside where energy becomes concentrated (the "Meissner effect"). The diameter of the self focusing energy filaments depends on the polarization density, or ordering of biological water. Del Guidice's group calculated a self focusing diameter of about 15 nanometers, *precisely the inner diameter of microtubules!* Del Guidice and colleagues feel the cytoskeleton is the material consequence of dynamic self focusing of polarization waves in the cytoplasm. The observed diameters of self focused optical beams in simple *nonbiological* liquids are of the order of microns; correlation among components is created by propagation of waves rather than as a specific property of the material itself. The Milan group concludes that focusing occurs in cytoplasm of eukaryotic cells due to the spatial coherence and ordering imparted by cytoskeletal electret behavior.

The self-focusing predicted by the Milan group would have interesting capabilities. Energy is refracted into beams which become surrounded by cylindrical waveguides: the Indian rope trick. Coherency imparted to the refracted energy by either a Fröhlich-type mechanism or periodic structure of a cytoskeletal waveguide biomolecule could lead to holographic mechanisms. A rudimentary theory of waveguide/holographic behavior in microtubules has been described (Hameroff, 1974). Photorefractive crystals can be used to generate dynamic, real time holography (Gower, 1985) and MT could be projecting dynamic cytoplasmic holograms.

Models of cooperative protein dynamics described by Davydov solitons or Fröhlich coherent oscillations may be different perspectives of the same phenomena. Tuszynski and co-workers (1984) have compared the two approaches and their respective emphasis. They observed that Fröhlich's model concentrates on time-independent effects, or stable states, to explain the establishment of order, whereas Davydov's model highlights time dependent propagation of order via solitons.

The overlapping cooperative models of protein conformational dynamics (coherence, resonance, solitons, electrets, self focusing) are of interest when applied to specific structural elements with relevant properties. For example, Davydov's model of soliton propagation was originally applied to contractile coupling between actin and myosin. The cytoskeleton appears uniquely suited to take advantage of cooperative dynamics related to information processing. Chapter 8 reviews evidence and models of cytoskeletal information processing based on cooperative dynamics. The next chapter describes anesthesia—the result of inhibition of collective, cooperative protein conformational dynamics in the brain.

# Chapter 7

# Anesthesia: Another Side of Consciousness

The net collective effect of protein dynamics and other factors in the human brain is consciousness. Since the time of ancient Sumerians, Egyptians, Assyrians and Greeks, opiate derivatives of the poppy (Papaver somniferum) and other drugs have been used to obtund or ablate consciousness in the face of pain, or for surgical procedures. In the 19th century, anesthetic gases including nitrous oxide ("laughing gas") and diethyl ether were developed. These inhalation anesthetics, when administered properly in a narrow range of concentration, were able to cause a reversible cessation of consciousness. As early anesthetists tragically discovered, excess anesthetic caused inadequate breathing, cardiovascular failure and death. However, when properly used, anesthesia became a boon to mankind. In addition to alleviating the suffering of countless surgical patients, anesthetics also became an important tool in the investigation of consciousness. A trail of experimentation and observation has led to the conclusion that anesthetics inhibit collective dynamic conformational changes in brain proteins.

## 7.1 Levels of Anesthesia/Consciousness

With the widespread advent of diethyl ether, anesthesia became somewhat standardized and it became useful to delineate "stages" in the continuum from the awake state through anesthesia to respiratory paralysis, cardiovascular collapse, and death. In 1847 John Snow published his pioneer monograph, *On the Inhalation of the Vapour of Ether* in which he described five

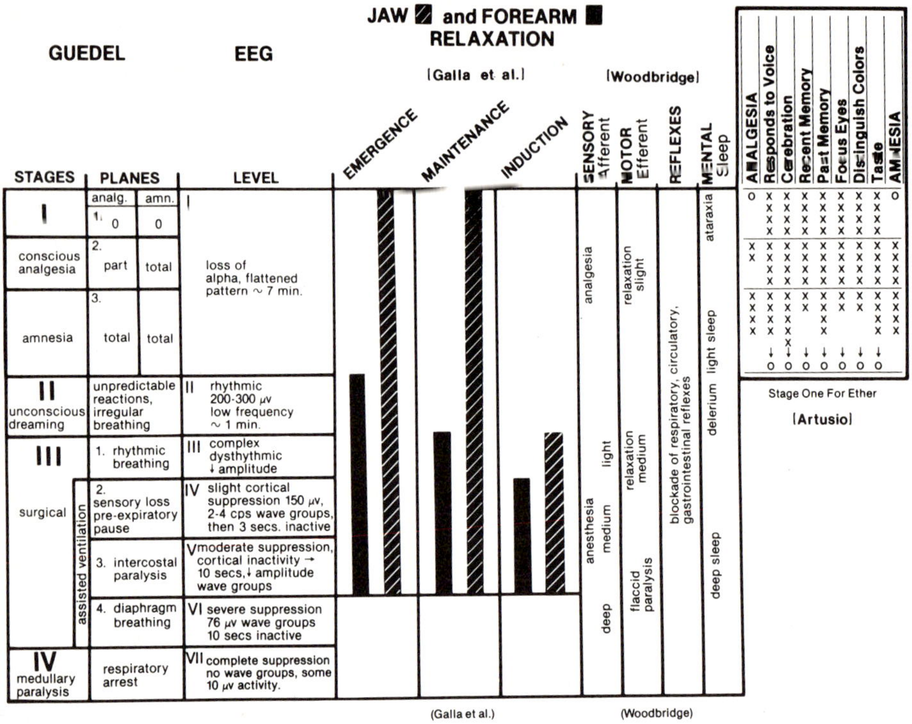


Figure 7.1: Levels of anesthetic depth from different systems displayed on a common axis. Guedel's (1937) stages and planes serve as a standard. EEG levels (Courtin, Bickford and Faulconer, 1950), jaw and forearm relaxation (Galla, Rocco and Vandam, 1958), Woodbridge's (1957) classification of nothria, and stage one for ether (Artusio, 1955) are also represented. With permission from Grantham and Hameroff (1985).

empirical levels through which an anesthetized patient progressed from consciousness to respiratory paralysis. Surgery could be performed in "stage three" characterized by analgesia (pain relief) and amnesia (lack of memory storage), or in "stage four" characterized by muscular relaxation and regular, automatic breathing. Snow's scheme was expanded in 1920 by Arthur Guedel who published codified stages and signs of anesthesia, later detailed in his 1937 monograph, *On Inhalation Anesthesia—A Fundamental Guide.* Guedel predicated his stages on obvious physical signs involving muscle tone, respiratory patterns, and eye signs. He enumerated four levels: stage of analgesia, stage of delirium, surgical stage (subdivided into four planes), and stage of respiratory paralysis. Others have added a fifth stage, anesthetic overdose: lack of tissue oxygen, convulsions, and imminent death.

In Guedel's stage one (Figure 7.1), the patient progresses from alertness and sensibility to pain to total amnesia, analgesia, and sedation. Breathing is slow and regular, with use of both diaphragmatic and rib muscles; the eyelid reflex (twitching of the eyelids in response to gentle brushing) is intact. Stage two has been variously termed the stage of delirium, excitement, unconsciousness, or the "dream state." The patient may pass through this stage sedately or may manifest wild, uninhibited activity. Breathing is irregular and unpredictable, pupils of the eye may be dilated, ocular muscles are active, and eyelid reflex active. In this stage the patient is at risk for unwanted reflex activity such as vomiting, spasm of the airways and cardiac dysrhythmias. The stage two excitement phase caused by anesthetics is somewhat puzzling. Early explanations described inhibition of neocortical "inhibitory" brain circuits, leading to unchecked primitive activity. However, anesthetic effects on single neurons may also show an excitatory phase, thus anesthetic induced "excitation" may be a molecular level effect at low concentrations. Guedel's stage three has four different "planes" of progressively deeper anesthesia during which surgery may be performed. Muscle relaxation is slight in plane one, but progresses to complete abdominal muscle relaxation in planes three and four. Breathing is very irregular and periodic in plane one, as in normal sleep. With plane two, a pause develops between inhalation and exhalation, and inhalation becomes shorter relative to exhalation. Paralysis of rib muscles begins in plane three, and diaphragmatic breathing is prominent. In plane four, paradoxical movement of the rib cage occurs with inhalation, breathing is irregular, and if anesthetic depth is not lightened breathing will stop altogether. Eye muscles are active initially in plane one, however the eyes become immobile when plane two is reached. The eyelid reflex disappears in plane three. The pupils tend to become dilated with progression to plane four. The fourth stage of anesthesia is respiratory arrest and ensuing cardiovascular collapse. Muscles are flaccid, and pupils are widely dilated. Death will occur if anesthetic depth is not decreased from stage four (Grantham and Hameroff, 1985).

Because of the risks of anesthetic overdose and the variable requirements of individual patients, anesthesiologists have sought methods to judge anesthetic depth in addition to the clinical signs enumerated by Snow and Guedel. Since the early 20th century, electroencephalography (EEG) has been recorded from anesthetized patients. EEG waves recorded at the scalp are thought to emanate from summated dendritic synaptic potentials ("dipole fields") generated by pyramidal cells of the cerebral cortex. Scalp potentials generally range from 10 to 200 microvolts, and in frequencies from several Hz to about 50 Hz. The frequency spectrum of the EEG is usually divided into 4 major classifications, delta: less than 4 Hz, theta: 4–8 Hz, alpha: 8–13 Hz, and beta: greater than 13 Hz. Generally, power at higher frequency decreases as anesthetic depth increases. Accordingly, efforts to derive an appropriate index of anesthetic depth have included attempts to quantify an EEG frequency below which most EEG power occurs. This involves a Fourier transform of raw EEG data into a frequency/power spectrum. One example is "spectral edge," the frequency below which 95 percent of EEG power occurs (Rampil, 1981). As anesthetic depth increases, spectral edge roughly decreases. Other attempts to quantitatively assess EEG and anesthetic depth have included the plotting of EEG voltage in "phase space," yielding phase portraits and chaotic attractors. Phase space plotting involves choosing an arbitrary "phase lag," and plotting EEG amplitude at any point in time against the amplitude at that particular time plus the phase lag. This results in geometric phase portraits which may be analyzed for complexity, or "dimensionality" on a continuum of order and chaos. As patients become more anesthetized, the dimensionality of their EEG becomes more ordered, and less chaotic (Figures 7.2 and 7.3, Watt and Hameroff, 1987). Consciousness may thus be described as a manifestation of deterministic chaos somewhere in the brain/mind.

In addition to observing activities of the brain, anesthesiologists must closely monitor other physiological functions of their patients during and immediately after surgery. Cardiovascular, pulmonary, neuromuscular, kidney, blood clotting, and other factors are carefully followed. A variety of technological advances in monitoring permit safer care of sicker patients for more complex procedures. Future monitoring may utilize even more advanced technologies. One possible example is the "biosensor": field effect transistors which are tiny membrane covered chips which can detect a variety of ions, molecules, drugs or hormones. Small enough to fit on the tip of a small catheter harmlessly inserted into a blood vessel or tissue, these biosensors connected to a computer can yield "on-line" monitoring of blood chemistry and many other functions. With the advent of nanotechnology, even tinier, more profoundly sensitive biosensors will materialize. The capability for monitoring nanoscale conformational dynamics of neural proteins (telemetrically or via very small implants) will lead, not only to more sensitive observation of brain function during anesthesia and surgery, but perhaps eventually to the manipulation of consciousness.

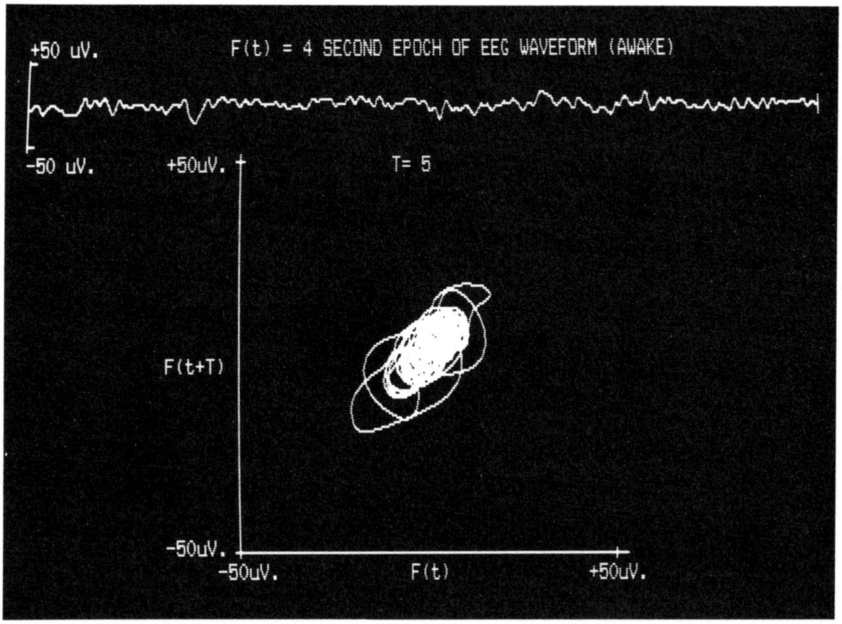


Figure 7.2: EEG from awake patient prior to anesthesia. Top: 4 seconds of "conventional" EEG which shows low voltage, high frequency waves. Bottom: same 4 seconds plotted as phase space trajectory with digitization of 300 Hz and phase lag of 5/300 sec. Phase portrait of awake state is densely centered. From Watt and Hameroff (1987).

## 7.2 Memory

Occasionally patients have reported awareness or recall during anesthesia, an abhorrent event to both patient and anesthetist. Research into this area has illuminated the mechanisms of memory consolidation and led to specific amnesia producing anesthetic drugs. Accordingly, awareness and recall during anesthesia is currently exceedingly rare. In an excellent review of anesthesia and memory processes, Cherkin and Harroun (1971) described a two-stage theory of memory which postulated that information is first perceived and stored in an unstable dynamic form (short-term memory), which may then be consolidated into a stable physical memory trace (long-term memory).

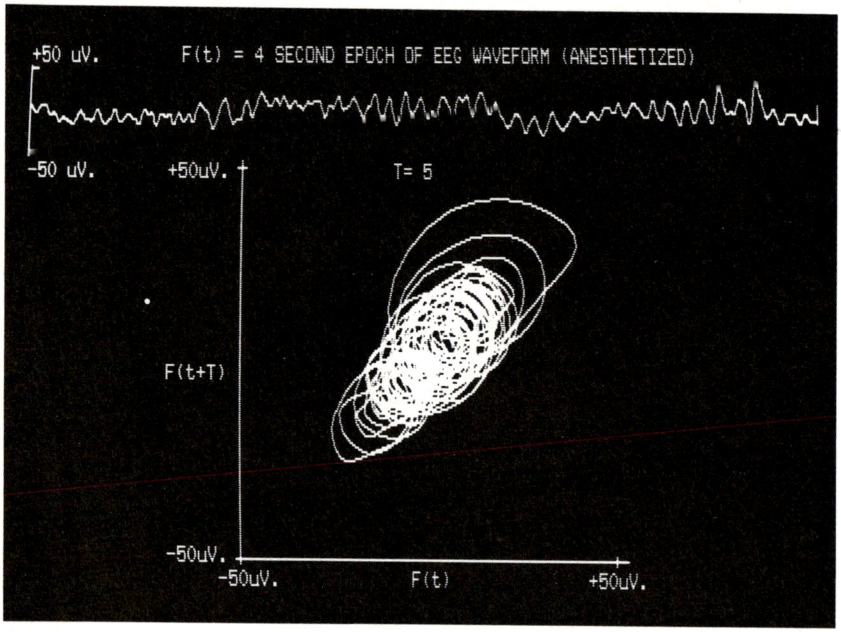


Figure 7.3: EEG from patient after induction of general anesthesia with thiopental and isoflurane. Top: 4 seconds of conventional EEG show higher voltage, lower frequency waves. Bottom: same 4 seconds plotted as phase trajectory as in Figure 7.2. Phase portrait of anesthetized patient is unraveled with "looser," geometric contour resembling an "attractor" Dimensional analysis reveals a less "chaotic," more "ordered" phase portrait in the anesthetized state compared to the awake state. From Watt and Hameroff (1987).

Perception and short-term memory storage may constitute awareness, but do not necessarily cause a physiological response or become long-term memory to produce postoperative recall.

Consolidation of short term memory to long term storage appears to take a finite period of time, on the order of 45 seconds (Figure 7.4). The establishment of long-term memory traces depends to a great extent on the initial impact of the information input. Many of us who are old enough

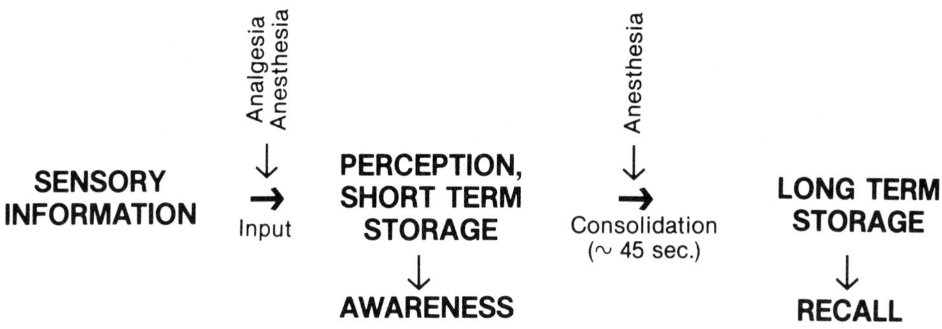


Figure 7.4: Sensory information may be perceived and register short term memory and awareness. A second step which takes a finite period of time results in long term memory storage and recall. Modified from Cherkin and Harroun (1971).

can remember exactly where we were when President Kennedy was shot. Similarly, threatening or derrogatory remarks are more likely to be heard and consolidated to long-term memory and recall in partially anesthetized patients (Bitnar, 1983). Amnesia (lack of recall) results from prevention of memory consolidation at any point from input to long-term memory storage. Anesthetic drugs can act at various stages of this process: narcotic painkillers prevent noxious input, amnestic tranquilizers block consolidation to long term memory, the general anesthetics inhibit these and all other cognitive functions. The consolidation process appears to occur in the brain's hippocampal region but memory storage and awareness are more diffusely distributed.

## 7.3 Mechanisms of Anesthesia

A variety of quite different molecular structures have anesthetic activity. Other molecules, quite similar in structure to anesthetics, may have opposite effects and cause convulsions. Despite these paradoxes, imaginative attempts have been presented to unify a single mechanism for anesthesia and represent clues to the mechanism of consciousness.

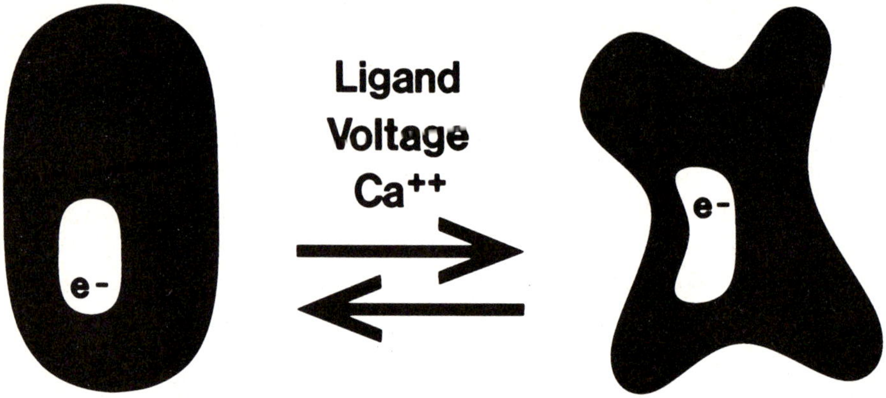


Figure 7.5: Anesthesia results from prevention of protein switching between two or more different conformational states induced by binding of ligand, calcium ion, or voltage change. Anesthetic gas molecules bind by Van der Waals forces within hydrophobic pockets within which electron dipole oscillations may be a trigger, or switching mechanism for protein conformation.

Several neural functions and structures are inhibited by anesthetics, although some are inhibited at concentrations higher than that required for anesthesia. Propagation of action potentials resulting in nerve conduction as well as microtubule dependent axoplasmic transport are blocked by anesthetics such as halothane, which also causes depolymerization of MT. However, the measurable neural function which is sensitive to anesthetics at concentrations barely sufficient to erase consciousness is synaptic transmission. Neurotransmitter release and/or neurotransmitter receptor activation thus appear to be the specific actions whose inhibition results in anesthesia. Inhibition of collective communicative activities within the cytoskeleton and attached membrane proteins related to synaptic transmission are presently undetectable with current technologies and their inhibition may still account for anesthesia. Anatomical localization of sites of anesthetic action has focused on regions with many synapses. Accordingly, the reticular activating system which is involved with regulation of wakefulness and attention and is "polysynaptic" was originally viewed as a major site of anesthetic effect. However, many other brain regions are inhibited by anesthetics at concentrations relevant to anesthesia. Further, different anesthetics which have the same end result have dominant actions at differing areas of the brain, and have differing characteristics and side effects.

The mechanism of anesthesia, the "other side" of consciousness, is a beguiling enigma. Efforts to understand the actions of anesthetics began in mid-nineteenth century; Claude Bernard observed that the anesthetic chloroform inhibited "protoplasmic streaming" in slime mold. Around the turn of the twentieth century Meyer (1899) and Overton (1901) discovered that the anesthetic potency of a group of compounds directly correlated with their solubility in a lipid environment, specifically olive oil. Because membranes are largely lipid, the natural conclusion was that anesthetics exerted their effects through actions on lipids in membranes. Much later it was determined that critical, dynamic effects in membranes occurred via proteins, and that hydrophobic regions *within* proteins were "lipid-like" and hence able to bind anesthetic gas molecules by weak Van der Waals forces. Most contemporary theories agree that anesthesia results from alteration of dynamic, conformational functions of important brain neural proteins: membrane ion channels, synaptic receptors, cytoskeleton, neurotransmitter releasing mechanisms, and/or enzymes (Koplin and Eger 1979; Kaufman, 1977; Eyring, Woodbury and D'Arrigo, 1973). Opinions and theories disagree as to whether anesthetics alter these dynamic protein functions *directly* via intra-protein hydrophobic pockets, or *indirectly* via membrane lipids surrounding proteins, or at lipid protein interfaces (Franks and Lieb, 1982). There is further disagreement as to whether a single, unitary mechanism can explain actions of a wide range of anesthetic compounds acting on a presumably wide range of neural proteins, as well as explain reversal of anesthesia by increased pressure, and the paradoxical relationship between anesthetics and convulsants (Halsey, 1976).

Correlations of anesthetic potency with solubility in "lipid-like" solvents (olive oil, octanol, lecithin) are consistent with anesthetic effects occurring either in lipids or in "lipid-like" hydrophobic pockets within proteins (Eger, Lundgren, Miller and Stevens, 1969). Wulf and Featherstone (1957) were among the first to demonstrate binding of anesthetics within hydrophobic regions of whale myoglobin and other proteins. They suggested that anesthetic binding caused a protein conformational change with a resultant increase in exposure of charged groups at protein surfaces which sufficiently altered protein function to cause anesthesia. Altered charge groups can change protein binding of surface water and can explain the findings which led Pauling (1965) and Miller (1969) to propose that anesthetic induced water-protein complexes ("clathrates") were the cause of anesthesia. Other research including Brammall, Beard and Hulands (1974) demonstrated that anesthetics can cause conformational changes in functional proteins, but at concentrations significantly higher than that required to cause reversible clinical anesthesia.

A logical inference is that anesthetics, rather than causing protein conformational changes, *prevented* those that were necessary for normal function. Looking for model systems of functional dynamic activity in proteins, a number of researchers have focused on a seemingly obscure group of systems which, on the surface, are distantly removed from any semblance of cognitive function: anesthetic inhibition of luminescence from photoproteins in firefly and a variety of bacteria (Harvey, 1915; Halsey and Smith, 1970; White and Dundas, 1970). Firefly luciferase is a protein dimer of about 100 kilodaltons (not unlike tubulin). When combined with a 280 dalton hydrophobic molecule called luciferin, and in the presence of ATP and oxygen, an excited electron state is induced which results in photon emission. The light emission from the luciferase/luciferin complex is exquisitely sensitive to anesthetic gases in proportion to their anesthetic potency. Studying the light emission from firefly luciferase, Ueda (Ueda, 1965; Ueda and Kamaya, 1973) proposed that anesthetic binding at a hydrophobic site in the luciferase protein induced an inactive conformational state which prevented the ATP activated excited state which would normally result in luminescence. In more recent anesthetic studies on firefly luminescence, Franks and Lieb (1984) corroborated earlier work that anesthetic potency correlated with luciferase inhibition potency, much like the earlier correlations with solubility in "lipid-like" hydrophobic environments. They also reported that anesthetic binding did not *cause* a luciferase conformational change, and that the anesthetic effects occurred by competitive binding with luciferin for a hydrophobic site in luciferase. Franks and Lieb suggested that anesthesia occurs due to displacement of endogenous ligands from brain neural protein hydrophobic pockets which bear some resemblance to the luciferin site within firefly luciferase. Displacement of endogenous ligands would not explain, however, anesthetic inhibition of functional protein conformational changes which occur in response to stimuli other than hydrophobic ligands (i.e. voltage change, calcium, other ions).

Ultimately, anesthesia results from impairment of protein conformational dynamics. As described in Chapter 6, mechanisms which regulate protein states are not totally understood, but collective, cooperative mechanisms appear important. The conformational regulation of anesthetic sensitive neural proteins is schematically summarized in Figure 7.5. Under normal, non-anesthetic conditions, there is a class of neural proteins which undergoes functional conformational change in response to appropriate stimuli. These proteins may be membrane bound ion channels which become permeable to sodium, potassium, chloride, or calcium ions; membrane bound receptors for acetylcholine, gamma-amino butyric acid (GABA), glycine, aspartate, glutamate or other neurotransmitters which are coupled to ionic channels; cytoskeletal proteins which perform a wide range of dynamic intracellular functions; or various enzymes. Stimuli which induce functional conformational change in these proteins are thought to include binding of ligands (i.e. neurotransmitters), changes in voltage, or ionic flux (i.e. calcium ions). These functional conformational changes may either facilitate neuronal excitatory activity or have inhibitory effects. The common anesthetic sensitive link for both excitatory and inhibitory effects appears to be a protein conformational change.

The coupling of conformational states to appropriate stimuli is not clearly understood; possible explanations were discussed in Chapter 6 and appear to involve collective, nonlinear coupling of electronic mobility to mechanical movements. Conformationally coupled electron mobility within neural protein hydrophobic pockets may thus be essential to highest cognitive function and sensitive to the effects of anesthetic molecules. Evidence from gas chromatography (McNair and Bonelli, 1969) and corona discharge experiments (Hameroff and Watt, 1983) suggest that anesthetic gases can directly alter electron energy, capturing, retarding, or enhancing their mobility by Van der Waals-London forces (instantaneous dipole coupling). Thus regulation of protein conformational state, as proposed by Fröhlich's theory of coherence, electret behavior or Davydov's soliton model, would be directly inhibited by anesthetic occupancy of protein hydrophobic pockets because the environmental medium for electron mobility would be significantly altered. Hydrophobic pockets vary in size and shape among different proteins which would account for the variability among different anesthetic gas molecules. The weak, reversible Van der Waals interactions by which anesthetics bind within hydrophobic regions explain the reversibility of anesthesia. The flip side of anesthesia, consciousness, may be inferred to occur due to dynamic protein conformational effects among a distributed class of brain neural proteins. Because consciousness is the brain function most susceptible to anesthetics (which can spare breathing control and many reflexes at appropriate concentrations) it appears related to *collective* effects of protein conformational dynamics. Inhibition of a subset of the collective dynamics would therefore inhibit consciousness.

Conversely, *excitation* of a specific subset involved in collective protein dynamics could lead to alteration or enhancement of consciousness. Psychoactive drugs such as amphetamines and LSD can exert profound effects on the brain/mind at extremely low concentrations. They are thought to bind and to exert their effects at membrane protein receptors such as the serotonin receptor in the case of LSD. Kang and Green (1970) correlated the potency of psychoactive compounds with their electronic bond structure and molecular orbitals. They found that the drug molecules' electronic orbital energy available to the receptors (and connected membrane and cytoskeleton) was an index of drug potency. Thus collective effects related to electron mobility within key proteins in a cooperative assembly may be important to consciousness.

# Chapter 8

# Models of Cytoskeletal Computing

Cooperative, collective effects of dynamic protein conformational states are a likely substrate for biological intelligence ranging from cytoplasmic probing to human consciousness. The activities, functions, and structures of microtubules and other cytoskeletal components appear suited to information processing and have led at least a dozen author groups to publish theoretical models of rudimentary cognition within MT and the cytoskeleton. The concepts range from passive MT signal transduction, to descriptive patterns among MT subunit states, to dynamic cooperative "automaton" effects among coherent oscillations of centrioles and the cytoskeleton, to cytoplasmic/cytoskeletal "sol-gel field" effects utilizing holographic imagery. If correct, these proposals could explain many aspects of information representation and dynamic organization in biological systems. Hopeful metaphors, these models are non-exclusive and may be overlapping and complementary.

Cytoskeletal information processing would require some mechanism of cooperativity, long range order, coherence, and/or energy transfer among cytoskeletal components and their subunits. Such possible mechanisms were discussed in the previous chapter. Specific evidence which supports transfer of energy and information within microtubules will be discussed here followed by thirteen models of cytoskeletal information processing.

## 8.1 Energy and Information in Microtubules

Direct support for the propagation of signals in MT has been generated by Vassilev, Kanazirska, and Tien (1985) who reconstituted bilayer membranes from brain lipids and studied their electrical excitability. They suspended membranes as parallel unconnected plane surfaces separated several millimeters apart in a buffer solution which contained depolymerized tubulin, GTP, and other physiological components. Each membrane was monitored electrically and baseline recording of the two membranes showed no electrical coupling; when one membrane was electrically stimulated it depolarized, but the other membrane remained silent. When the tubulin was caused to polymerize into MT (by lowering calcium ion concentration) MT bridges formed between the two membranes and electrical coupling between the two membranes was observed. Electrical stimulation of one membrane then resulted in depolarization in both membranes. The addition of the MT destabilizing drug colchicine prevented coupling, demonstrating that intact microtubules were necessary. The authors concluded that intermembrane signaling occurred by electrically induced polarization and conformational changes of MT components which linked the two membranes. They suggested that similar communication functions occurred routinely within the cytoskeleton.

Another series of experiments which supports the notion of MT mediated signaling is fluorescence resonance transfer among MT and membrane components. Becker, Oliver and Berlin (1975) developed a technique to study energy transfer among fluorescent groups separately attached to different MT subunits or to membranes. Resonance energy transfer occurs when a fluorescent portion of a molecule ("chromophore") which is electronically excited by the absorption of light energy transmits that energy to another "acceptor" chromophore some distance away. This transmission requires the overlap of the emission spectrum of the "donor" chromophore with the absorption spectrum of the acceptor, without involving the actual reabsorption of light by the acceptor. The process is therefore referred to as "nonradiative" resonance energy transfer and occurs if the distance between the chromophores is relatively close, not to exceed about 10 nanometers. In the study by Becker, Oliver, and Berlin, fluorescein isothiocyanate (FITC) was used to fluorescently label one population of unpolymerized MT subunits or membranes, and another fluorescent label, rhodamine isothiocyanate (RITC) was used to label a second population of tubulin. They chose these chromophores because they bind covalently to tubulin or membranes, and because the emission spectrum of FITC "donors" extensively overlaps the absorption spectrum of RITC acceptors. Recordings of fluorescence spectra reveal the "resonance transfer" when it occurs. When MT were depolymerized, a mixture of donor labeled and acceptor labeled tubulin did not show resonance transfer. With polymerization or aggregation of MT subunits, fluorescent excitation of fluorescein labeled tubulin resulted in fluorescent emission by rhodamine labeled tubulin, as the chromophores were brought sufficiently close together in a common lattice to permit resonance energy transfer. The energy transfer occurred not only among tubulin subunits in MT, but among MT subunits and membrane components.

Evidence for another mode with communicative implications in microtubules is suggested by the parallel alignment of MT in applied electric and magnetic fields (Vassilev, Dronzine, Vassileva, and Georgiev, 1982). They cite the postulated existence of low intensity electric fields (Jaffe and Nuccitelli, 1977; Adey, 1975) in the range of 20 to 500 millivolts per centimeter within cells (one millionth of the field strength across polarized membranes). Vassilev and colleagues isolated rat brain tubulin and created polymerizing conditions in the presence of pulsed electric fields of about 25 millivolts per centimeter. Electron micrographs showed that the MT polymerized in perfect parallel alignment with the applied field. Similar results were obtained when low intensity (0.02 Tesla) magnetic fields were applied. If assemblies of MT can also *generate* electric and/or magnetic fields of similar intensity via an electret effect, then a cooperative communication comparable to the "Indian rope trick" may be utilized in cellular growth, differentiation, and synaptic plasticity. MT could then generate their own pathways for cytoplasmic movement.

Other data (Matsumoto and Sakai, 1979; Alvarez and Ramirez, 1979) suggest that the intraneuronal cytoskeleton is necessary for nerve membrane excitability and synaptic transmission. Nerve membrane proteins including ion channels and receptors which are anchored to the cytoskeleton may be the "tips of an iceberg" of a cytoskeletal communicative medium which could utilize a number of possible modes to achieve collective cooperativity and intelligent cellular behavior. The following models suggest some possible strategies.

## 8.2 Cytoskeletal Information Processing

The following models have been roughly grouped by authors and significant concepts.

### 8.2.1 MT Sensory Transduction/Atema

Several authors have discussed the transduction of sensory information by the mechanical distortion of cilia: membrane covered centriole-like structures which protude from cells. Lowenstein, Osborne, and Warshall (1964) suggested that the "kinocilium" of the hair cells in the inner ear served as motile cilia in reverse. They reasoned that motile cilia produced movement using chemical energy provided by ATP hydrolysis, so mechanoreceptor cilia should transduce mechanical deformation caused by environmental stimuli to provide the cell with "patterns of chemical energy representing information." Lettvin and Gestalind (1965) also proposed that receptor cilia operated "by virtue of transmitting mechanical signals to the base of the cilium."

Biologist Jelle Atema (1973) of the Wood's Hole Oceanographic Institute, whose work had focused on acoustical perception at the cellular level, also linked microtubules and sensory transduction. Noting common cilia-like structure in sensory receptor organelles among a wide variety of organisms, Atema proposed that sensory cilia conveyed environmental information to the rest of the cell. He suggested that transduction occurred by propagated conformational changes in the microtubule subunits which constituted these cilia. He argued that microtubules were active functional units in reception and transduction of sensory information.

Atema reviewed conformational changes in MT tubulin subunit dimers observed by a variety of authors and suggested that these occurred under physiological conditions. For example, oscillations of sperm flagella are apparently not controlled by their cell membrane but rather are direct properties of their microtubules in the presence of ATP (Lindemann and Rikmensboel, 1972). Atema concluded that a sequence of subunit conformational changes is likely to occur in microtubules. In motor cilia and flagella, the distortion would originate at the base of the structure and be propagated distally, resulting in wavelike or whiplike motions which propel the organism. In sensory cilia, the distortion apparently originates near the distal end of the ciliary MT and propagates in either direction or at least proximally towards the cell body. There the signal may propagate via either an excitable membrane or through the anchoring basal body and cytoskeleton. Because mechanical deformation and/or local chemical distortion are sufficient stimuli to start a propagating signal in sensory cilia, Atema's microtubule theory assumed that distortion of tubulin conformation by any number of sources was sufficient to propagate a conformational wave. Thus light energy, chemical bond energy, and mechanical forces could be transduced by sensory cilia.

Atema's view of MT information processing was an "all or none" propagation by allosteric conformational changes along tubulin protofilaments. His was the first theory to look beyond the global behavior of cilia to consider conformational effects in tubulin components. Subsequent theories became more elaborate to consider localized analog functions, switching, and collective neighbor interactions among MT subunits.

### 8.2.2 MT Mechano-Ionic Transducers/Moran and Varela

Biological sensory systems ranging from human inner ears to honey bee hairplate receptors use sensory cilia to transduce mechanical deformation into cognitive information like position in space ("proprioception"). Harvard biologists Moran and Varela (1971) studied the proprioceptive transduction of mechanical deformation in the legs of cockroaches. There, tactile spines contain mechanoreceptor structures called "campaniform sensilla" which consist of a single bipolar neuron from whose dendritic tip extends a modified cilium containing 350 to 1000 microtubules. Moran and Varela determined that this parallel bundle of MT was directly involved in the transduction of mechanical deformation to the neuron.

All neurons are mechanical receptors, being stretch sensitive to cause ionic currents. In the cockroach campaniform sensilla, the cilium is the only bridge and transduction does not occur when MT are incapacitated with tubulin binding drugs like colchicine and vinblastine (Moran and Varela, 1971). MT are thus directly involved in proprioceptive determination of the system's spatial relation to its environment. Moran and Varela saw two possible roles for MT in their mechanoreceptor function: as passive translation rods, or as generators of ionic/electrical currents. They suggested that MT behaved as "mechanochemical engines driven backward." Compression or bending of MT would cause release of bound ions from MT subunits, resulting in an ion flux. Moran and Varela saw these ionic currents capable of generating membrane depolarization, being perhaps the first to suggest that MT can regulate membranes. Each tubulin dimer reversibly binds 16 calcium ions so ionic fluxes of significant current could result from an active assembly of parallel MT. This process may be analogous to the coordinated release of calcium ion waves by sarcoplasmic reticulum in muscle cells which triggers actin-myosin contractile interactions. Calcium waves are also thought to regulate the bending and waving of cilia and flagella, and can regulate the cytoskeletal "ground substance" by coupling to sol-gel states. Moran and Varela's contribution was to observe that MT could release ions such as calcium in a controlled and modifiable manner useful for intracellular communication.

### 8.2.3 Cytomolecular Computing/Conrad and Liberman

Wayne State University computer scientist Michael Conrad and Soviet information scientist E. A. Liberman have collaborated to consider aspects of biomolecular computing including the cytoskeleton. They consider that the "computing power of the brain is primarily based on intracellular processes." They propose that the cyclic nucleotide system (energy rich molecules such as cyclic AMP) sculpts three dimensional dynamic patterns in the cytoplasm of neurons and other cells which are the analog texture of real time information processing. Conrad and Liberman view reaction diffusion patterns of cyclic AMP, regulated and perceived by both cell membranes and the cytoskeleton, as a link between macroscopic neural activities and molecular scale computing. Conrad, known for his conceptualization of protein enzymes as possible computer components (Chapter 1), has also advanced the notion of molecular automata within cells (Conrad, 1973). He and Liberman suggest that the intracellular cytoskeleton perceives mechanical stretch or distortion subsequent to membrane events and accordingly regulates cyclic AMP and other biomessengers.

Conrad and Liberman (1982) suggest:

> in neurons, mechanical stimulation appears on movement of the intraneuron tubule skeleton and micromuscle ... details of the skeleton and micromuscles (are) suitable for constructing the molecular analog ... of the real physical field in which this system moves.

Conrad and Liberman view a molecular analog within the cytoskeleton as a representation of the external world. Extrapolated to complex systems like the brain, such a cytoskeletal analog could suffice as a medium of cognition.

Conrad (1985) notes that

> highly parallel signal processing and vibratory behavior on the part of microtubules and other cytoskeletal elements could play a significant role.

### 8.2.4 MT Signal Processing/DeBrabander

Cell biologist Marc DeBrabander has extensively studied activities of the cytoskeleton in mitosis and cellular organization in general. He and his colleagues at Belgium's Janssen Pharmaceutica Research Laboratories (DeBrabander, DeMey, VandeVeire, Aerts and Geuens, 1975, 1986) have examined the distribution and movement of cell membrane proteins and gathered evidence that they are linked to the cortical actin filament system. In turn, the ordered activity of the filaments appears to be controlled by the microtubule system, analogous to its function in cell movement and cell shape. DeBrabander (1977) observes that in these phenomena MT act beyond their capacity as skeletal support elements and perform as signal transducers from the cell center to the periphery, and vice versa. Interactive signaling in the cytoskeleton fits with observations that local events at one site in the cell often lead to a global cellular rearrangement. DeBrabander raises several MT features which could favor signaling: MT intrinsic polarity, the existence of "centrifugal and centripetal microtubules," the conformational propagation mechanism proposed by Atema (1973), directional organelle movement and topography, and transport of ions such as calcium along microtubules.

DeBrabander and his associates (1986) have also innovated new techniques in the examination of cytoskeletal structure and dynamics. These include "Nanovid Microscopy," a nanometer scale video microscopy in which transport of labelled particles along individual MT may be visualized, and a method of labelling individual tubulin subunits within MT which are either tyrosinated or glutamated (Geuens, Gundersen, Nuydens, Cornelissen, Bulinski, DeBrabander, 1987). As described in Chapter 5, polymerized MT may be detyrosinated in the cytoplasm subsequent to DNA directed genetic input. Tyrosine, the terminal amino acid in tubulin, may be removed to expose glutamate as the new terminal amino acid. Thus all tubulins within polymerized MT are either tyrosinated, or glutamated. A variety of cytoplasmic factors determine whether or not a particular tubulin is tyrosinated. The precise significance or function of tyrosination/glutamation is unknown but could serve as a convenient programming or memory function. DeBrabander and colleagues developed a double labeling technique in which immunogold particles bind to tubulin subunits. Large immunogold particles (10 nanometers) identify glutamated tubulin and smaller particles (5 nanometers) bind to tyrosinated tubulin. Patterns of tyrosinated/glutamated tubulin within MT show heterogenous distribution, and suggest the potential for a coding mechanism. MAP attachment, tubulin conformation, calcium binding

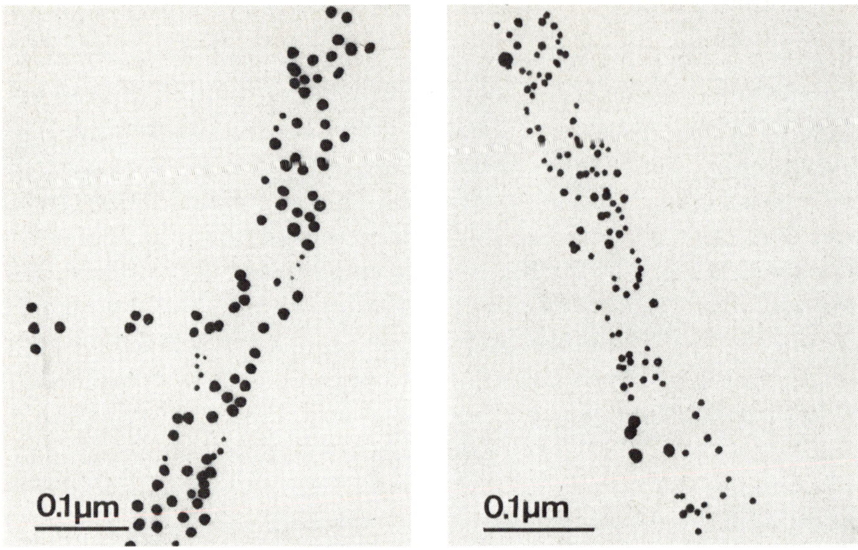


Figure 8.1: Microtubules from PtK2 cells double labeled with antibody and immunogold under high magnification. Large circles (10 nanometer gold particles, arrows) tag glutamated tubulin subunits, small circles (5 nanometer gold particles) label tyrosinated tubulin subunits. Left: interphase microtubule, Right: spindle microtubule. With permission from Geuens, Gundersen, Nuydens, Cornellisen, Bulinski, and DeBrabander (1986), courtesy of Marc DeBrabander and Janssen Pharmaceutica Research Laboratories.

and other factors could be coupled to MT function via such patterns. Although predisposition to detyrosination may be genetically linked to tubulin isozymes, the actual patterns of tubulin detyrosination are determined by "real time" cytoplasmic factors. DeBrabander and colleagues have provided the first direct evidence of modifiable patterns of tubulin variability in intact microtubules. Consequently, MT appear capable of signal *processing* in addition to signal transduction.

### 8.2.5 Cytoskeletal String Processors/Barnett

Brooklyn College computer and information scientist Michael P. Barnett has pursued the design of computer components suited to molecular scale

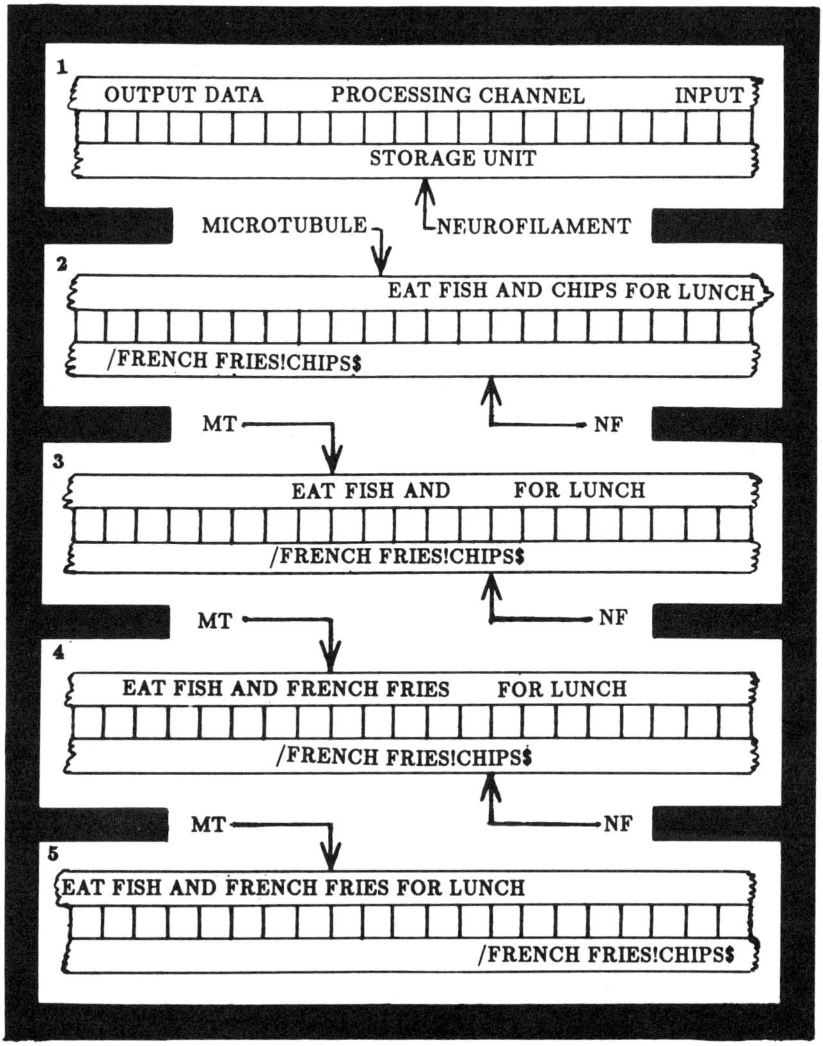


Figure 8.2: Barnett's string processor model of microtubule-neurofilament memory storage. 1) a processing channel (microtubule-MT) is laterally connected to a parallel storage unit (neurofilament-NF), 2) information traverses the processing channel from right to left; on the storage channel, synonym pairs move from left to right, 3) as the word "chips" moves by its own self in storage, it is deleted, 4) "chips" is replaced by its synonym "french fries" and 5) the replacement is complete. By Paul Jablonka.

devices. In addition to the development of tiny binary switches to be assembled into circuits on digital architecture, Barnett (1987) has also sought an associative memory which can analyze imprecise analog data as well as conventional digital information. Fabricated systems with these features could be versatile, powerful and a boon to the goals of artificial intelligence. Like many AI oriented researchers, Barnett turned to biology for clues. However, unlike most AI researchers, Barnett scoured the *subcellular* biological realm in search of molecular scale information processing concepts. His fancy was captured by cytoskeletal microtubules and neurofilaments! He has proposed information representation as patterns in the subunits of cytoskeletal polymer subunits. He proposes that specific subunit states may be characterized by "electrons transferred from delocalizable molecular orbitals," but his basic premise would also be supported by other causes of conformational state variability among MT and neurofilament polymer subunits. Barnett suggests that filamentous cytoskeletal structures operate like information strings analogous to word processors.

In Barnett's conceptualization (Figure 8.2), information strings move from right to left along processing channels, which run parallel to one-dimensional memory channels in which character strings can be stored. Barnett's string transformers can perform global replacements on sequences of characters like common word-processors. His model assumes the existence of processing channels (MT) along which strings of information can move, and memory channels (neurofilaments) which consist of a succession of locations, each of which can hold a single character. Parallel array and lateral interconnectedness of MT and neurofilaments could qualify these cytoskeletal elements as string processors, assuming that information may be represented in the polymers. Barnett's model is thus compatible and complementary with other models of conformational *patterns* within MT and the cytoskeleton.

### 8.2.6 Microtubule "Gradions"/Roth, Pihlaja, Shigenaka

Roth, Pihlaja, and Shigenaka (1970), and Roth and Pihlaja (1977) have considered information processing in two types of biomolecular assemblies: membrane rosettes and microtubules. Citing cooperative allosteric effects among adjacent proteins or protein subunits, they propose that *conformational gradients* in protein arrays represent information by patterns of conformational states among near neighbors in protein lattices.

Rosettes are ordered rings which consist of from seven to twelve membrane embedded proteins important in membrane fusion in lower organisms.

One example of rosette function is in "suctorian feeding", in which certain protozoa affix themselves to a host cell membrane and feast upon its cytoplasm by sucking its contents through a rosette common to both membranes. Bardele (1976) suggested that rosettes utilize concepts of cooperativity and allosterism because the triggering of conformational dynamics within the complex does not require contact with all particles. Stimuli at only one or a few subunits followed by allosteric changes in the rest can cause activation of the total complex. Satir (1973) showed that rosettes in membranes of tetrahymena have rosette pairs in register with each other: internal and external rings of eight proteins each. Induction of a conformational change by binding of an effector molecule at the external rosette causes allosteric cooperative changes in the conformation and functional states of all sixteen subunits. In mutants with aberrant rosettes, a minimum of 5 subunits is required before rosettes functionally respond.

Roth, Pihlaja, and Shigenaka considered the next level of complexity in protein assemblies to be exemplified by microtubules. They viewed MT as highly oriented patterns of tubulin dimers and anticipated at least three different conformational states of each tubulin dimer, based on studies of tubulin binding to vinblastine and GTP (Luduena, Shooter, and Wilson, 1976). Biologists Roth, Pihlaja, and Shigenaka (1970) studied the patterns of linkage among MT within the axopod of a simple organism called Echinosphaerium, a complex spiral assembly of hundreds of MT. The precisely patterned, interwoven spiral arrays of MT which comprise the axopod are explained by inter MT bridges of two different lengths. The authors considered that these two types of linkages can not only account for the complex structure of the axopod, but also indicate two different conformational states of the tubulin dimers to which the bridges attach. Along the lengths of the axopod, microtubule linkages are found at intervals that reflect binding at approximately every fourth dimer. These precise linkage sites represented a code of some sort to Roth, Pihlaja, and Shigenaka who proposed in 1970 that recurrent patterns, or "gradients" existed in microtubules to localize these precise binding patterns. They defined these patterns as "gradions": repeating conformational sequences or fields in which multiple varying patterns can exist in the same molecular architecture simultaneously.

Conformational states of individual tubulin subunits were thought to be controlled by factors Roth, Pihlaja, and Shigenaka termed "gradionators." These could include ligand induced conformation, tubulin isozyme, detyrosination, or binding of MAPs including inter-microtubule bridges. They defined discrete gradion fields within microtubule surfaces as the neighbor-

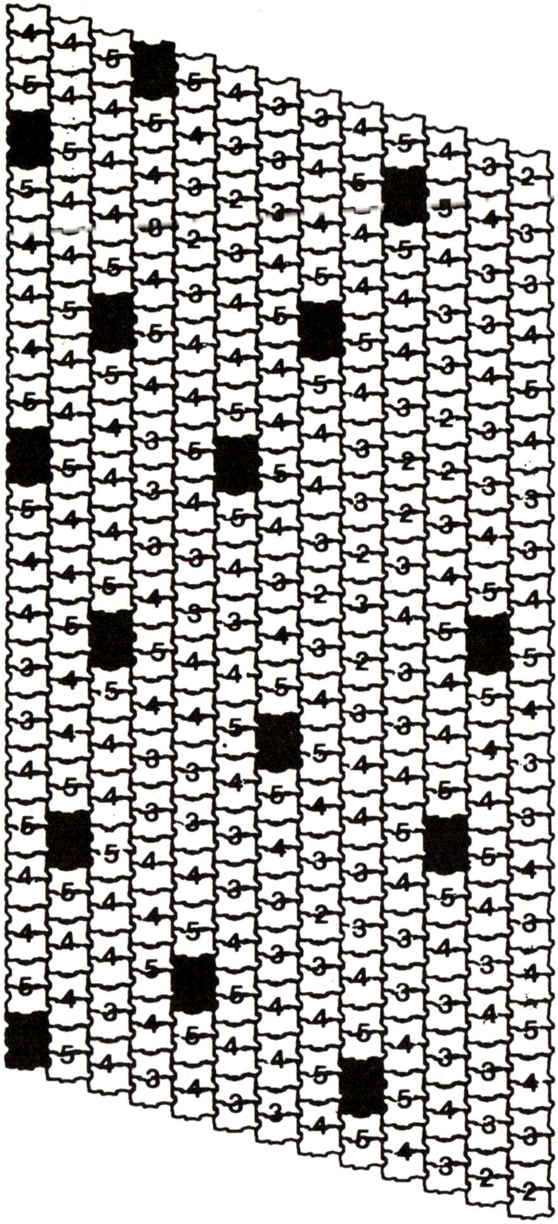

Figure 8.3: "Gradions" within microtubule lattice. Dark shaded dimers are MAP attachment patterns; numbered dimers are determined by proximity to MAP attachment sites. MAP binding and other "gradionators" are thought to induce coded patterns representing information. From Roth, Pihlaja and Shigenaka (1970) by Paul Jablonka.

hood areas around intermicrotubule MAP bridges. They further assumed five possible conformational states for each dimer determined by inter-MT linkage sites and their cooperative allosteric effects, as well as dimer binding by a number of different native substrates or foreign molecules. The pattern of tubulin dimer conformational states in a region of MT lattice "governed" by attachment of inter-MT linkages (or other MAPs) is defined as a "gradion," or conformational field which includes about seventeen dimers. Consequently, $5^{17}$ different gradion patterns could exist if the conformation of each subunit were independent of all others. Cooperativity and allosteric effects preclude independence, so the number of possible MT "gradions" is somewhat less, but still substantial. Roth and Pihlaja (1977) see formation of many "gradions" within axopod MT as a mechanism for position determination and coding for connections. Applied to neuronal cytoskeleton, such deterministic patterns would be generally useful for mental processes in the brain and be a viable candidate for "grain of the engram."

The gradion theory may be applied to any large field of allosteric particles. Allosterism and cooperativity in membranes, cell junctions, and protein particles in many cells could explain aspects of cooperative communication, however microtubules appear especially well suited for information functions. The MT gradion model of Roth, Pihlaja, and Shigenaka is an early specific theory of information processing in protein assemblies in general, and microtubules in particular.

### 8.2.7 Gyroscopic Centrioles/Bornens

French molecular biologist Michel Bornens (1979) has investigated cellular mechanisms of organization and argues for a dynamic stability based on organizational properties of centrioles and the cytoskeleton. Specifically, Bornens proposes that centrioles are animated by rapid oscillatory rotation about their longitudinal axis which results in a dynamic stability and inertia analogous to a spinning top or gyroscope. Peripheral movements throughout the cell are interconnected by the cytoskeleton, with the gyroscopic centrioles an inert point of reference which provides cellular gravity. Centrioles and microtubule organizing centers (MTOC) are the origin for the cell's spatial coordinates and cytoplasmic movement appears to occur relative to the MTOC. Bornens likens movement in each cylinder to a stepwise electric motor in which the central "cartwheel" (or "pinwheel") is equivalent to an ATP-dynein "stator," and the centriolar wall the moving "rotor." Continuous torque rotation of centrioles could serve to propel them through cytoplasm, like an "Archimedes screw" (Record, 1986). Bornens also considers the possibility of "back and forth" oscillation with utility for scanning the cell environment.

Reviewing Bornens' gyroscopic centriole model, Albrecht-Buehler (1981) questioned the high rate of rotation required for significant inertia. Albrecht-Buehler calculated that centrioles would have to rotate at frequencies of 2.3 million revolutions per minute before their kinetic energy matches one kT (the thermal energy of one molecular degree of freedom). Consequently, "much greater rotational frequencies would be required before the centriole could withstand the impact of thermally moving molecules around them and maintain stable axial orientation" (Albrecht-Buelhler, 1981). Bornens responded by suggesting a submicroscopic mechanism "allowing more independence of the centriole with respect to surrounding material." Factors which could support his contention include ordered water coupled to centriolar oscillation, some unknown property of the "pericentriolar material," ionic charge layer, or even superconductivity as suggested by Fröhlich and Del Giudice's group (Chapter 6). Rotatory oscillations in the range of $10^7$ per second would be consistent with the findings of several researchers (Chapter 9) who found collective energy absorption in the range of $10^7$ per second by protein assemblies such as virus coats.

Bornens views centrioles as the center of a dynamic cytoskeleton which communicates and integrates cellular information. In Bornens' view of cytoskeletal organization, microtubules are conductors of spatial centrifugal information, rigid organizers of cell space in which physical or electrical signals propagate as conformational modifications of MT subunits. Intermediate filaments and the microtrabecular lattice also participate in Bornens' vision of a dynamic network of pulsating polymers. He also suggests that ATP generated centriolar rotation triggers propagating impulses along microtubules by transitory contact/stimulus of the nine rotating MT doublet/triplets in the centriole wall with their surrounding satellite bodies. This would lead to rhythmic signals through the cytoskeleton with a frequency nine times greater than that of the centriole's rotation. Rhythmic, propagating signals are compatible with coherency, solitons, and other models. Their occurrence throughout the cytoskeleton would be mechanisms close to the nature of life itself.

### 8.2.8 Centriole-MT Signaling/Albrecht-Buehler

Northwestern University biologist Guenter Albrecht-Buehler has considered the general question of intelligence in cytoplasm (Chapter 5) as well as two specific models of cytoskeletal information processing: centriole signal detection, and propagation of MT impulses. The walls of centrioles are composed of nine MT doublets or triplets arrayed in a cylinder. Each MT triplet can consist of two incomplete C-shaped MT and a complete O-shaped one fused longitudinally. The triplets are arranged at an angle of 30–45 degrees from the main cylinder, pitched to form a blade which advances to a final increment of one ninth of the perimeter just below the position of the next blade. A line drawn through any blade connects with the inner edge of the preceding blade (Figure 8.4).

Collective behavior of cytoplasm would seem to require some communicative format, and Albrecht-Buehler suggests that centrioles are perfectly designed to detect both intensity and direction of linear signals. One *possible* example of a highly specific, yet ubiquitous signal propagated in a straight line in the cellular environment is infrared radiation of the molecules inside and around cells (Albrecht-Buehler, 1981). As discussed in Chapter 6, transmission of biomolecular infrared energy would require shielding from bulk water, and perhaps nonlinear coupling to structural conformational states. Both of these requirements may be met by the ordered water and ions surrounding the cytoskeleton and the electron-dense pericentriolar material.

Cellular navigators, centrioles are involved in directional orientation of moving cells, establishment of cell architecture in cell growth and differentation, and all dynamic rearrangements of cytoplasm. In an attempt to understand how centrioles could navigate and orient, Albrecht-Buehler asks how an optimally designed, technological spatial signal detector would appear. Instruments such as radar scanners can determine direction of signals by scanning different directions sequentially. However, such scanners miss signals which arrive from one direction while another is being scanned. A properly designed nonscanning instrument can listen simultaneously to all directions with no moving parts. Albrecht-Buehler points out some geometric features of an optimally designed, nonscanning "angular" detector. With signals arriving from arbitrary directions, a detector designed as a circle with a number of regularly spaced marks around its circumference would be accessible and capable of identifying direction. Nine fold symmetry provides small size with sufficient angular resolution. To *locate* a signal source, a detector must prevent an emitted signal from arriving at more than one

Figure 8.4: Centriole in cross section is comprised of nine triplet MT angled like "Venetian blinds" from centriole axis. Centriole pairs consist of perpendicular cylinders. Albrecht-Buehler (1985) contends these features are ideally suited for signal detection. Bornens suggests rotatory oscillations of centrioles leading to gyroscopic function. By Paul Jablonka.

receptor. A simple circular arrangement would fail to meet this requirement because about half the receptors are accessible to a signal emitted from any source. A simple way to improve the design is to attach blinds or "blades" to one side of each receptor to absorb or deflect a signal wherever it interacts with them. A radial arrangement of straight blinds would be inadequate because at least two receptors would remain accessible to signals from the same source. However, if the blinds are bent circumferentially, an optimal angle is achieved when the blinds prevent access to all but one receptor without producing "blank spots," areas from which signal sources cannot reach any of the receptors. Blinds that restrict access to single receptors are even better if their shape is concaved. Used by manufacturers of "Venetian blinds," this curvature averts the possibility that a signal located directly in line with a straight blind could reach two adjacent receptors. Consequently, a signal is received *not* by the receptor closest to the signal source, but by one located at a fixed angle off the incident direction. Centriolar "blades" extend above and below the centriolar cylinder and are pitched, or twisted. Albrecht-Buehler sees this torque as further angular resolution, but the "propeller-like" arrangement could also serve to screw centrioles through the cytoplasm, assuming a centriolar rotation as suggested by Bornens (Record, 1986). Two detectors are best placed at right angles to each other so that one of them can locate the "longitude" while the other locates the "latitude" of the signal source. Centriole-like basal bodies fixed perpendicularly into a two dimensional cell surface are usually not accompanied by a second basal body at right angles. Thus the right angle paired cylinder formation permits global exposure and three dimensional reckoning unnecessary in basal bodies. Albrecht-Buehler's view of centrioles as perfectly designed signal detectors is complementary with Bornens' concept of a gyroscopic oscillator and signaling center. Combination of the two models results in a dynamic cell center capable of piloting cytoplasmic activities.

Albrecht-Buehler (1985) has also considered a mechanism for signal propagation along microtubules. He considers that each MT protofilament is a chain of alpha-beta tubulin dimers: AB, AB, AB, ..., AB. Within the wall of a microtubule each monomer is in contact with other monomers of the same protofilament and with those of adjacent protofilaments. Each monomer is consequently subject to attractive Van der Waals forces from surrounding tubulin monomers which hold together the protofilaments and MT cylinders. Albrecht-Buehler proposed that each of these interactions must weaken the A-B dimer bond; consequently the wall of a microtubule exists in a

> state of resonance as to the relative strengths of the intermonomeric bonds. For example, (A-B) (A-B) ... (A-B) (A-B) could resonate with (A) (B-A) (B-A) ... (B-A) (B). Such a resonating chain could propagate information at close to the speed of light.

Albrecht-Buehler is suggesting a coherent communicative resonance among protein conformational states within MT assemblies.

### 8.2.9 Dynamic Tensegrity/Heidemann and Jarosch

Buckminister Fuller proposed an interesting architecture constructed from components which may have a recursive or fractal structure (Fuller, 1975). Its macro level structure, a "tensegrity mast," is a *rigid* structure constructed from an assembly of tension and compression members. The compression members of solid struts are isolated from each other, held together by the tension members. In one of Fuller's variations, he notes that in the macro tensegrity mast, each individual *solid* strut may be replaced by a miniaturized version of the macro tensegrity mast. And then each one of the miniature solid struts may itself be replaced by a still smaller subminiature tensegrity mast, and so on down to the atomic level. Thus tensegrity structures may have a fractal substructure.

Joshi, Chu, Buxbaum and Heidemann (1985) have shown that cytoplasm has both compressive and tensile elements. Semi-rigid microtubules are under compression presumably due to tension generated by actin filaments and the microtrabecular lattice or "cytomusculature" (Chapter 5). In general, MT do not contact each other so that the self-supporting capability of cytoplasm may stem from tensegrity.

Robert Jarosch (1986) has published a series of papers describing actin-MT interactions which suggest that 1) contractile actin filaments are spirally wound around microtubules, 2) coordinated contraction of the actin filaments imparts a rotational torque to MT, somewhat like a spinning top, 3) actin filaments wound in opposite directions on the same MT can cause rotational oscillations of the MT. These two models fit together to provide a picture of a dynamic cytoplasmic tensegrity network in which the cytoskeleton may be twisting back and forth, even "rockin' and rollin'!" Perturbation of any part of such a tensegrity network could have dynamic consequences throughout its domain. Transient changes in tension, compression, or oscillatory rhythm caused by a variety of factors would be detectable and possibly amplified throughout the cytoskeleton.

### 8.2.10 Dynamic MT Probing/Kirschner and Mitchison

Strong evidence supports the concept of dynamic instability in microtubule assembly (Chapter 5). Many MT exist in either growing or shrinking phases and MT stabilized at merely one end by centriole based microtubule organizing centers (MTOC) tend to predominate. Selective retention of MTOC based MT establishes cell polarity important in extension of axons and den-

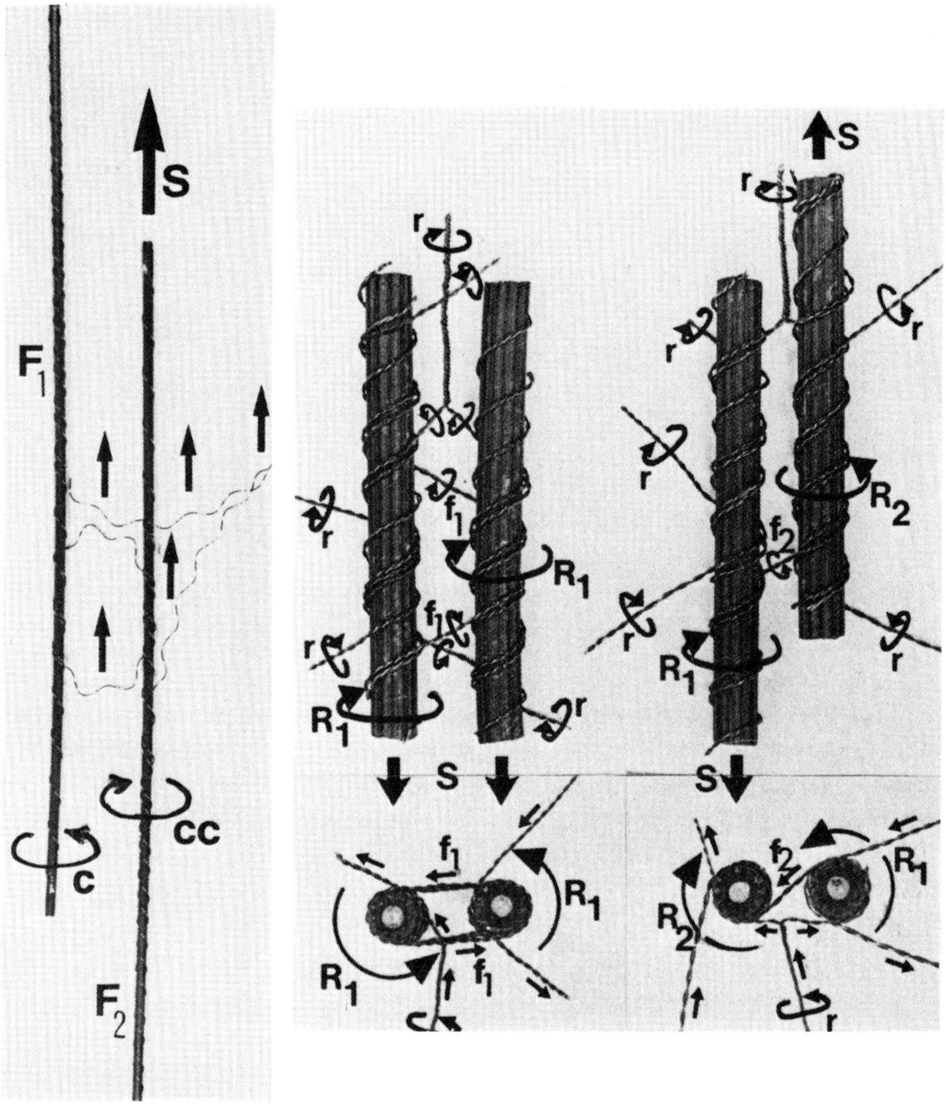


Figure 8.5: Robert Jarosch has proposed that actin filaments wind around microtubules and contract, causing MT to rotate and oscillate around their long axis. With permission from Robert Jarosch (1986).

drites, elongation of cells in embryological development, and formation of lamellipodia and filopodia in locomotory cells. All are examples of the "Indian rope trick" in which cells somehow choose their direction of growth, and then grow in that direction. A cue presented at the cell periphery can lead to rearrangements of cell symmetry and polarity. Kirschner and Mitchison (1986) ask: "how can a peripheral clue lead to reorganization deep within a cell?" One possibility is that a signal is relayed to the microtubule organizing center leading to a change in its structure, orientation, and directed nucleation of microtubules. This would be consistent with a primary role for information integration and decision making within the MTOC. A simpler "non-hierarchical" idea is that a signal at the periphery affects distribution directly. Since the whole cytoskeletal array is very dynamic, it would only be necessary to stabilize or destabilize a particular subset of microtubules for the entire cytoskeleton to rapidly transform. Preferential stabilization of a microtubule (MTOC, GTP capping, binding to membrane related structures etc.) could be mediated by interactions at the cell periphery close to the site receiving environmental information. Kirschner and Mitchison (1986) have proposed that the dynamics of the microtubule array results in probing many regions at random and, by stabilizing certain conformations as they arise, the cell "can arrive at a structure that is not precisely defined by genetic information but fulfills a particular functional role as dictated by environmental factors." Dynamically unstable MT offer many possibilities for controlling the distribution of MT by selective stabilization. Kirschner and Mitchison suggest that the tendency for probing and transformation is a fundamental advantage which favored the evolution of dynamically active microtubules.

### 8.2.11 Sphere Packing Screw Symmetry/Koruga

There are 32 possible symmetry arrangements of packed spheres in a cylindrical crystal. Erickson (1973) used hexagonal packing of protein monomers to explain the form and patterns of viruses, flagella and microtubules. Djuro Koruga (1986) of the University of Belgrade's Molecular Machines Research Unit has analyzed the symmetry laws which describe cylindrical sphere packing and the structure of microtubules. Koruga has used both hexagonal packing and face centered cubic packing of spheres to explain microtubule organization. Koruga (1986):

> The particular symmetry group which represents the packing of spheres in microtubules is 'Oh(6/4).' Hexagonal packing may be

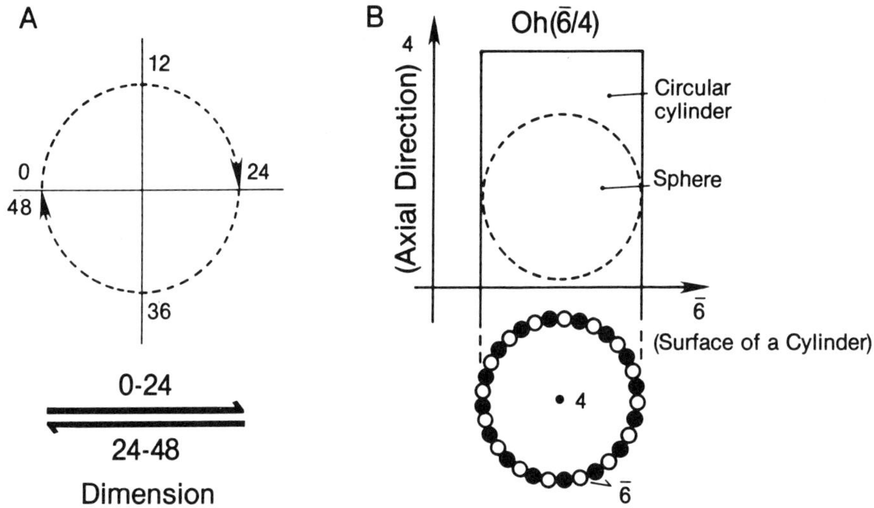


Figure 8.6: Koruga's derivation of the symmetry of microtubules. With permission from Djuro Koruga (1986).

> described by using fixed conditions if the centers of the spheres lie on the surface of the cylinder and if the sphere values in the long axis of the cylinder are the same as in the dimension of face centered cubic packing. The six fold symmetry and dimer configuration lead to screw symmetry on the cylinder: a domain may repeat by translocating it in a spiral fashion on the cylinder. From coding theory, the symmetry laws of tubulin subunits suggest that 13 protofilaments are optimal for the best known binary error correcting codes with 64 code words. Symmetry theory further suggests that a code must contain about 24 monomer subunits or 12 dimers.

Koruga's symmetry arguments may be compared with the "gradion" concept of Roth, Pihlaja, and Shigenaka in which a field of about 17 monomers is thought to represent a basic information unit. Koruga also concludes that microtubule symmetry and structure are optimal for information processing.

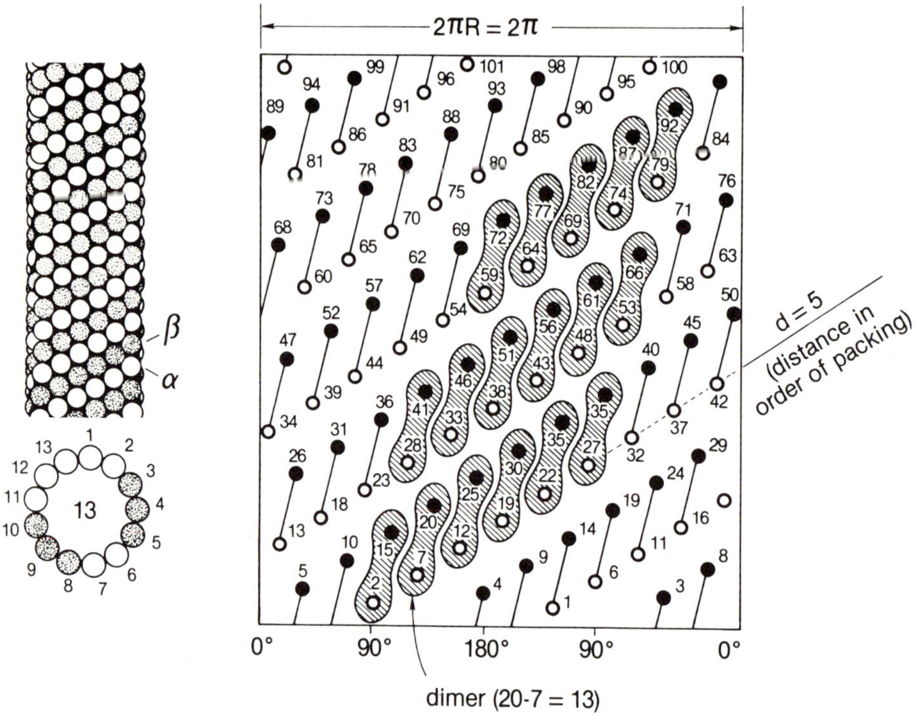


Figure 8.7: Koruga's screw symmetry and optimal information unit in lattice wall of microtubule. With permission from Djuro Koruga (1986).

### 8.2.12 Cytoskeletal Self-Focusing/Del Giudice

As described in Chapter 6, a group of scientists from the University of Milan have applied the mathematical tools of "many body problems" to the activities of biomolecular dipoles. Del Giudice, Doglia, Milani and Vitiello (1985, 1982) have used quantum field theory to describe the electret state of biological systems (ordered water surrounding linear biomolecules) and determined that there exists a strong likelihood for the propagation of particle-like waves in biomolecules. Further, the ordering of water should lead to self-focusing of electromagnetic energy into filamentous beams excluded by the ordered symmetry. For ordered cytoplasm, they calculate the diameter for the confinement and propagation of particle-like waves (massless bosons, or solitons) in biomolecules to be about 15 nanometers, exactly the inner diameter of microtubules.

The proposal by the Milan group has a number of implications. Confinement within filamentous regions excluded from water would favor the propagation of electromagnetic energy in biological systems, and provide a mechanism for alignment and communication (the "Indian rope trick"). Further, cytoskeletal polymers may be capable of capturing and utilizing ambient or biologically generated electromagnetic energy. One possible example is infrared energy which is routinely generated by dipoles in biological molecules. This energy is generally believed to be dissipated into heat within the aqueous cytoplasm, however "self-focusing" could utilize this energy productively in a communicative medium. The Milan model also includes lateral force generation by focused energy within cytoskeletal filaments which would be useful in biomolecular maneuvering and communication.

### 8.2.13 MT Automata, Holography/Hameroff, Watt, Smith

The self-focusing of electromagnetic energy described by the Milan group is thought to occur by an electret induced increase in the refractive index of cytoplasm. A similar concept was proposed (Hameroff, 1974) in which microtubules were thought to act like "dielectric waveguides" for electromagnetic photons. Living tissue does transmit light more readily than nonliving material. Van Brunt, Shepherd, Wall, Ganong and Clegg (1964) measured penetration of sunlight into mammalian brain by routes other than the visual system. Stereotactically placed photoreceptors recorded intensities of $10^{-3}$ lumens in sheep hypothalamus when surface intensity was 0.4 lumens, with a logarithmic diminution. The most light permeable areas were in the temporal regions of the skull, lateral to the orbits; the brain's temporal poles and hippocampus received maximum light intensity. When the animals were sacrificed, light penetration to the hypothalamus remained constant for about 30 minutes following which the brain opacity rose sharply. This suggests that some property of living brain tissue is relatively translucent to optical photons; polymerized MT acting as waveguides may be such a property.

Hameroff (1974) also proposed that the periodic array of MT subunits "leaked," or diffracted energy with 8 nanometer periodicity, resulting in a source of "coherent" energy (or calcium ions) from each MT. Cytoplasmic interference of the coherent sources from among multiple MT would lead to holographic imaging in cytoplasm. Coupling of calcium concentrations to cytoplasmic sol-gel states could "hardwire" holographic patterns into the microtrabecular lattice. In parallel arrays of MT within nerve fibers, graded potentials or traveling action potentials were thought to collectively acti-

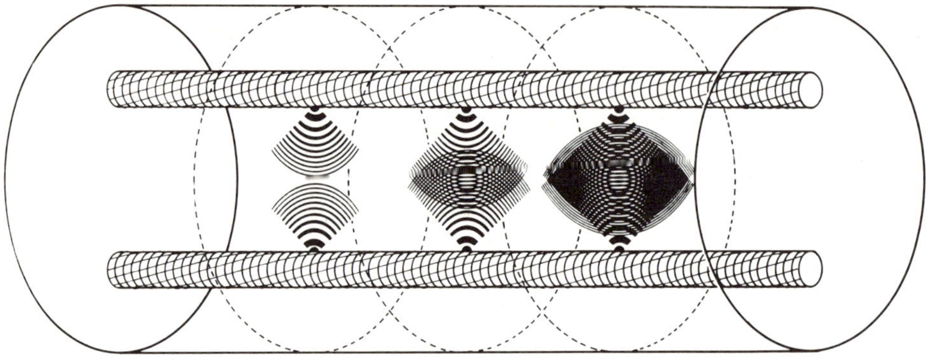


Figure 8.8: Interference patterns in cytoplasm caused by coherent waves (e.g. $Ca^{++}$, sol-gel state, MTL) generated by dynamic activities in microtubules may be a basis for holographic information imagery.

vate "planes" of cytoplasm perpendicular to the long axis of the MT and nerve fibers. These traveling planes may be likened to image screens as in TV sets. In a TV picture tube, the screen is motionless and electron beams move to create a picture by their intersection with the screen. Perhaps imaging within neurons occurs on traveling screens generated by action potentials moving through parallel MT arrays. The content of such images would depend on programming mechanisms in the conformation of tubulin subunits which comprise the MT walls and which update with each successive action potential. Hameroff and Watt (1982) described a method of MT tubulin programming in which charge carriers (calcium ions, electrosolitons) or conformational waves such as phonons or solitons were steered through MT lattices by genetically or cytoplasmically programmed tubulins and specific MAP binding sites. MAP bridges to other MT, cytoskeleton, or other organelles were thought to act as "sinks" or "sources" which conveyed pulse trains of charge/conformation among MT throughout the cytoskeleton as a regulatory and communicative medium. Hameroff and Watt (1982, 1983) likened MT to microprocessors in which switching in a "Boolean matrix" was determined by programming factors intrinsic to the tubulin subunits (Figure 8.9).

In subsequent papers, Hameroff, Smith and Watt (1984, 1986) utilized

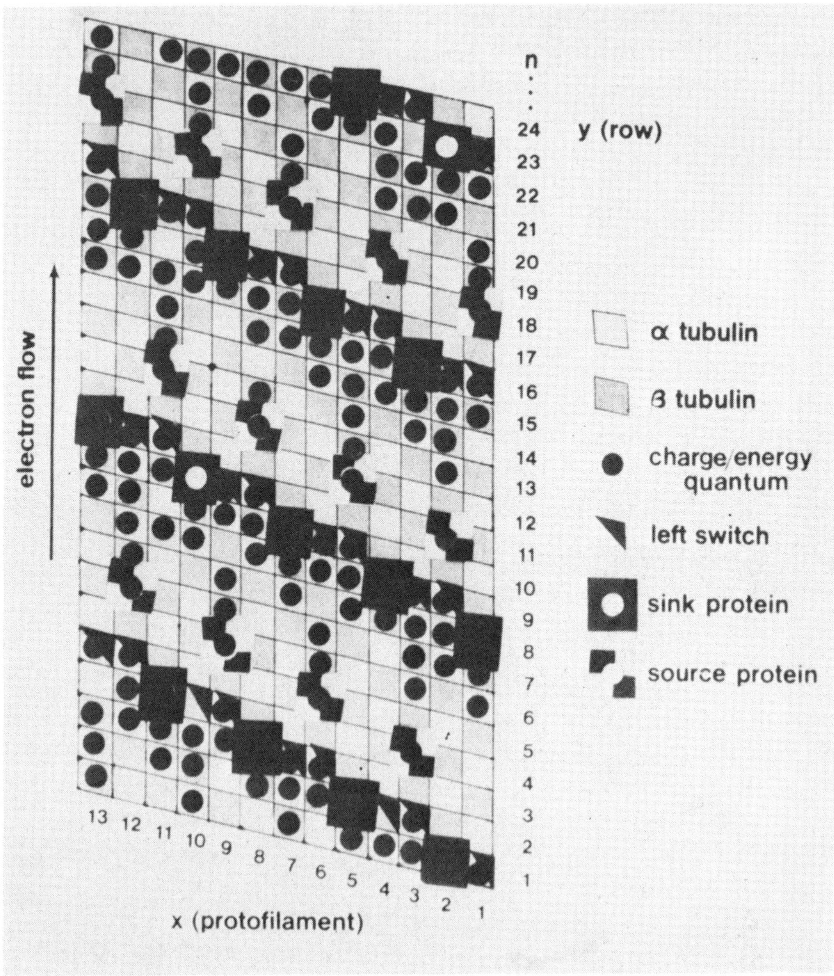


Figure 8.9. Boolean switching matrix in microtubule lattice from Hameroff and Watt (1982). Alpha and beta tubulin subunits in the MT lattice wall were considered to be transiently occupied by quanta of charge/energy/conformation. MAPs behaved as "left switches" or "sink" or "source" proteins. Sequences and patterns of conformational states traveled through the MT lattice and via "sink" and "source" inter-MT bridges throughout the cytoskeleton somewhat like a pinball machine. The model demonstrated a capability for tubulin-MAP "programming" of dynamic activities related to biological intelligence.

principles of cellular automata to explain information processing in MT. As described in Chapter 1, cellular automata are dynamical systems which can generate and process patterns and information, and are capable of computing. Cellular automata require a lattice structure of "like" neighbors with discrete states and neighbor rules, and a universal "clock" to which all neighbors are timed. Adopting Fröhlich's model of coherent nanosecond dipole oscillations coupled to conformational states as a clocking mechanism, the authors calculated MT lattice neighbor Van der Waals dipole interactions as rules for an MT-automaton computer simulation. Each tubulin dimer was considered to be in one of two possible states at each nanosecond "generation." The two states were related to Fröhlich's concept of dipole oscillation so that the dipole can be oriented either toward the alpha tubulin end ("a," Figure 8.2.13) or toward the beta tubulin end (represented by a dot in the Figure). The polarity and electret behavior of MT indicate that in the resting state, tubulin dimer dipoles should be oriented toward the beta monomer.

Dimer states at each "clock tick," or generation were determined by neighbor states at the previous generation.

$$\text{state} = \alpha, \ \text{if} \sum_{i=1}^{n} f(y) > 0, \ \text{state} = \beta, \ \text{if} \sum_{i=1}^{n} f(y) < 0,$$

where n = 7 as the number of neighbors, and f(y) the force from the "$i$th" neighbor in the y direction.

$$f(y) \propto \frac{\sin\theta}{r^2}, \ \sin\theta = \frac{y}{r}; \ f(y) \propto \frac{y}{r^3}.$$

The dipole state of any particular dimer at each clock tick thus depends on the summation of the dimer's neighbor dipole states (including its own) at the previous clock tick. The neighbor influences are unequal because of the screw symmetry of the MT lattice. Distant dimers (more than one neighbor away) would be expected to have little influence because of the dropoff in force intensity ($y/r^3$) with distance. However, collective influences from many "like" oriented dimers could lead to long range cooperativity. Using only near neighbor influences, computer simulation of an MT automaton yielded interesting patterns and behavior of dipole/conformational states. These included both stable and traveling interactive patterns capable of computing and regulation of cytoskeletal activities. For example, Figure 9.13 shows a "kink-like" pattern traveling through an MT lattice, leaving an altered "wake," or memory. Assuming nanosecond generations, these

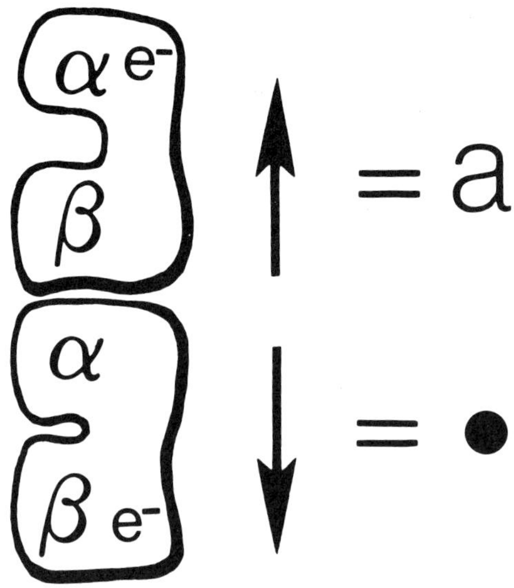


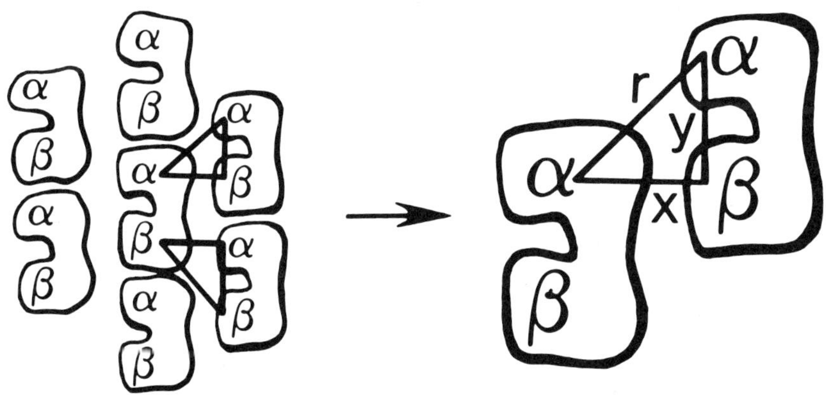


Figure 8.10: Top: MT-tubulin dimer subunits comprised of a and b monomers. Relative electron occupancy of either monomer may correlate with coherent dipole oscillations in the nanosecond time domain. Bottom: screw symmetry hexagonal packing leads to unequal neighbor rules based on lattice distances and Van der Waals dipole interactions. For dimers shown at right, x = 5 nm, y = 4 nm, r = 6.4 nm. The relative strength of each neighbor dimer dipole state is $y/r^3$.

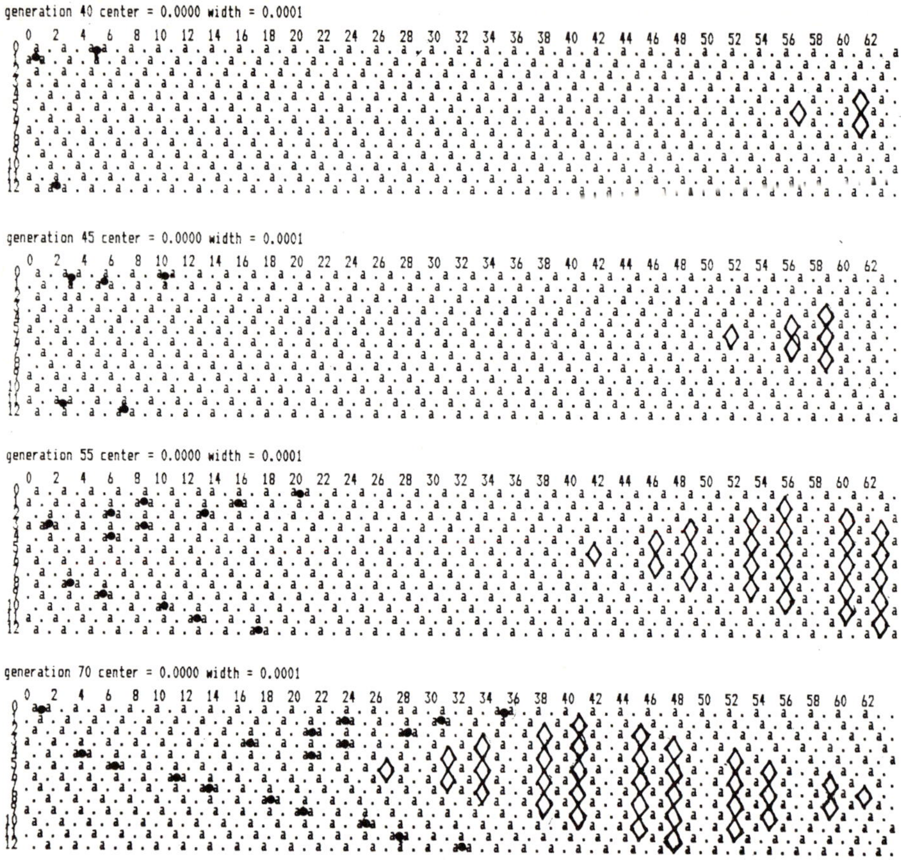


Figure 8.11: 30 "nanoseconds" of computer simulation of MT cellular automaton based on Fröhlich coherent oscillations and tubulin subunit neighbor rules based on dipole-dipole interactions. 13 protofilaments are arranged horizontally, 64 rows are shown vertically yielding a flat MT lattice. "Alpha" states of tubulin (Figure 8.10) are shown by "a," beta states by a dot. 4 contiguous beta states are highlighted by diamond patterns, 4 contiguous alpha states by darkened dots. The simulation sequence (generations 40, 45, 55 and 70) shows movement of patterns at velocity of 80 nanometers per nanosecond (80 meters per second). With permission from Hameroff, Smith, and Watt (1984).

traveling patterns would travel at 8 nanometers per nanosecond (80 meters per second), a velocity consistent with propagating action potentials, or solitons. Variability in individual tubulin dimer isozymes, ligand bind-

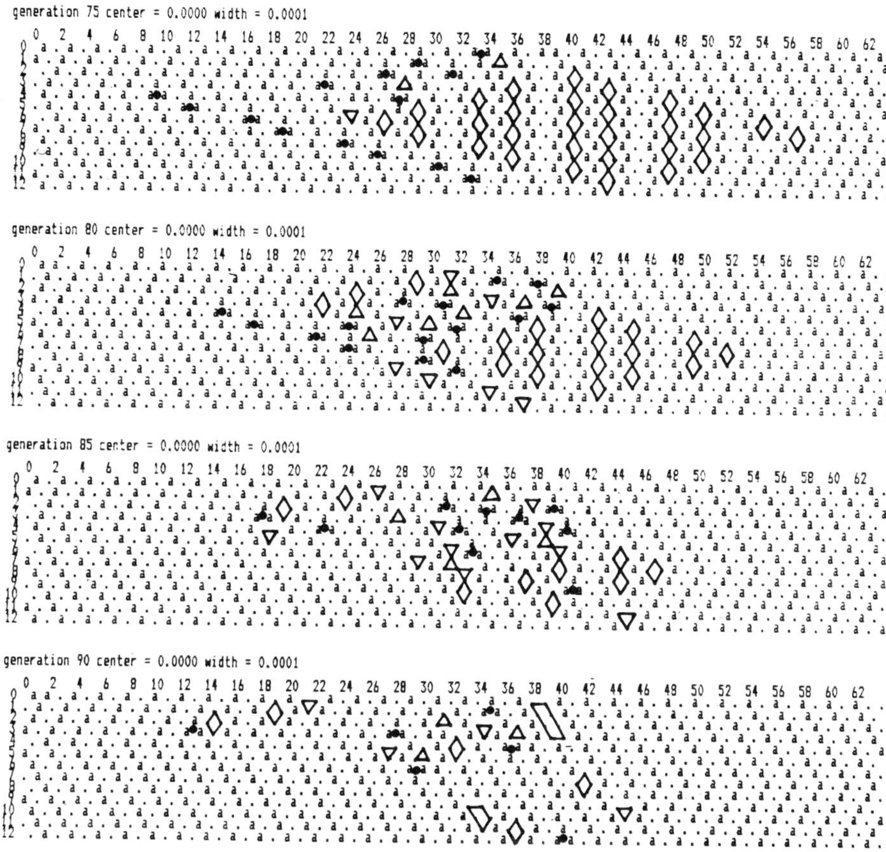

Figure 8.12: Continuation of MT cellular automaton sequence of Figure 8.11. Alpha and beta patterns collide and generate bizarre shapes, leaving configuration altered from initial state. Such interactions could represent rudimentary "computing" and "memory." With permission from Hameroff, Smith and Watt (1984).

ing, or MAP attachments could "program" and "read out" information in routine cellular functions. As one example, propagation of MT conformational patterns could coordinate the activities of contractile MAP sidearms in axoplasmic transport. In a general sense, MT automata may be the information substrate for biological activities ranging from ciliary bending to human consciousness. Specific automata patterns distributed throughout wide volumes of cytoskeletal arrays within the brain could lead to cooperative resonance and collective effects resulting in a thought or idea similar

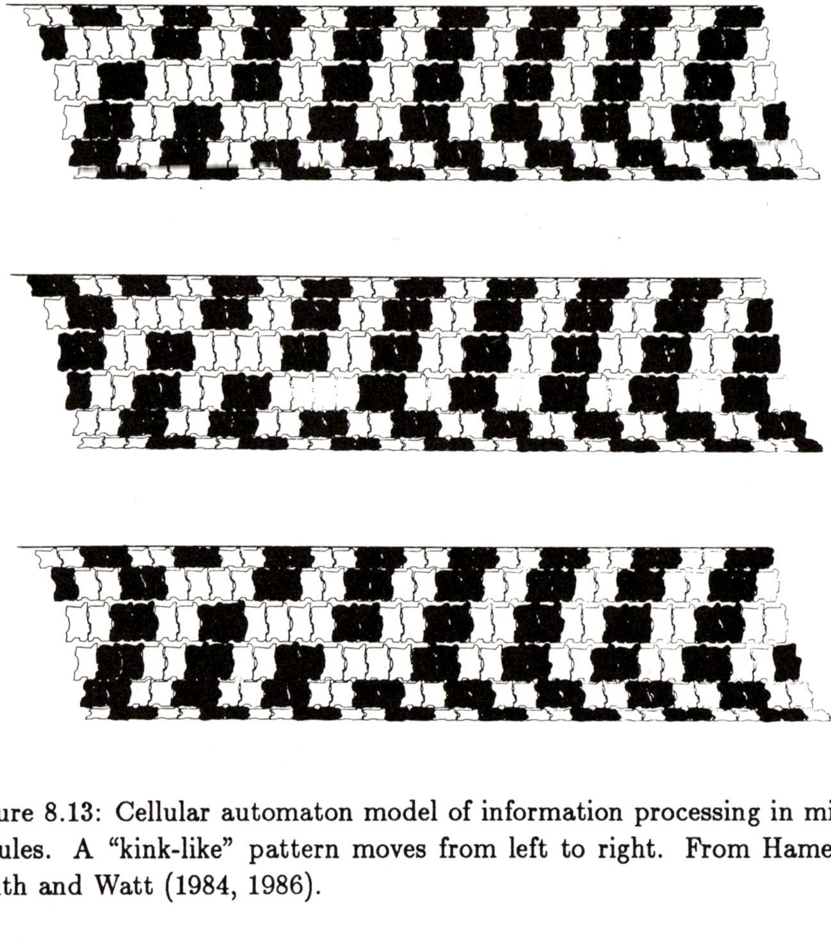

Figure 8.13: Cellular automaton model of information processing in microtubules. A "kink-like" pattern moves from left to right. From Hameroff, Smith and Watt (1984, 1986).

to the manner in which coherence induced phase transitions in metals yield emergent collective properties such as superconductivity.

## 8.3 The Cytoskeletal Connection

Microtubules, centrioles, and the cytoskeleton evolved to take command of biology. By being at the right size scale and moving in the right time scale, they appear capable of utilizing many forms of available energy. MT polarity, periodic structure, and helical lattice array of conformationally programmable subunits qualify them as "ultimate computers" and nanoengines as well, generating purposeful force and movement. They may capture and

Figure 8.14: "Kink-like" pattern continues left to right through MT automaton. Such dynamic patterns can represent and compute information, serve as binding sites for transported molecules, designate MAP and inter MT bridge attachment sites, or generate coherent waves of calcium coupled sol-gel states.

self focus electromagnetic radiation, induce coherency, and propagate energy quanta with minimal loss. Propagation may manifest as coherent polarization waves as described by Fröhlich, massless bosons as described by Del Giudice or as solitons as described by Davydov. Charge density waves or proton transfer at MT surface layers of ordered water, and positively charged counter ions such as calcium are other possible modes available for cytoskeletal based information and energy transfer.

MT electron surpluses occur due to an abundance of acidic amino acids. With the intrinsic polarity of each tubulin dimer subunit, MT reside in a strong polar field or "electret" which possesses piezoelectric properties. Each MT subunit can thus "integrate," being capable of absorbing and sensing input in the form of acoustical energy, optical photons, chemical ATP, altered potential changes, fluxes of calcium and other ions, and responding with conformational changes accordingly. Each subunit within a microtubule lattice can not only represent information, but can input and output information into an ongoing automaton. Coherent oscillations in MT tubulin subunit states among regional forests of microtubules can provide a communicative medium in which any subunits which are out of phase induce waves of phase uncoupling along MT structure. Alterations which disturb coherency could be amplified by collective oscillations among cytoskeletal macroassemblies like rotatory centrioles or tensegrity structures of MT wound by contractile actin filaments.

Several other possible modes of information management present themselves in the structure of MT. Could the gaps between tubulin subunits act like pores? The relative abundance of electronegative charges in the outer surface of cylindrical microtubules may create gradients across tubule walls if the MT interior is neutral or positive. Presently unmeasurable, even a small gradient across 4 nanometer MT walls would comprise a significant voltage field. Thus a propagating soliton, charge density wave, or transient osmotic swelling of nerve processes could open "cracks" among the tubulin subunits to allow for ion flux or radiation of electromagnetic energy focused and trapped within MT. Alternatively, an electronegative MT interior core might support Del Giudice's suggestion of cytoskeletal superconductivity. Self-focused energy exerting force perpendicular to the long axis of MT could account for mysterious effects such as the perpendicular birth of daughter centrioles and inter-MT MAP bridges. Energy or ion fluxes radiated from MT due to propagating waves would be coherent due to spatial periodicity of the gaps between the tubulin subunits, and could thus form the basis of coherent wave interference in a three dimensional hologram. Coupled to calcium/sol-gel states, holographic cytoplasm may be an "infoplasmic" canvas for dynamic biological information.

In the brain, highly parallel arrangements of cytoskeletal proteins within asymmetrical axons and dendrites suggest analogies to parallel computing. Propagation of action potentials or dendritic depolarization waves could induce sequences of alterations of MT subunit states due to influx of calcium, alteration of electrical fields, or direct mechanical effects as sodium and water transiently swell the nerve process. Signals may cooperatively reverberate throughout the cytoskeleton by waves of calcium release and uptake, mechanical/electrosolitons propagating through the cytoskeletal lattice, and/or lateral propagation through sidearms and other components of the cytoskeletal network. Transient fluxes of calcium resulting from conformational waves propagating down cytoskeletal lattices in parallel with action potentials or dendritic current waves can result in traveling frames of dynamic imagery. Sequences of image frames traveling through holographic cytoplasm and changing with each nerve firing may collectively be the "Mind's Eye," the "grain of the engram."

The current prevalent model of brain function is the "neural network," a collective dynamical system whose output is determined by the states of component neuronal synapses. In turn, the state of each synapse is determined by other neurons modulated by collective dynamic effects of the interneuronal cytoskeleton. In turn, the cytoskeleton is a collective dynamical system whose net activity is determined by the states of its component subunits. Their states, in turn, are determined by another level of organization including cytoplasmic factors, ordered water and ions, genetic isozymes of individual subunits, and lower level cytoskeletal elements such as the microtrabecular lattice. Each cytoskeletal subunit is a computer capable of integrating multiple inputs to a specific output state. Cognitive processes which have been ascribed to a neural net concept may thus be more accurately interpreted as a fractal hierarchy of dynamical systems which are highly parallel, highly interconnected, and of increasing capacity as they become more microscopic. Collective effects may occur at each level and at successively more macroscopic levels. The net effect may be consciousness.

Microtubules and the cytoskeleton created their place in evolutionary history by being problem solvers, organelle movers, cellular organizers, and intelligence circuits. Where do they go from here?

# Chapter 9

# Viruses/Ambiguous Life Forms

## 9.1 What Is the Essence of Living Matter?

Oparin (1938) proposed that living matter be defined as having the following properties: metabolism, self-reproduction, and mutability. Eigen (1971) observed that these criteria could be met by decidedly non-biological entities such as von Neumann's "self-reproducing automata," or various robots. Even some simple biological systems are somewhat ambiguous regarding their "life-like" status. Primitive systems which blur the distinction between animate and inanimate include viruses, prions (protein crystals which may be causative agents in some human diseases) and proteinoids (self organizing protein assemblies implicated in life's origins (Fox, 1972). Independent cytoskeletal elements also have characteristics of being "alive." The slithering microtubules which R. D. Allen and colleagues (1985) isolated from squid axoplasm ambulate, avoid collisions, and have a sense of direction. Are they alive? If future technologies lead to nanoscale replicating automata, the distinction between living and nonliving entities will be further clouded.

## 9.2 Virus (Mis)Behavior

Among the more basic varieties of life are viruses. Their simple structure and activities have prompted questions as to whether they are actually "alive," or are merely chemical robots which multiply and reside within truly living organisms. Viruses form spontaneously as their simple subunits assemble

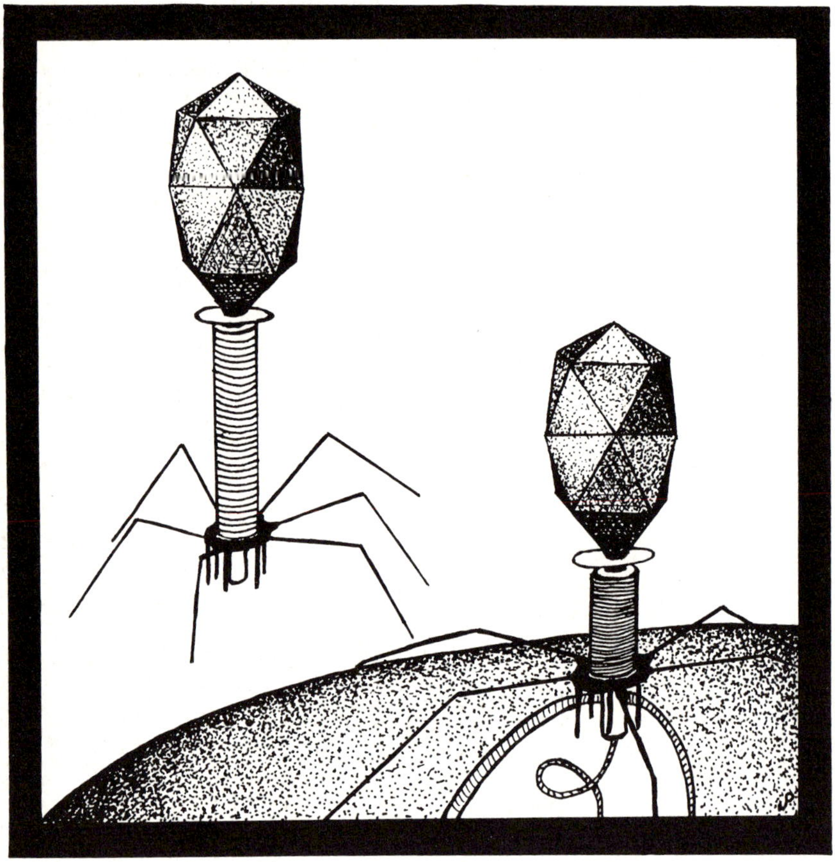

Figure 9.1: Phage virus landing and injecting genetic material into host cell. The cylindrical collar undergoes a collective conformational contraction associated with the injection. By Paul Jablonka.

into complex three dimensional structures. The assembly process, driven by the increased stability or lowered energy of the completed virus, flows against the second law of thermodynamics which dictates that order proceeds to disorder. Data from the tobacco mosaic virus (Chapter 6) show that hydrophobic interactions which exclude water from unassembled subunits offset the increase in order (negative entropy) of the assembled virus.

Generally, viruses enter cells and pirate their genetic machinery to cause virus multiplication; they then escape to begin another cycle. Some viruses quietly and harmlessly reside within host cells for extremely long periods of time whereas the sojourn of other viruses involves the total commandeering of cell machinery and can have extremely damaging consequences for the host. Virus induced human diseases range from trivial common colds and influenza to chicken pox, measles, hepatitis, polio, and deadly diseases such as smallpox, rabies, yellow fever, cancer, and acquired immunodeficiency syndrome ("AIDS"). Some viral diseases such as smallpox have been controlled by modern medicine but others such as AIDS remain significant threats to human health.

Once assembled inside a host cell, viruses can escape in several ways. One route involves causing the death and disintegration of the host cell, allowing the newly formed viruses to spill out and carry their infection elsewhere. Viruses which are covered by lipid membranes can also escape by "budding" or reverse endocytosis ("pinocytosis"), a mechanism similar to secretion or extrusion of neurotransmitter vesicles from synaptic boutons.

An important subclass of viruses are retroviruses which contain RNA without DNA. When arriving within a host cell, the retrovirus brings an enzyme known as reverse transcriptase which converts the virus single stranded RNA into a double stranded DNA copy. This allows the virus to commandeer the host cell genetic machinery. Among the human retroviruses is HTLV-III (or "HIV"), carrier of AIDS.

## 9.3 Virus Structure and Collective Oscillations

Virus structure includes up to three major components: genetic material (either DNA or RNA), a protein coat which surrounds the genetic material, and a membrane (Figures 9.1 thru 9.3). One example of a viral protein coat is the cylindrical lattice similar to microtubule structure found in the tobacco mosaic virus, the first virus to be isolated (by Russian botanist Dimitri Ivanovski in the late 19th century). In adenoviruses, the protein coat is an icosahedron: a 20 sided polygon with 12 vertices. Influenza virus protein coats are spherical and may include outward extruding glycoproteins surrounded by lipid membranes. Another interesting viral structure is that of bacteriophages which actively inject their DNA into host bacterial cells. The bacteriophage looks somewhat like a nanoscale lunar lander and has an icosahedron head mounted by a collar onto a cylindrical tail which in turn is supported by a base plate which contacts a bacterial cell surface. Upon

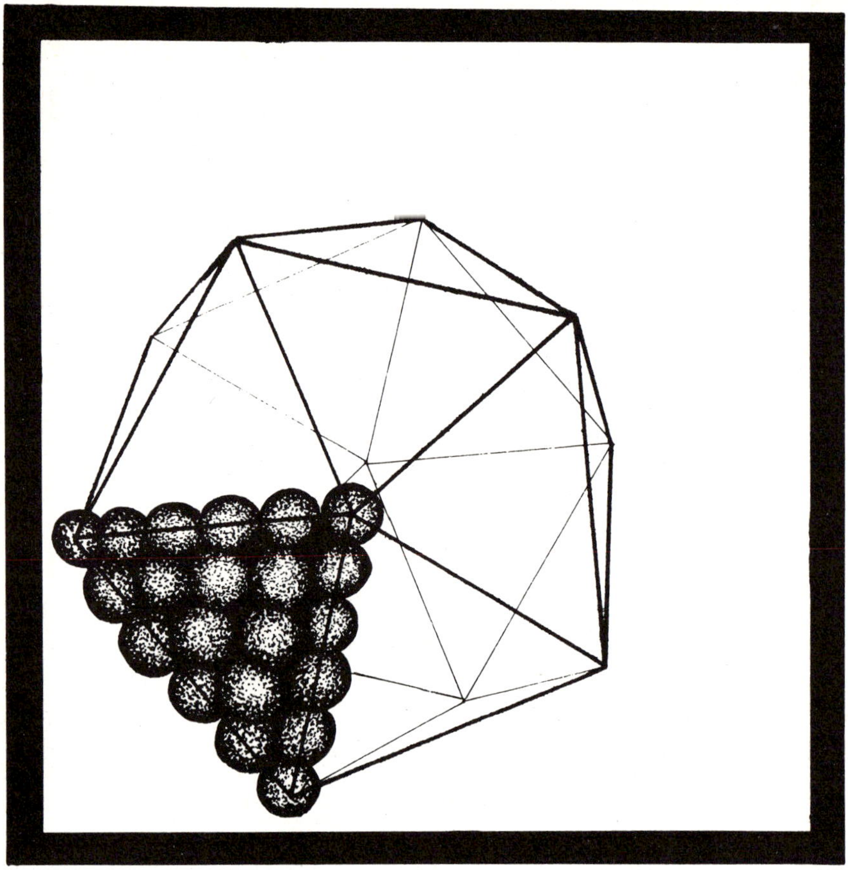

Figure 9.2: Icosahedral structure of an adenovirus protein coat. By Paul Jablonka.

an appropriate stimulus, the bacteriophage virus protein coat undergoes a collective conformational event; contraction of the entire tail assembly results in the active injection of viral DNA through the bacterial cell wall and into the bacterial cell cytoplasm. This sequence of events requires a coordinated communication and active response and may be construed as a rudimentary intelligence.

How do viruses perform cognitive acts such as sensing host cell surfaces and injecting into them their nucleic acids? One possibility is by perturbations of collective vibrations of their protein shells. Michels, Schulz, Witz, Pfieffer, and Hirth (1979) measured ultrasonic absorption of protein assemblies over a range of $10^5$ to $10^9$ oscillations per second. They found the

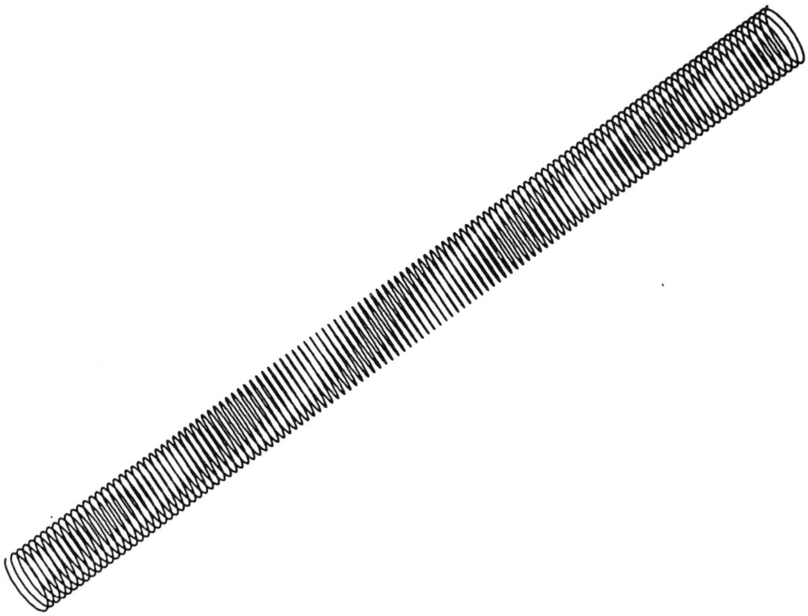

Figure 9.3: Longitudinal view of the DNA path in a tobacco mosaic virus structure. The path takes about 120 complete turns. By Paul Jablonka.

assembled structures absorb much greater ultrasonic energy than do the individual unpolymerized subunits, suggesting a *collective* oscillation of the protein assembly. For example, the intact icosahedral Brome mosaic virus absorbed energy with a mean frequency of $5 \times 10^7$ per second. Robach, Michels, Cerf, Braunwald and Tripier-Darcy (1983) found similar results with frog virus, and determined that the displacement amplitude of the oscillations were on the order of a tenth nanometer. Most virus research has focused on the genetic capabilities of viral DNA or RNA in coopting host cell activities. The protein coats have been considered as simple environments for DNA or RNA transport, however they perform functions analogous to organized cytoplasm. For example, the glycoprotein spikes which extend outward from some viruses while remaining anchored to the protein coat interact with the host membrane and cell wall surfaces to facilitate contact and entry. This general activity involves information signaled through the protein coat enacting a mechanical conformational change. Michels, Dormoy, Cerf and Schulz (1985) used similar techniques to study two strains of tobacco mosaic virus as well as other protein assemblies including microtubules. They observed that the fluctuations depend on specific organization and complexity of the assembly, and that certain assemblies oscillate more than others. They speculate that collective oscillations of virus coats could function in communicative functions such as the injection of viral nucleic acids into host cells. Viruses perform limited functions devoted to their own replication and survival, but their collective oscillations could be a clue to the essence of living matter. Collective oscillations of cytoskeletal proteins could serve a more versatile communicative role in all eukaryotic cells.

## 9.4 Nature and Origin of Viruses

Deciding whether or not viruses are alive depends on a definition of life. In his book *Pirates of the Cell,* Scott (1985) concludes that life is a spectrum of animate to inanimate material, and that viruses are somewhere in the middle. Some vital characteristics of living things include locomotion, nutrition, growth, respiration, excretion, sensitivity and reproduction. By these criteria viruses are not alive, however "inanimate" materials like crystals do manifest life-like growth, nutrition, reproduction, and locomotion. Defining life by the apparent ability to defy the second law of thermodynamics (creating order from disorder) by assembling complex organized biological structure from the disordered nonliving world ignores the fact that the thermodynamic law remains inviolate overall. Molecular level cytoskeletal protein subunits or viral particles assemble into more highly ordered structures, however the hydrophobic exclusion of water from protein subunits counteracts the change in entropy and the net effect remains order proceeding to disorder.

There are three general theories as to the origin of viruses (Scott, 1985). The first is that viruses originated very early in evolution before the advent of eukaryotic cells. According to this view, modern viruses are direct descendants of primitive early molecules floating in the primordial soup or mud (Chapter 3). The second idea is that they are derived from parasites which invaded other cells and gradually became simpler, de-evolving to be totally concerned only with survival and multiplication. The third notion, which dominates current beliefs, is that viruses evolved from genetic material of cellular life: the "escaped gene" hypothesis.

## 9.5 Domesticated Viruses

The capabilities of viruses may be harnessed (Scott, 1985). Two hundred years ago Edward Jenner began to develop safe and effective anti-viral vaccines, a technique which amplifies the body's immune system. Slopek and co-workers (1983) have treated patients afflicated with drug resistant bacterial infection by using viral bacteriophages selected for their effectiveness against the resistant organism. British scientists (Williams, Smith and Huggins, 1983) have used bacteriophages to treat intestinal infections in animals and direct use of viruses to combat bacterial infections in humans have also been attempted. Bacteriophages have potential advantages over modern drugs: they are highly specific, can leave the host cells unharmed with minimal side effects, and could be produced inexpensively. However, bacteria could develop resistances to bacteriophages as they do to some drugs.

Other researchers have used exploited viruses to mass produce natural proteins in "genetic engineering." For example, the "Ebstein-Barr" virus has been used to transform selected immune cells which then multiply and produce large quantities of specific antibodies useful in medicine and industry. In some cases viruses are changed into novel forms for use as live vaccines. Genes for influenza and hepatitis B organisms can be combined in a vaccine virus genome to protect us against diseases such as hepatitis B and the flu. Genes encoding proteins of parasites such as the protozoan that causes malaria are being added to suitable viral genomes (Smith, 1984). Viruses can transfer foreign genes into bacterial cells. Gene coding for any wanted protein can be linked up to the genetic material of a virus and, when the virus infects suitable bacterial cells, the foreign gene is carried in with it. Once inside, the transferred gene may then begin to direct the manufacture of plentiful supplies of the protein it codes for. Viruses are thus being turned into versatile ferrymen which can carry a whole range of proteins into humans and livestock.

Other exciting possibilities include transferring new genes into human cells. Many research groups including Richard Mulligan and cohorts at MIT (Kolata, et al., 1984) are trying to construct viruses that will carry new genes to human cells to cure genetic diseases. A number of hereditary diseases could thus be treated by genetic manipulation using bacteriophages. A first step would be having a retrovirus insert a therapeutic gene into the DNA of cells deficient in that particular gene. A promising area might be diseases of blood cells which proliferate continuously. Related ideas are to use membranes, virus protein coats, or other proteins to transport drugs to immuno-targeted intracellular destinations ("magic bullet").

Unfortunately, viruses may also be utilized as diabolical weapons, capable of infecting large populations. However like other technological "double-edged swords," their potential benefit is even greater. Nanotechnology combined with genetic engineering could lead to virus-like entities which could stalk and destroy lethal infectious agents, malignant cells, atherosclerotic plaques which obstruct blood vessels, scar tissue which limits nerve regeneration, neurofibrillary tangles associated with senile dementia, and perhaps other diseases. Future virus-like "nano-doctors" may be making cellular house calls.

# Chapter 10

# NanoTechnology

Functional communication among biomolecules and technological devices will require dramatic advances in our abilities to fashion logic devices from matter. The attainable limit appears to be computer components precisely structured at the atomic level, on the order of 0.1–0.3 nanometers. Nanoscale fabrication of information devices capable of biological interfacing would also enable construction of valuable nanoscale robots, sensors, and machines (Schneiker, 1986). Ultimate computing may be but a single facet of a wide range of applications: "nanotechnology."

## 10.1 Early NanoTechnologists

The American Physical Society held its 1959 annual meeting at the California Institute of Technology. Scheduled to speak was a man who would win the 1965 Nobel physics prize for his historic work in quantum electrodynamics. Still later he would serve on a Presidential Commission and find the fault in the Challenger disaster in a disturbingly brief period of time. In 1959 however, he spoke of other work that may be even more important. In his talk entitled *There's Plenty of Room at the Bottom,* physicist Richard Feynman (1961) proposed a simple and straightforward strategy for constructing useful structures ranging in size down to the atomic scale! He suggested using machine tools to make many more sets of much smaller machine tools, which would in turn make many times that number of other, even smaller machine tools, and so on. At the lowest level, he noted the possibility of mechanically assembling molecules, ***an atom at a time.***

Feynman proposed the construction of nanomachines, nanotools, tiny computers, molecular scale robots, new materials and other exotic products which would have far reaching applications and benefits. Considering the relevance of nanotechnology to living molecules, Feynman (1961) noted,

> A biological system can be exceedingly small ... . Consider the possibility that we too can make a thing very small, which does what we want—that we can manufacture an object that maneuvers at that level!

Machines able to directly manipulate matter on the submicron to nanometer size scale are referred to as "Feynman Machines" (FMs). The only existing FMs currently known are biological, however computer controlled or teleoperated FMs may in the future implement a broad range of nanotech applications.

Following Feynman's lead, other scientists delved into nanotechnology. Von Hippel (1962) predicted dramatic material science possibilities if new advances in "molecular designing" and "molecular engineering" of materials could be achieved. Noting the eventual possibility for repairing human tissue (molecule by molecule if necessary) for life extension, Ettinger (1964) suggested repair machinery for modification and interaction with existing organisms; later he proposed development of nanorobotics. Ettinger envisioned nanoscale scavenger and guardian organisms designed to emulate and surpass the actions of white blood cells which might hunt down and clean out hostile or damaging invaders (Ettinger, 1972). Such nanorobots might be useful for fighting AIDS or cancer, excavating blocked blood vessels, or straightening neurofibrillary tangles in senile neurons.

Shoulders (1965) reported the actual operation of micromanipulators able to position tiny items with 10 nm accuracy while under direct observation by field ion microscopy. Ellis (1962) also developed similar (but much larger) micromanipulators and proposed the construction of "microteleoperators": remote controlled nanodevices! Drexler (1981, 1986) has described some advantages and hypothetical dangers of nanotechnology. Capabilities for atom-by-atom assembly and nanoengineering could lead to new materials and pathways (Feynman, 1961). One such material is "diamond-like carbon" films which are "transparent, insulating, chemically inert, have a high dielectric strength, good adhesion and are relatively hard" (Aisenberg, 1984). Drexler suggests that rotary hammers a few molecules long might be used to hit carbon atoms in graphite at just the right angle and force to create lightweight diamond films and fibers useful in a variety of material applications. Drexler warns of two Frankenstein aspects to nanotechnology:

nanosensor surveillance, and uncontrolled replication of nanodevices with consumption of biosphere resources. However, the existing, dramatic developments in microsensor technology and microelectronics render worries about nanosensor surveillance superfluous (Schneiker, 1986). Schneiker also discounts Drexler's extreme "end of the world" worries about nanoreplicators, pointing out that the Feynman machine (top-down) approach to nanotechnology, *micro*replicators and other factors obviate the problem. Nonetheless, like genetic engineering, nuclear power, the automobile, and junk food, nanotechnology may well be a "double edged sword" which demands responsible management.

Potential benefits from nanotechnology attainable in perhaps a decade or two might include (Schneiker, 1986):

> ... vastly faster, much more powerful and numerous computers with extremely large capacity memories, ultrastrong composite materials, greatly improved scientific instrumentation, microscopic mobile robots, and automated flexible manufacturing systems, replicating systems, and achieving the practical miniaturization limits and maximum performance in virtually every area of technology.

Despite these lofty hopes and predictions, nanotechnology and molecular computing have remained mere dreams. Obstacles to their implementation center on the absence of available Feynman machines. A feasible solution has been advanced by a present day nanotechnologist whose contributions may eventually eclipse all others. Conrad Schneiker (1986) may have found the bridge to the nanoscale. He predicts that atomic level manipulative capabilities embodied in a 1981 invention, the scanning tunneling microscope (STM), can implement nanoscale Feynman machines (Hansma and Tersoff, 1987). Schneiker had noted that even without the STM, and before its invention, silicon micromechanics combined with other technologies could have been used. Not content with two approaches, he also proposed another route based on augmented machine tool technology originally developed for single crystal diamond machinery. But STMs are much more convenient and much less expensive.

## 10.2 Scanning Tunneling Microscopes (STMs)

Figure 10.1: Scanning tunneling microscopy (STM) depends on ultraprecise, but rapid movement of a sharp, single atom tip over a surface. The movement is controlled by X-Y-Z direction piezoceramic arms which move fractions of nanometers in response to voltage changes. By Paul Jablonka (Schneiker and Hameroff, 1987).

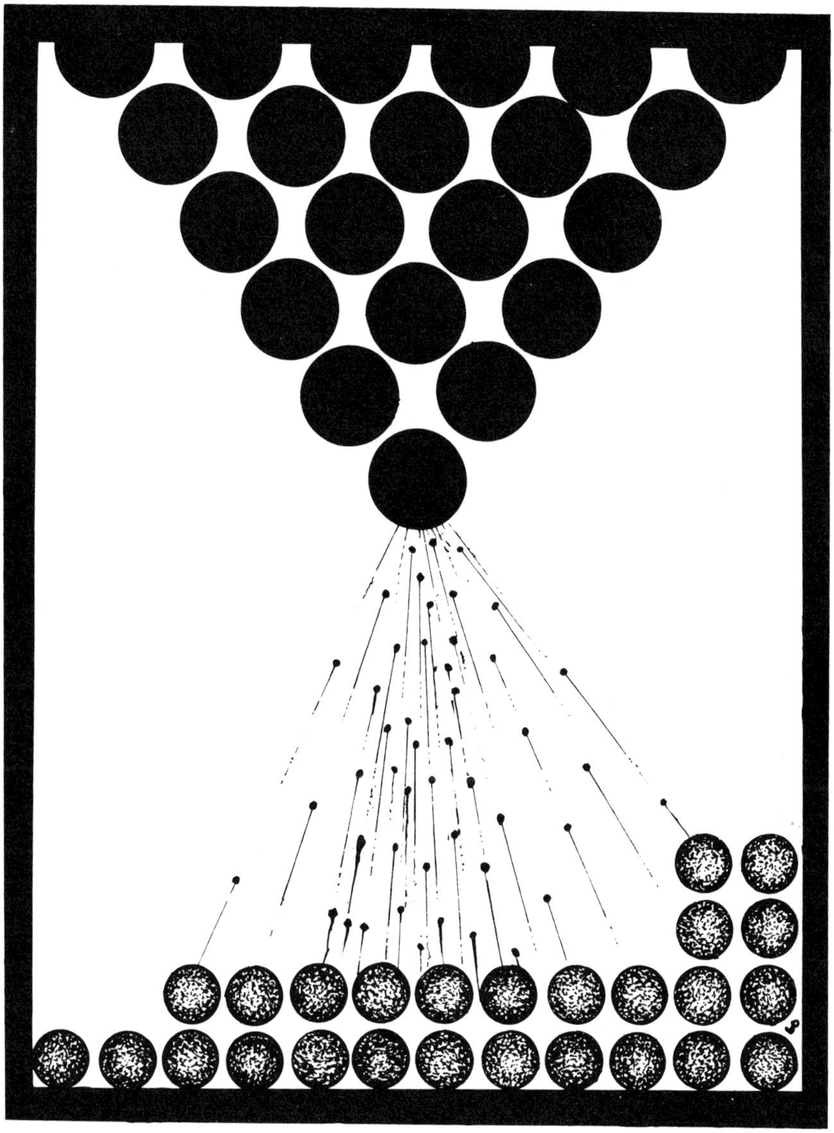

Figure 10.2: Nanoscale view of substrate to STM tip electron tunneling. By Paul Jablonka (Schneiker and Hameroff, 1987).

| Mode Number | I | II | III | IV |
|---|---|---|---|---|
| Quantity held constant | i, v | h, v | h, i | h, i, v |
| Quantity measured | h | i | v | i/v |

Table 10.1: STM operating modes; i, v, h are tunneling current, voltage, and height (tip to sample surface distance) respectively (Schneiker, 1986).

The 1986 Nobel Prize for Physics was split between the 50 year old cornucopia of microknowledge, the electron microscope, and a brand new, unfulfilled technology, scanning tunneling microscopy (STM). The STM half of the Prize was awarded to Gerd Binnig and Heinrich Rohrer of the IBM—Zurich Research Laboratories where STM was invented in 1981 (Binnig, Rohrer, Gerber and Weibel, 1982). The principle used in STM is extremely simple (Figure 10.1). Piezoceramic materials expand or shrink extremely small distances (i.e. angstroms, or tenth nanometers) in response to applied voltage. In STM, piezoceramic holders scan an ultrasharp conducting tip (such as tungsten) over a conducting or semiconducting surface (Figure 10.1). When the STM tip is within a few angstroms of a surface, a small voltage applied between the two gives rise to a tunneling current of electrons (Figures 10.2, 10.3 and 10.4). The tunneling current depends exponentially on the tip-to-substrate separation (about an order of magnitude per angstrom or tenth nanometer). Depending on the substrate, typical tunneling currents and voltages are on the order of nano-amperes and millivolts, respectively. A servo system uses a feedback control that keeps the tip to substrate separation constant by modulating the voltage across a piezoelectric positioning system. As the tip is scanned across the surface, variations in this voltage, when plotted, correspond to surface topography. Detailed surface topography maps of various materials have been obtained which demonstrate individual atoms like cobblestones. By varying and measuring different combinations of tunneling current, voltage, and distance, different types of information may be obtained about the surface being probed.

The basic modes of operation of an STM, described by Hansma and Tersoff (1987), are summarized in Table 10.1. Here i, v, and h are the tunneling current, the voltage across the gap, and the gap size respectively. Mode I is used to measure the topography of the surface of a metal or semiconductor and is the slowest mode since the electro-mechanical servo system must follow the shape of the surface during the scanning operation.

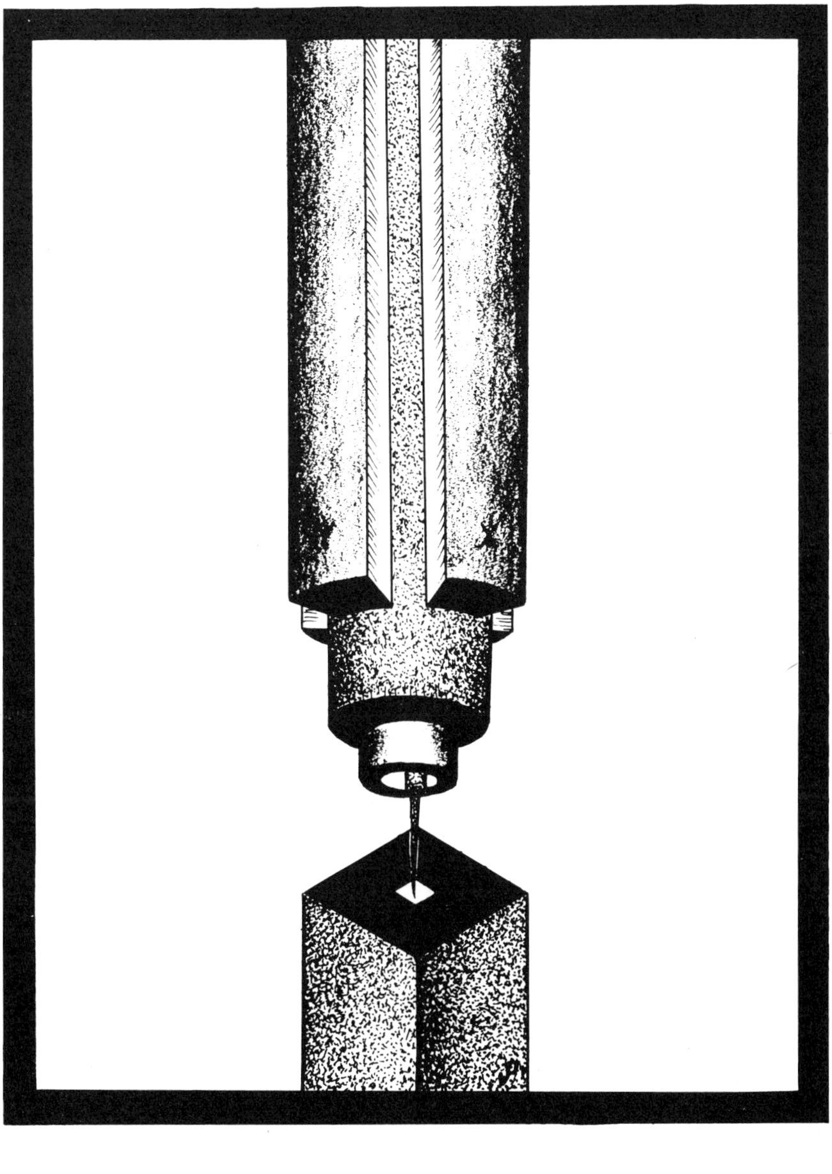

Figure 10.3: Multi-directional movement of STM tip can be controlled by single piezo-tube scanner. By Paul Jablonka (Schneiker and Hameroff, 1987).

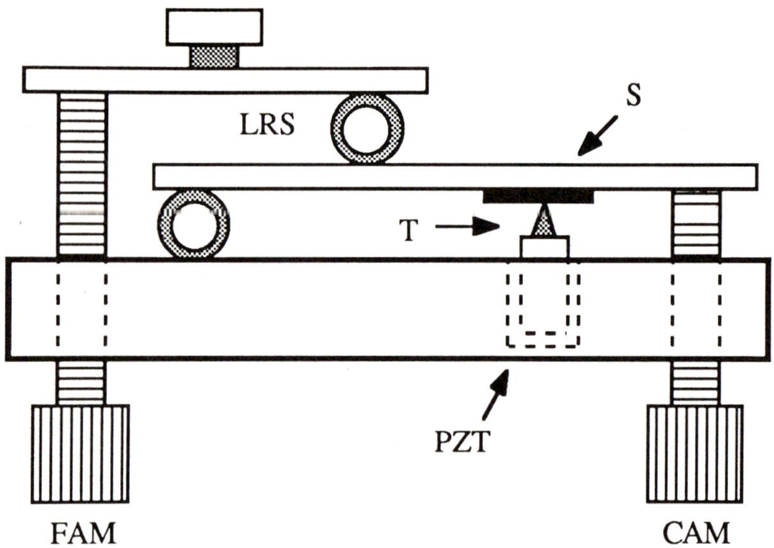


Figure 10.4: Schematic of STM from Tunneling Microscope Corporation (Doug Smith). T = tip, S = sample stage, PZT = piezoceramic tube, CAM = coarse adjust micrometer, FAM = fine adjust micrometer, LRS = lever reduction system. By Conrad Schneiker (Schneiker, 1986).

The scanning speed in Mode I is determined by the response of the servo system. Modes II and III are faster since the tip maintains only a constant average height above the surface. The scanning speed in these modes is determined by the response of the preamplifier only. Mode IV measures the joint density of states which, for a small tip, is a map of the local distribution of electron states of the substrate. For this mode, one "dithers" the tip-to-substrate bias voltage with a small alternating current signal and monitors i/v. Using this mode, Smith and Quate (1986) have done spatially resolved tunneling spectroscopy and observed charge density waves. STM can thus be used to identify the elements comprising specific atoms and to monitor molecular dynamics (Figure 10.5).

Adaptability and versatility have been shown by STMs operated in air, water, ionic solution, oil and high vacuum (Drake, Sonnenfeld, Schneir, Hansma, Solugh and Coleman, 1986; Miranda, Garcia, Baro, Garcia, Pena and Rohrer, 1985). Their scanning speed may be pushed into the real-time

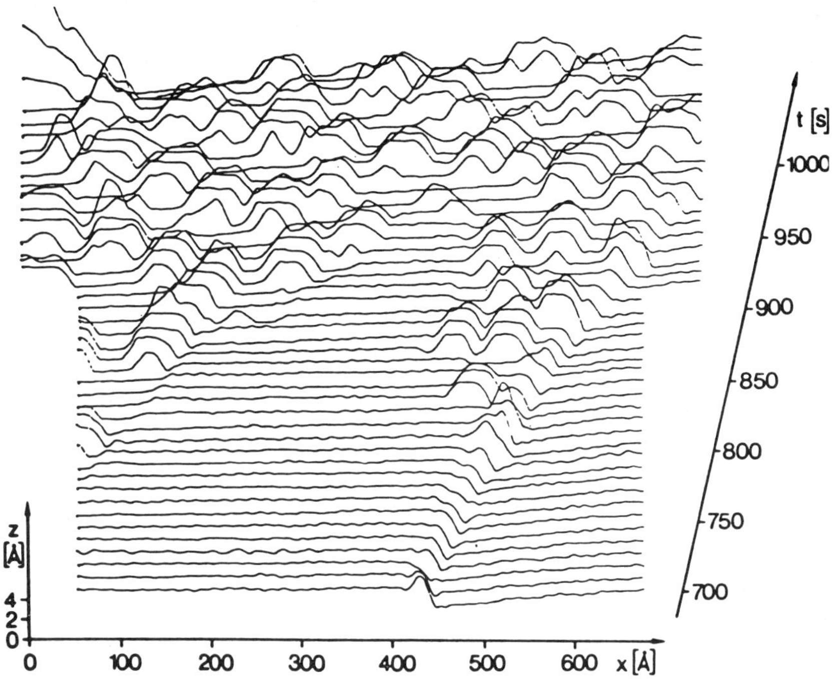


Figure 10.5: STM scan with t-axis showing evolution of a nanosystem over time. With permission from Ritter, Behm, Potschke and Wintterlin (1987), courtesy of Eckhart Ritter.

imaging domain (Bryant, Smith and Quate, 1986). One technique for machining STM tips is based on a simple ion milling process that can generate ultrasharp tips with single atom points; a similar technique can generate ultrasharp knife edges and other nanotool shapes as well (Dietrich, 1984). Dieter Pohl (1987) of IBM-Zurich considers STM one example of a group of "stylus" microscopic technologies. Tips are being developed as thermocouples of two different metals sensitive to extremely low changes of heat, and hollow pipettes able to administer or detect individual molecules in solution. Pohl refers to these STM capabilities as molecular "tasting and smelling." Another STM related function can "touch and feel." In "atomic force microscopy," a nanoscale lever positioned by STM piezo-technology is deflected by contact and/or movement of atoms, molecules or their surrounding ions. Van der Waals force induced lever deflection is monitored by an STM tip or by interferometry. Binnig, Quate, and Gerber (1986) propose to measure

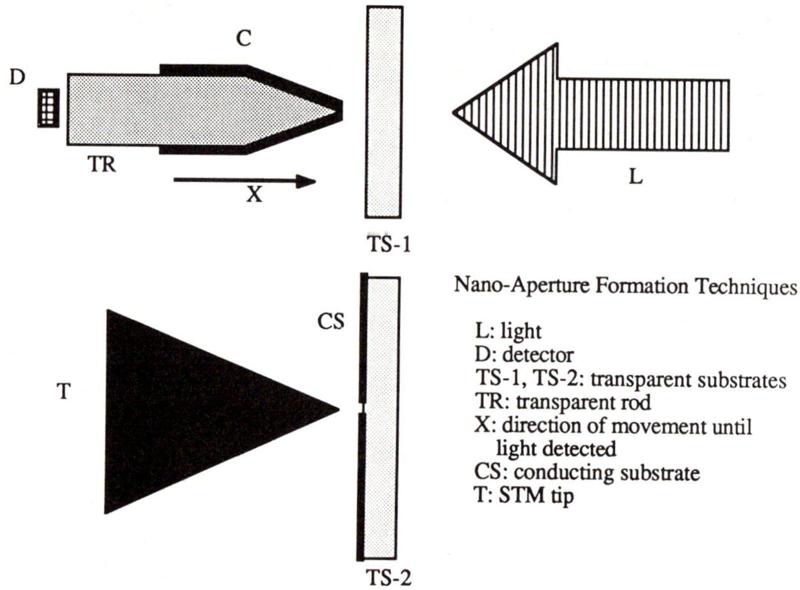


Figure 10.6: Formation of near-field optical nano-apertures. By Conrad Schneiker (Schneiker, 1986).

forces as small as $10^{-18}$ newtons using atomic force microscopy. Mechanical shape and structural dynamics can be probed without tunneling through the sample material using this mode, which may be particularly important in the study of biomolecules.

Another STM spinoff may lead to "nanovison." Resolution of "near-field microscopy" depends on the diameter of the aperture through which the reflected light passes. Binnig (1985) described how to use an STM to make 20 nm holes in opaque metal films which, when used in scanning light microscopes, become capable of resolving features much smaller than the wave-length of the light. In general, a cone of light cannot be focused to a spot which is significantly smaller than the light's wavelength. However, light passed through an aperture that is many times smaller in diameter than its wavelength results in a narrow beam of light that may be scanned mechanically close to a sample being examined. STM technology is capable of creating nanoapertures, and also the scanning mechanism required for imaging an area (Figure 10.6). Pohl, Denk and Lanz (1984) described their "optical stethoscope" in which "details of 25-nm size can be recognized using 488-nm radiation." A significant advantage of the optical stethoscope concept is "that it allows samples to be ***nondestructively*** investigated in their ***native*** environments" (Lewis, Isaacson, Harootunian and Muray, 1984). Betzig and colleagues (1986) conclude that this technology will be "able to follow the temporal evolution of macromolecular assemblies in living cells ... to provide both kinetic information and high spatial resolution." Scanning nanoapertures could also use ultraviolet, X-ray, gamma ray, synchotron, or particle beams to "snoop" in the nanoscale if suitable materials and configurations are found (Schneiker, 1986).

Schneiker's breakthrough was to realize that STM tips can be used as ultraminiature, ultraprecise robot fingers that can both "see" and be used to directly manipulate individual atoms and molecules along the lines suggested by Feynman. The scaling down process proposed by Feynman (machines building smaller machines, and so on) can be reduced to just one step! According to Schneiker's concept, STMs can directly link up to the nanoscale to implement, construct, and evaluate Feynman machines and other nanotechnologies. He notes that many natural biomolecules such as enzymes, t-RNAs, antibodies, cytoskeleton, and many synthetic compounds provide a wide variety of potential finger tips to realize Feymans vision of molecular manipulation. Rudimentary nanomanipulation has already been described. Ringger and colleagues (1985) have described STM nanolithography: atomic scale surface etching. Becker, Golovchenko and Swartzentruber (1987) of Bell Labs have reported a single atom modification of the surface of a nearly perfect germanium crystal by the tungsten tip of an STM. They conclude that manipulation of atomic sized structures on surfaces will lead to "new frontiers in high density memories, new devices whose electrical properties are dominated by quantum-size effects, and modification of single, self replicating molecules,"—supporting the earlier predictions of Schneiker. In its first few years, STM has imaged single atoms, probed atomic forces, manipulated atoms, and won a Nobel prize. Eventually STMs and their spinoffs may enable researchers to alter and communicate with genes, viruses, proteins, cytoskeletal assemblies, and perhaps biomolecular consciousness.

Schneiker and Hameroff (1987) have proposed an integrated system of STM and other technologies to best take advantage of STM capabilities. The "nanotechnology workstation" (Figures 10.7 and 10.9) proposes to combine multiple STM tips, optical microscopy, nanoapertures, and fiber optic waveguides in a hopeful attempt at synergism. For example, the STM is extremely "near-sighted;" it can have difficulty finding its target ("a needle in a haystack"). If integrated into a high powered light microscope,

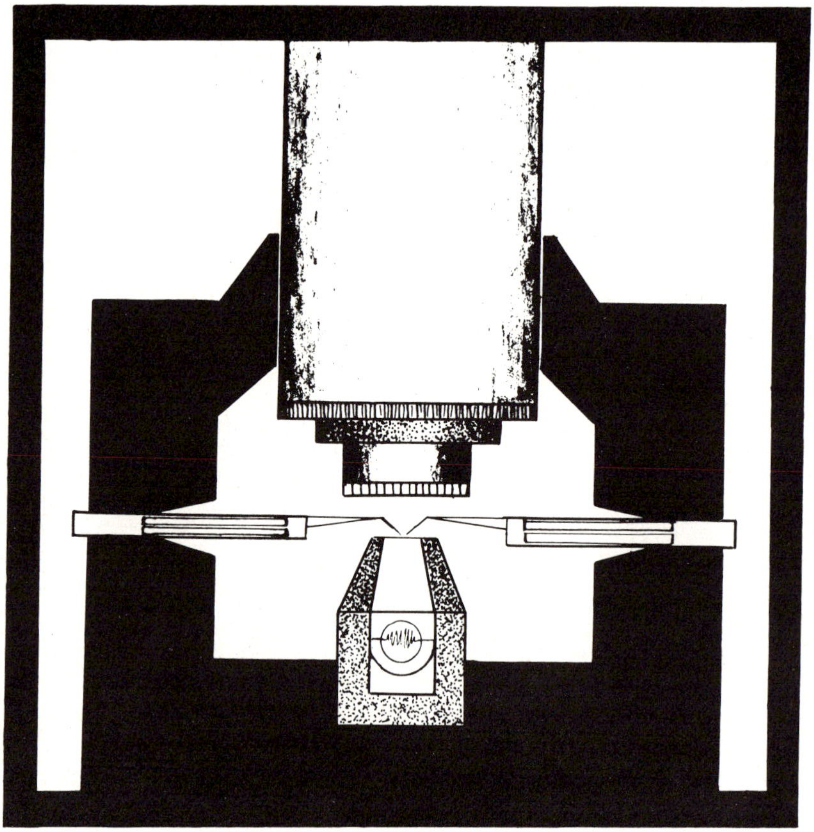

Figure 10.7: Nanotechnology workstation is envisioned as a synergistic combination of multiple STM tips mounted in an optical microscope. By Paul Jablonka (Schneiker and Hameroff, 1987).

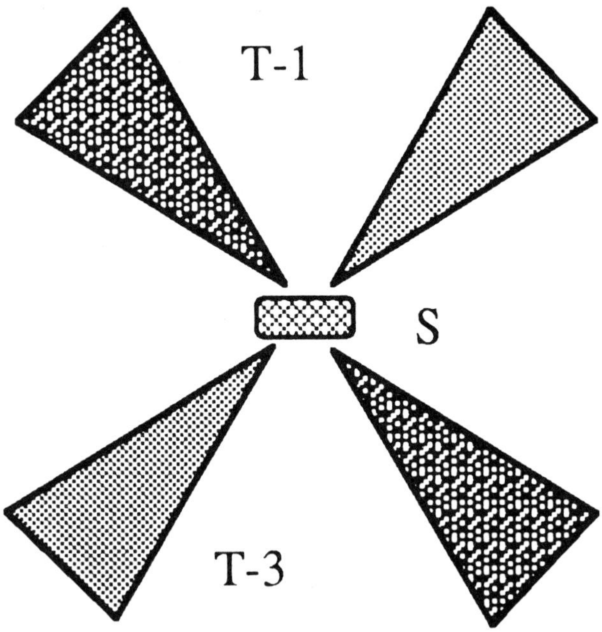


Figure 10.8: Multiple tip geometry for STM probing of nanoscale systems. By Conrad Schneiker (Schneiker, 1986).

the microscope can act as a coarse approach mode to bring the STM tip to the desired area. From there the STM can perform as a "cursor" mode for the microscope: able to expand the view of, and manipulate, atoms and molecules. In one configuration of the nanotech workstation, four STMs are placed symmetrically about a central STM and these each form a 45 degree angle to the sample plane; they may later be augmented with fiber optical interferometers for precise spatial coordinate determination and motion control (Figure 10.8). The central STM may be used for high speed scanning while the peripheral STMs are used as oblique STM scanners, nanoaperture scanners, electrodes, mechanical "fingers" or optical frequency antennas. In addition, four other micrometer-adjusted mini-arms can carry horizontal optical fibers: two for general illumination and two for experimental interfaces.

Reflecting optical microscope objectives may be positioned such that the optical axis is perpendicular to the plane of the opposite pair of STMs, and the focus is where their tips would meet when fully extended. The

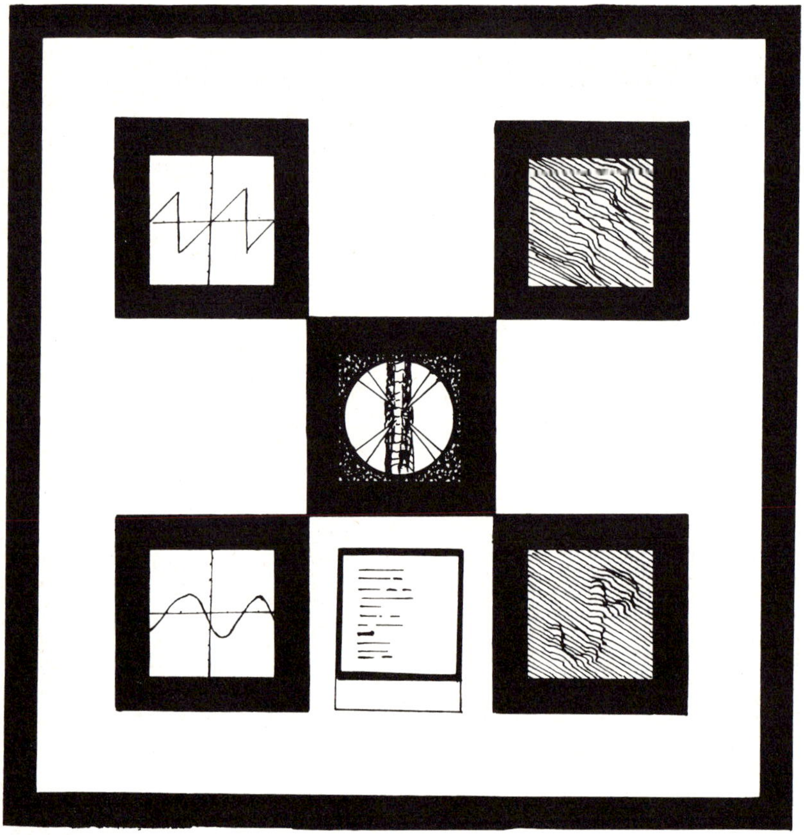

Figure 10.9: Graphic displays, monitors and control module for nanotechnology workstation. By Paul Jablonka (Schneiker and Hameroff, 1987).

optics of these microscopes would permit them to peer around the fiber optic illumination system mounted on the forward axis of each microscope. The optical system would be achromatic, and able to work in far infrared to ultraviolet which should lead to increased optical resolution. Fluorescent coating for STM tips (excluding the last few nm) can aid in visualizing their location. The final imaging system should consist of solid state detectors whose output may be displayed on high resolution color monitors: STM modes, optical microscope, nanoaperture "nanovision," etc. (Figure 10.9 and 10.10).

Figure 10.10: View of four STM tips approaching sample stage in nanotechnology workstation. By Paul Jablonka (Schneiker and Hameroff, 1987).

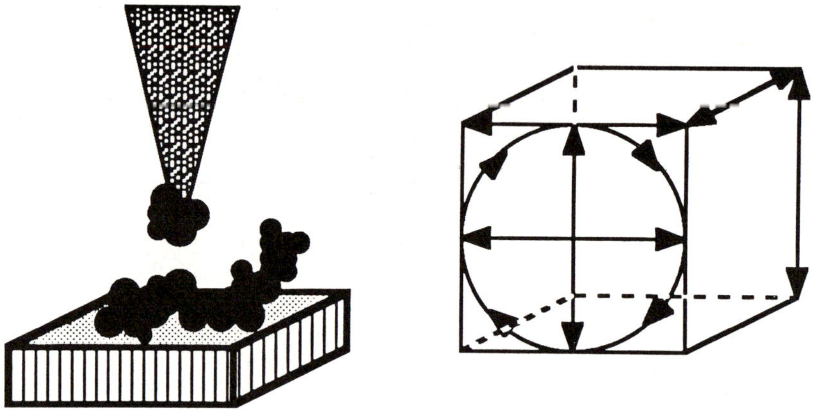

Figure 10.11: Molecular movement and construction with an STM. By Conrad Schneiker (Schneiker, 1986).

A system akin to the STM nanotech workstation may dynamically observe and manipulate nanoscale materials and systems, capabilities consistent with Feynman's notions for molecular machines.

## 10.3 STM/Feynman Machines(FMs)

STMs utilized to implement Feynman's ideas concerning atomic and molecular level fabrication and machining (Feynman machines: "FMs") have been dubbed by Schneiker as "STM/FMs."

Feynman had noted the possibility of doing chemical synthesis mechanically. In an effort to move in this direction, Schneiker (1986) proposed the following STM experiments (Figure 10.11). 1) Use an STM tip to move a pair of adatoms (or molecular fragments) on a substrate (or STM tip) together so as to induce chemical bonding. 2) Use an STM tip to cleave a chemical bond of a molecule adsorbed on a substrate. 3) Use an enzyme or synthetic catalyst adsorbed on an STM tip to repeat experiments (1) and (2).

Perhaps the first step in such ambitious proposals has already been accomplished. The Bell Labs group (Becker, Golovchenko and Swartzentruber, 1987) used an STM tip to place a single atom on a germanium surface. Further STM/FM modifications and applications elucidated by Schneiker (1986)

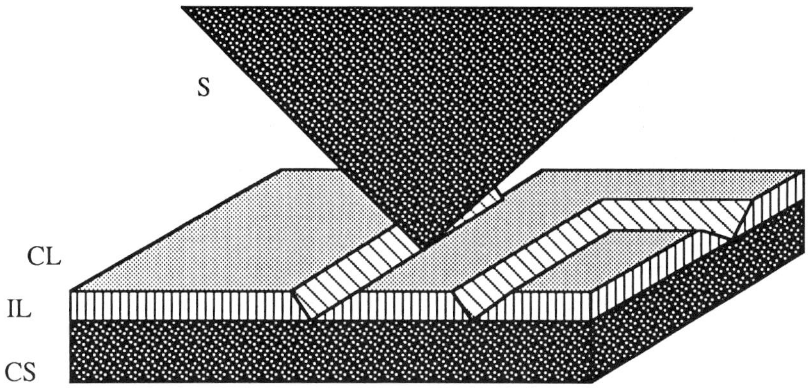


Figure 10.12: Nanomilling/nanolithography. By Conrad Schneiker (Schneiker, 1986).

include: 1) tip shape modifications (scalpel, chisel, cylindrical, and other configurations) for scanning, scribing, etching, milling, and polishing operations or for electrical interfaces, electrochemical synthesis or machining, 2) attached tip structures (enzymes, synthetic catalysts, shape selective crown ethers, transducer molecules, etc.) for molecular recognition with species selective (and perhaps electrostatic or electromagnetic assisted) pick, place, join, and cleave operations, or nano-environmental sensing, 3) multiple tip configurations (parallel or radial configurations) for use as ultraminiature tweezers, jigs, and arcs or interface electrodes, or to generate rapidly rotating electric fields, and 4) tip materials modification (insulating, semiconducting, ferroelectric or ferromagnetic) for electrostatic, electromagnetic, magnetic, kHz-GHz acoustic (longitudinal, transverse, or torsional) and optical modulation in mono-or multi-polar configurations (Figures 10.12 and 10.13).

In addition to the above modifications, STM/FMs or their tips could be augmented with a wide variety of sensors and transducers; the atomic force microscope of Binnig, Quate and Gerber (1986) is an excellent example. The augmentation of STM/FMs with fiber optic interferometers (or comparable techniques) could provide extremely accurate realtime calibration of absolute and relative STM/FM tip positioning, thus overcoming the problems of electrical noise, creep, ageing and hysteresis inherent in present STM piezo-positioning systems. The technique given in Dietrich, Lanz and

Figure 10.13: Nanoscale view of STM nanomilling/nanolithography. By Paul Jablonka (Schneiker and Hameroff, 1987).

Moore (1984) for making tips with uniform tip-to-base conical profiles would be useful for STM/FMs using closely spaced multiple tips. Schneiker conjectures that properly configured and instrumented sets of STM/FMs can operate as machine tools with effectively perfect lead screws and bearings. STM/FMs also can be used as sub-atomic resolution proximity detectors and coordinate measuring machines on conducting surfaces (AFMs would be used for insulators) to monitor nearly perfect superaccurate nanomachining operations (limited by the graininess of atoms and other materials science considerations). Many useful macroscopic mechanical structures and mechanisms may thus be duplicated at the submicron level, and many of these mechanisms may require no lubrication due to force/area scaling and very rapid heat dissipation (Feynman, 1961). For even smaller mechanisms, a switch to Feynman's mechanical chemistry approach would be needed to build up molecular devices in a series of joining and trimming operations. Many presently existing synthetic molecules could be utilized as building blocks; the molecular gear and bearing system of Yamamoto (1985) provides an interesting example and is thought to be capable of rotating about a billion times a second.

STM/FMs may execute many complex mechanical motions for driving such nanomechanisms (Schneiker, 1986). For instance, an essentially infinite series of three dimensional tip motions (single straight lines, circles, spirals, helices, etc.) can be made at speeds presently limited mainly by mechanical resonances in the driving system of current STMs—another motivation for miniaturizing them considerably. Thus it would be possible, for example, to turn a molecular scale mechanical crank (say of 2 nm diameter) at thousands of revolutions per minute. Similar to flagella and cilia of single cell organisms, these could be used for propulsion of nanoscale robots and other uses.

The capabilities of STM/FMs have been heretofore nonexistent; therefore applications may abound in the near future as scientists in many disciplines become aware of this versatile and inexpensive technology.

## 10.4 Micro/Nano STM Contest

Feynman's original (1961) proposal for molecular and atomic scale machines included prizes and competition to "get kids interested in this field." He offered \$1000 to the first person to a) reduce the information on the page of a book to an area 1/25,000 smaller in linear scale to be read by an electron microscope, b) construct an electric motor which is only 1/64 inch cube.

The latter prize was presented by Prof. Feynman on November 28, 1960 to William McLellan, who built an electric motor the size of a speck of dust (Gilbert, 1961). The former prize was awarded in 1985 to Tom Newman, a Stanford graduate student who used electron beam lithography to reduce a page from *A Tale of Two Cities* to 5.9 × 5.9 micrometers and magnified it back using an electron microscope.

Motivated by a desire to accelerate development of STM derived technology as a bridge to more powerful Feynman Machines, Schneiker (1985, 1986) has announced a series of construction challenges and prizes (Hansma and Tersoff, 1987). The challenges are to construct, operate, and publicly demonstrate STMs (including the active/moving part of the mechanical positioning and scanning systems) which can be controlled from the outside, and which (not counting lead-in wires, fiber optics, etc.) are of the following sizes or smaller: $(1\ \text{mm})^3$, $(100\ \mu\text{m})^3$, $(10\mu\text{m})^3$, $(1\ \mu\text{m})^3$, $(100\ \text{nm})^3$, and $(10\ \text{nm})^3$. The imaging capability should be comparable to that indicated by the picture on page 482, vol. 56 of the physics journal Helvetica Physica Acta, 1983. The two imaging tests will be: a) imaging a $(10\ \text{nm})^2$ square highly ordered pyrolytic graphite surface, and b) imaging the entire conical surface of another atomically sharp (nonrotating) tungsten STM tip (forming a 60° cone or less), extending 2 nm from that tip. Successive scans must be made to demonstrate that your device doesn't modify these test items. An additional pair of challenges are 1: to use an STM/FM to a) mechanically synthesize 10 molecules of either of the proposed spherical $C_{60}$ or $C_{180}$ molecules known as "Buckminster Fullerene" (Schneiker, 1986) and to b) demonstrate conclusively that the synthesis was successful, and 2: to use an STM/FM to a) mechanically synthesize a 10 nm × 10 nm layer of graphite, and to b) demonstrate conclusively that the synthesis was successful. The current prizes being offered, one per category, are $1000, with a maximum of one prize being paid out per year. Potential winners should contact the author. Additional sponsors and prizes are solicited. And as Feynman originally advised:

> ... *have some fun! Lets have a competition between laboratories.*

## 10.5 STM/FMs and Molecular Computing

The ability to use multi-tip STM/FMs for mechanical chemistry and as electrical interfaces might solve two formidable problems in the development of molecular computing devices as envisioned by Carter (1983): 1) the synthesis of prototype devices and 2) making individual, reliable, electrical connections to them for testing.

Arrays of nanoscale electrodes or STM tips comprising sub-micron scale assemblies may be extremely useful if their position can be precisely and rapidly regulated. Schneiker (1987) has suggested that optimal utilization of piezo-ceramic materials may be used for parallel scanning of nanoscale arrays. The piezo-ceramic "biomorph-bender" described by Dieter Pohl and colleagues (1986) may be used to manipulate such arrays for fast parallel reading and writing of nanoscale array chips for mass storage and processing of information (Figures 10.14 and 10.15, Schneiker, 1987).

STM/FMs could be used for more conventional ultraprecise circuit or component trimming and repair operations. The extremely accurate positioning systems of STM/FM could be exploited for minimal scale wirebonding systems. Other STM derived applications could include inspection of, and electrical and mechanical interfacing to, superlattice quantum well devices or "quantum dot" devices on the scale of cubic nanomenters. In 1959 Feynman noted that using his strategy it should eventually be possible to make what are now called superlattices and that they would probably have some very interesting properties (Feynman, 1961). Even though STM/FMs may not ultimately be used to mass produce such devices, they could be used to construct prototypes, help characterize test devices, and be used to optimize them. STM tip induced sputtering (Binnig, Gerber, Rohrer and Weibel, 1985) might be used to etch or mill ultrafine conductors for such devices. The availability of even small numbers of such extremely high performance devices would have great instrumentation value as interfaces and transducers (Figures 10.16 thru 10.19).

Computing components each consisting of a few atoms might use quantized energy levels or spin effects (Feynman, 1960); *if* such devices could be designed and assembled, they could be thousands of times faster than conventional devices used in today's computers. Other aspects of quantum computers are discussed in Maddox (1985), and other interesting references may be found in Schneiker (1986). Efficiently interfacing these and other such devices to the outside world in a manner that can effectively utilize their potential computing bandwidth presents some interesting problems in clocking, input/output, and other areas. Schneiker claims that the optical electronics technology suggested by Javan (1985, 1986) might be useful in these and other nanoapplications as well. Using micro and nano-antennas coupled to laser radiation, intense, localized high frequency electric fields

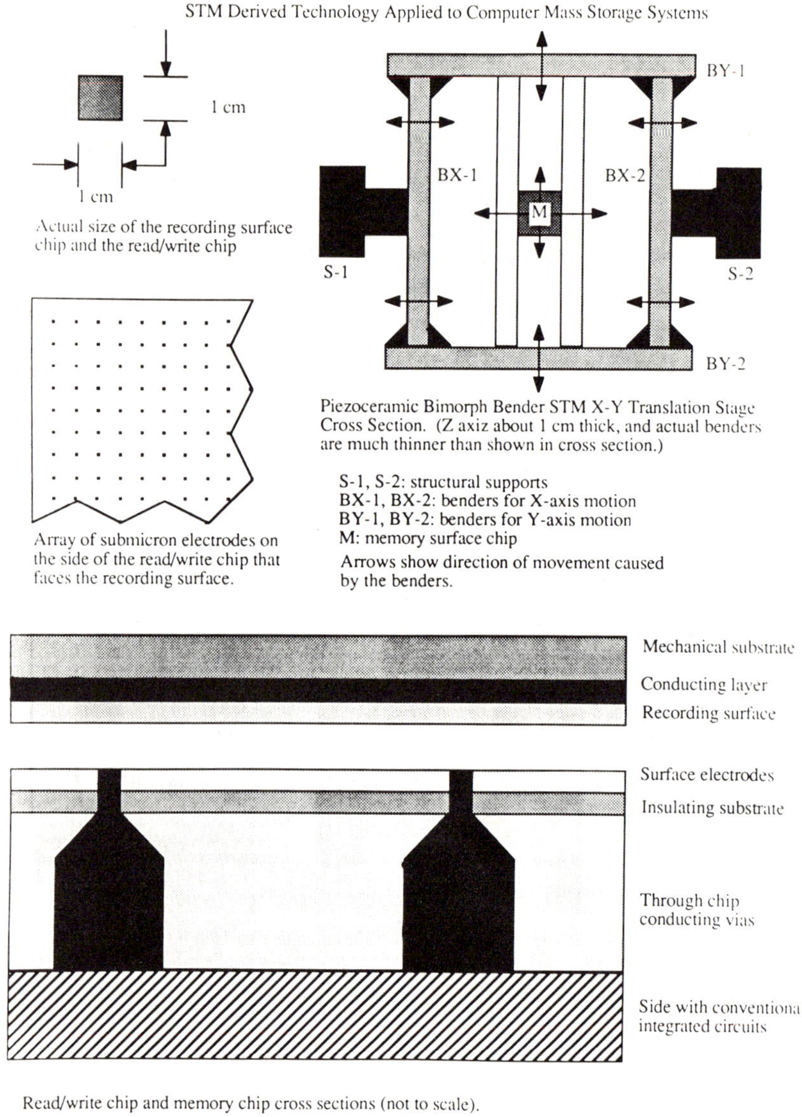


Figure 10.14: Computer mass storage for fast parallel reading and writing "nano-film" coated chips. Ultimately, electrodes could be replaced with micro-STMs (Figure 10.15) for accessing individual molecular devices. A simple system might also be configured for highly parallel nanolithography. By Conrad Schneiker (Schneiker, 1986).

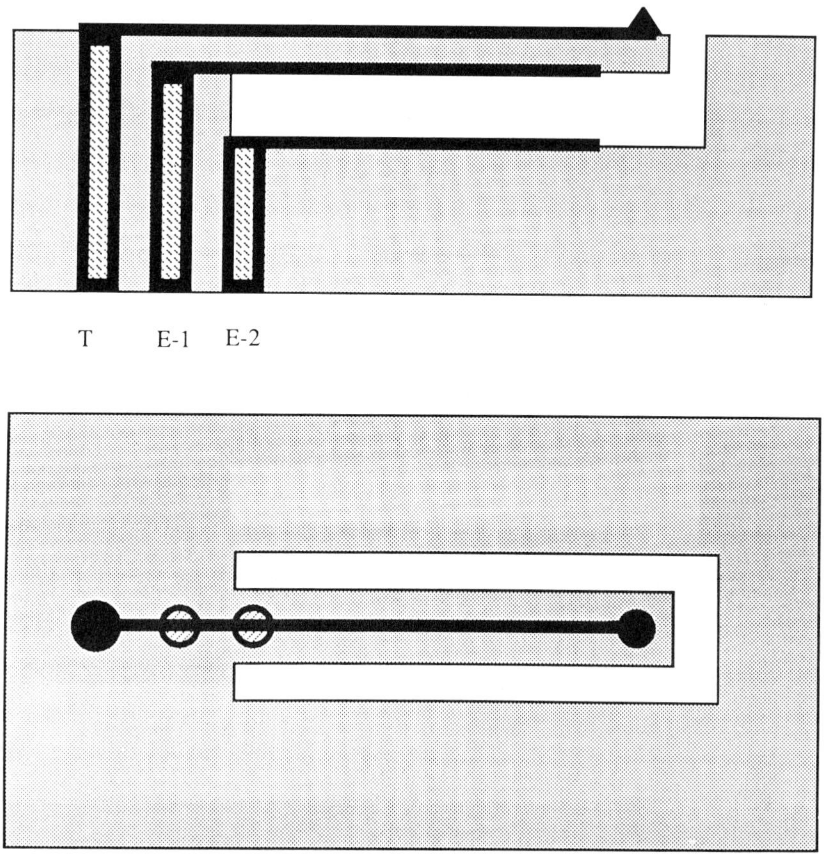


Figure 10.15: A micro-STM formed on a silicon substrate; thousands of these structures may be placed on a single silicon chip. By Conrad Schneiker (Schneiker, 1986).

Figure 10.16: Three STM tips in configuration for tunnel modulation experiment. By Paul Jablonka (Schneiker and Hameroff, 1987).

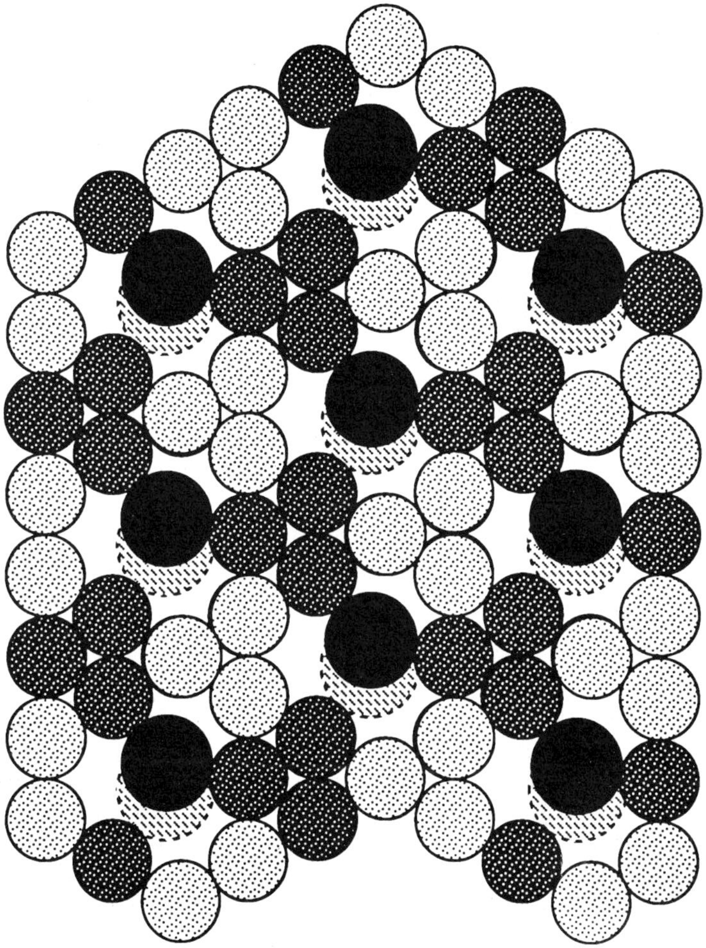

Figure 10.17: Electrochemical tunnel switch/atomic memory device. By Conrad Schneiker (Schneiker, 1986).

modulated by the laser beam's intensity, phase, and polarization could control and power arrays of STM/FM constructed nanoscale devices. This same idea could be used to control, power, and communicate with swarms of mobile nanorobots.

During recent talks on his quantum computing ideas, Feynman briefly speculated on a simple possibility for making nanocomputer components: use STM tips to make tiny holes in very thin metal sheets, thus forming

Figure 10.18: STM-induced/detected molecular conformation change. By Paul Jablonka (Schneiker and Hameroff, 1987).

grids for tunneling "nanovacuum tubes," perhaps around 3 to 10 nm in size, or smaller (Feynman, 1985, 1986). Analogous, but much larger (down to 100 nm scale) devices with calculated *picosecond* range switching speeds have been proposed by Shoulders (1965). Although he considered even smaller and faster devices, limitations of electron beam micromachining technology at that time prevented further size reduction (Shoulders, 1986). STM/FMs could solve that problem and many others. Indeed, the Naval Research Laboratory is now studying *subpicosecond* (thousandth nanosecond) submicron vacuum tubes (Robinson, 1986).

## 10.6 STM/FMs and Biomedical Applications

Nondestructive STM interactions with biological material have immense potential, assuming certain technical obstacles are overcome. Potential problems include the following:

1. Biomolecules, cells, and tissues are not conductors. Electron tunneling through these materials, if it occurs, may be damaging. Nevertheless, several groups including IBM Zurich have succeeded in imaging biomaterials in air such as protein coated DNA and virus structures. The tunneling is thought to occur from tip onto biomolecular surface followed by "low resistance electron transport to the conducting substrate" (Travaglini, Rohrer, Amrein and Gross, 1986). Simultaneous optical microscopy, as can occur with the nanotech workstation, may help this situation since photons can lower tunneling barriers. Appropriate choice of optical microscopy wavelengths may thus facilitate STM imaging by permitting non damaging tunneling currents.

An alternative approach is to utilize an atomic force microscope (AFM) mode of operation for biological materials. In this case a lever arrangement is adapted to, and mounted on, an STM which trails along the surface, or is held steady to observe mechanical dynamics of the material (i.e. protein conformational change). The movement of the lever is monitored by the STM, so that mapping and dynamics can be observed without direct tunneling through the biomaterial. Nanoscale thermocouple and pipette STM tips as described by IBM's Pohl (1987) may also be useful for biological materials.

2. Biomaterials should be studied in a stable environment as much like the aqueous medium within cytoplasm as possible. Temperature, pH, ionic concentrations, availability of high energy phosphate groups and numerous other parameters need to be closely regulated. Researchers at University of California-Santa Barbara have demonstrated that STM imaging can oc-

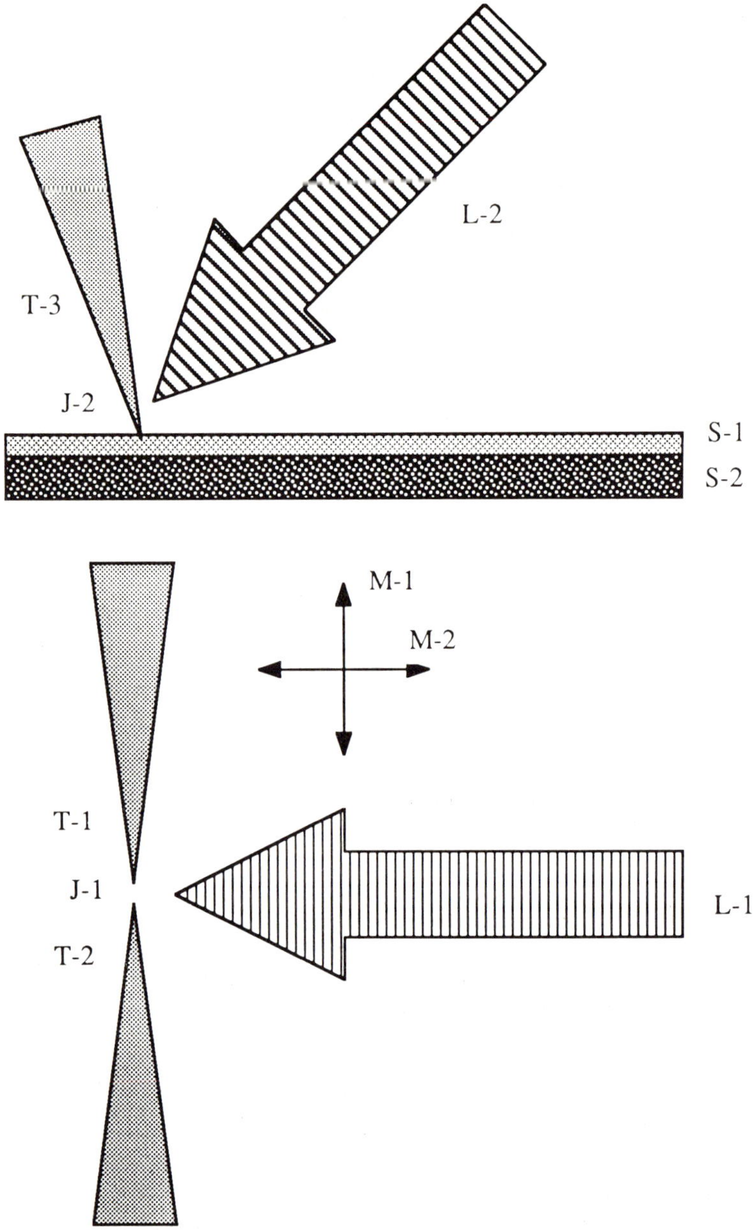


Figure 10.19: STM tunneling switches modulated at optical frequencies with lasers. By Conrad Schneiker (Schneiker, 1986).

cur at ionic liquid-solid interfaces. By insulating STM probes to very near their tips, significant leakage of current to the ionic aqueous environment is apparently avoided (Sonnenfeld and Hansma, 1986). Thus STM should be applicable to characterization of living biomolecules under stable physiological conditions.

3. A third problem has been the "near sightedness" of the STM. Hansma's Santa Barbara group and the IBM-Zurich team had difficulty in locating DNA molecules on the background substrate with the STM. This "needle in a haystack" problem is potentially solved by a nanotechnology workstation configuration in which the coarse approach of the STM is made simpler by visual observation. Further, immunofluorescence techniques may be utilized to identify specific biomolecular targets and the STM probes themselves. Electophoretic fields applied to the sample stage may attract and immobilize charged biomolecules rendering them easier to locate and study (Lindsay, 1987).

Potential applications of STM to biology and medicine include repetitive imaging and structural mapping of living material, nondestructive study of dynamical structural changes such as protein receptors (via alterations in the AFM mode), detection of possible propagating phonons or solitons (using the multiple STM configurations, one tip can perturb and another detect perturbation at a second position on a macromolecule or polymer), spectroscopic analysis (i.e. DNA base pair reading, cytoskeletal lattice communication), and real time imaging (via high speed scanning) of living structures can expand the horizons of experimental biosciences. Further, a capability for nanoscale manipulation of biomaterials and organelles offers a host of imaginative possibilities. For instance, a few dozen multitip STM/FMs might be developed to directly extract, manipulate, analyze, and modify DNA or other cellular components, thereby greatly assisting some subfields of genetic engineering and medical research. Synthesis or modulation of important biological molecules which are "topologically complex" (e.g. receptors, enzymes, antibodies, cytoskeleton) which may be difficult and relatively expensive using present day synthetic chemistry, could ultimately be feasible with STM/FMs. Feynman (1961) commented that great progress may be expected in biology when "you can simply look at, and work with individual molecules." STM/FMs could help materialize some even grander biomedical applications when used in combination with genetic engineering: to create, modify, or program micro-organisms for therapeutic uses.

Ettinger (1972):

> If we can design sufficiently complex behavior patterns into microscopically small organisms, there are obvious and endless possibilities, some of the most important in the medical area. Perhaps we can create guardian and scavenger organisms in the blood, superior to the leukocytes and other agents of our human heritage, that will efficiently hunt down and clean out a wide variety of hostile or damaging invaders.

Asimov (1981) suggests that we:

> Consider the bacteria. These are tiny living things made up of single cells far smaller than the cells in plants and animals .... [We] can, by properly designing these tiniest slaves of ours, use them to reshape the world itself and build it close to our hearts' desire .... .

White (1969) proposed a similar notion for programmable cell repair machines:

> appropriate genetic information can be introduced by means of artificially constructed virus particles into a congenitally defective cell for remedy (and) ... repair. The repair program must use (the cells own) protein synthesis and metabolic pathways to diagnose and repair any damage.

Approaches to biomedical nanotechnology relying solely on genetic/protein engineering and self assembly (Asimov, 1981; Aridane, 1983; Drexler, 1981, 1986) are severely constrained by the lack of direct control and observation: the potential attributes of STM/FMs. Hybrids and components of organisms, cytoskeletal structures, viruses, and synthetic structures may evolve, facilitated by STM/FM capabilities, to fabricate, examine, and modify nanostructures. Ettinger (1972) anticipated "nanominiaturized robots to serve us." Feynman (1961) reported that a friend, Albert R. Hibbs suggested:

> a very interesting possibility for relatively small machines. He says that, although it is a very wild idea, it would be interesting in surgery if you could swallow the surgeon. You put the mechanical surgeon inside the blood vessel .... Other small machines might be permanently incorporated inside the body .... .

Feinberg (1985) sees

> ... the construction of nanosensors that could be implanted into the human body. They could ... [monitor] various physiological functions from subcellular molecules up through tissues and organs ... essential in determining some of the mechanisms involved in growth and aging.

Feinberg (1985) also proposed the use of laser energy to power and communicate with implanted sensors, and to direct activities of nanorobots.

> ... using coherent, short-wavelength radiation, we too will be able to match the accomplishments of the cells that compose us, and do molecular engineering.
>
> These molecular structures will be much more complex than anything that human technology has thus far achieved ... . It may be possible to extend some of the methods using ... short-wavelength coherent radiation ... to nanofabrication as well as to seeing what we are doing in the nanoworld, since microholography can be used to demagnify patterns as well as magnify them.

There exist rationale and applicability for mobile nanodevices whose missions might be to stalk and destroy lethal viruses or malignancies, excavate clogged blood vessels, correct genetic expression and differentiation, repair or remove the processes of ageing including senile tangles of cytoskeletal proteins, and perhaps augment natural capabilities. Virtually every disease might be amenable to intracellular house calls by such structures, if they ever exist. Such nanodevices might be patterned after, or directly utilize, cytoskeletal elements (centrioles, microtubules, etc.), viruses (bacteriophages) programmed by specifically engineered genes, and/or controlled by telemetry. Synthetic materials including nano STM/FMs composed of piezo-materials may be incorporated.

Ellis (1962) discussed the possibility of microteleoperators (and suggested that protein enzymes may function in this manner). Nanoantennas have been proposed by Marks (1985) who described two orthogonal layers (about 200 nanometers in length) of metallic dipole antenna arrays which can convert photons to direct current with 75 percent efficiency. Javan (1985, 1986) has studied metal whiskers which can form tunnel junctions and may be suitable as nanoantennas. Feinberg's (1985) notion of communication by short wavelength coherent radiation, and Pohl's nanoaperture concept could be utilized for remote nanovision feedback and observation of nanoscale activities (Schneiker, 1986).

STM/FMs combined with genetic engineering and immunological techniques could be used to create, observe, and evaluate such mobile nanodevices whose capabilities would be optimized as replicating automata.

## 10.7 Replicating Automata

Ideas of automatic machines and robots have existed for centuries, but not till the work of John von Neumann in the mid twentieth century was there mathematical proof of the *possible* existence of self-replicating automata in a real physical sense. Assuming the proper material could be found, and using relatively simple neighbor logic, subunits could spontaneously assemble into complex structures, dynamically disassemble, multiply, or rearrange into successive configurations. Freeman Dyson (1979), Schneiker (1986) and others have considered the profound usefulness of such hypothetical robotic replicators in uses and sizes ranging from outer space exploration to household furniture to nanoscale "Hibb's machines," or nanodoctors. Dyson (1979) foresaw self-replicators which, after being launched from earth to an asteroid or planet, could mine their own raw materials and form symbiotic relationships with other "life" forms. Shoulders (1961) extended the concept of replicators to the nanoscale where their existence could have wide ranges of scientific and medical applications.

Implementation of real-world, useful replicators faces many obstacles. A NASA study group (Bekey and Naugle, 1980) found that von Neumann's famous proof made major assumptions and avoided significant problems. Schneiker (1987) has summarized an approach to simplifying the development of replicative automata by minimizing the different types of materials/parts needed, the number of scale-dependent factors, part tolerance and wear, and assembly complexity. One other requirement is to increase reliability (i.e. by intrinsic error detection, repair, regeneration). Presently, these traits are fulfilled only by computer simulations or purely biological structures such as cytoskeletal proteins and viruses.

Schneiker (1986) notes that simple *micro*replicators, augmented with STM/FMs could mass produce nanotechnology products in virtually unlimited quantities. Nanotechnology applied to new superconductive materials (and vice versa) may help to implement useful replicative micro-automata which in turn could turn out nanodevices in vast quantities. Until recently superconductivity has been thought limited to near absolute zero temperatures, however some materials have been shown to have superconductive transitions at much warmer, easily attainable temperatures (Robinson, 1987).

The dramatic increase in superconducting transition temperatures in certain materials, and the availability of lossless current, magnets, and motors may herald a wave of scientific, technological, and economic advances. Schneiker contends that among these will be the facilitation of STM technology (superconductive tunneling), Josephson junctions, magnetic field sensors, as well as ***superconducting replicators***. Superconducting materials offer scaling benefits, design simplicity, low power needs, and high component density. Superconducting magnets and coils could form switchable "glue" or "clamps" to hold subassemblies, generate switchable magnetic field patterns, direct assembly, positioning, and dynamic functions, drive rotary/linear motors or form superconducting channels or guideways for material transport (Schneiker, 1987).

Collective intelligence occurring in parallel replicating automata has been studied by Christopher Langton (1986) of Los Alamos National Laboratories. Langton is the organizer of a Los Alamos conference on "Artificial Life," the study of computer systems that exhibit behavior characteristic of natural life. By exploring models of cellular automata and self-replicators, Langton (1987) hopes to "extract the molecular logic of living systems."

Langton (1987) states:

> Microelectronic technology and genetic engineering will soon give us the capability to create new life forms *in silico* as well as *in vitro*. This capacity will present humanity with the most far-reaching technical, theoretical, and ethical challenges it has ever confronted.

# Chapter 11

# The Future of Consciousness

Nanotechnology may enable the dream of Mind/Tech merger to materialize. At long last, debates about the nature of consciousness will move from the domain of philosophy to large scale experiments. The visions of consciousness interfacing with, or existing within, computers or mind piloted robots expressed by Moravec, Margulis, Sagan and Max Headroom could be realized. Symbiotic association of replicative nanodevices and cytoskeletal networks within living cells could not only counter disease processes, but lead to exchange of information encoded in the collective dynamic patterns of cytoskeletal subunit states. If these are indeed the roots of consciousness, a science fiction-like deciphering and transfer of mind content may become possible. One possible scenario could utilize a small window in a specific brain region. Hippocampal temporal lobe, a site where memories enter and where electromagnetic radiation from outside the skull penetrates most readily and harmlessly, is one possible area where information distributed throughout the brain may perhaps be accessed and manipulated. Techniques such as laser interferometry, electroacoustical probes scanned over brain surfaces, or replicative nanoprobes immunotargeted to key hippocampal tubulins, MAPs, and other cytoskeletal components might be developed to perceive and transmit the content of consciousness.

What technological device would be capable of receiving and housing the information emanating from some $10^{15}$ tubulin subunits changing state some $10^9$ times per second? One possibility is a customized array of nanoscale automata, perhaps utilizing superconducting materials. Another possibility is a genetically engineered array of some $10^{15}$ tubulin subunits (or many more) assembled into parallel tensegrity arrays of interconnected microtubules, and other cytoskeletal structures. Current and near future genetic engineering capabilities should enable isolation of genes responsible for a specific individual's brain cytoskeletal proteins, and reconstitution in an appropriate medium. Thus the two evident sources of mind content (heredity and experience) may be eventually reunited in an artificial consciousness environment. A polymerized cytoskeletal array would be highly unstable and dependent on biochemical, hormonal, and pharmacological maintenance of its medium. Precise monitoring and control of cytoskeletal consciousness environments may become an important new branch of anesthesiology. Polymerization of cell-free cytoskeletal lattices would be limited in size (and potential intellect) due to gravitational collapse. Possible remedies might include hybridizing the cytoskeletal array by metal deposition, symbiosis with synthetic nanoreplicators, or placement of the cytoskeletal array in a zero gravity environment. Perhaps future consciousness vaults will be constructed in orbiting space stations or satellites. People with terminal illnesses may choose to deposit their mind in such a place, where their consciousness can exist indefinitely, and (because of enhanced cooperative resonance) in a far greater magnitude. Perhaps many minds can comingle in a single large array, obviating loneliness, but raising new sociopolitical issues. Entertainment, earth communication, and biochemical mood and maintenance can be supplied by robotics, perhaps leading to the next symbiosis—robotic space voyagers (shaped like centrioles?) whose intelligence is derived from cytoskeletal consciousness.

Yes, this is science fiction. Will it become reality like so much previous science fiction has? Probably not precisely as suggested; but if past events are valid indicators, the future of consciousness may be even more outrageous.

Figure 11.1: Multiple STM tip probing microtubule. By Paul Jablonka (Schneiker and Hameroff, 1987).

# Chapter 12

# Bibliography

Aamodt, E. J. and J. G. Culotti (1986). *Microtubules and Microtubule Associated Proteins from the Nematode Caenorhabditis Elegans: Periodic Crosslinks Connect Microtubules In Vitro.* J. Cell Biol., 103, 23–31.

Abraham, R. H. (1976). *Vibrations and the Realization of Form.* In: *Evolution and Consciousness.* E. Jantsch and C. H. Waddington eds., Redding, Massachusetts, Addison Wesley.

Abraham, R. H. (1985). *On Morphodynamics.* Santa Cruz, Aerial Press.

Abraham, R. H. (1985). *Dynamic Models for Thought.* J. Social Biol. Struct., 8, 13–26.

Adey, W. R. (1977). *Models of Membranes of Cerebral Cells as Substrates for Information Storage.* Biosystems, 8, 163–178.

Adey, W. R. and A. F. Lawrence, Eds. (1984). *Nonlinear Electrodynamics in Biological Systems.* New York, Plenum Press.

Adey, W. R. (1975). *Evidence For Cooperative Mechanisms in the Susceptibility of Cerebral Tissue to Environmental and Intrinsic Electric Fields.* In: *Functional Linkage in Biomolecular Systems.* F. O. Schmitt, D. M. Schneider and D. M. Crothers eds., New York, Raven Press, 325–339.

Adler, J. (1969). *Chemoreceptors in Bacteria.* Science, 166, 1588.

Adrian, E. D. (1947). *The Physical Background of Perception.* Oxford, Clarendon Press.

Afzelius, B. (1979). *Immotile Cilia Syndrome and Other Ciliary Diseases.* Int. Rev. Exp. Pathol., 19, 1-43.

Aizawa, M., A. Morioka and S. Suzuki (1978). *II. Electrochemical Determination of IgG with an Antibody-Bound Membrane.* J. Membr. Sci., 4, 221–228.

Alberts, B., D. Bray, J. Lewis, M. Raff, K. Roberts and J. D. Watson (1983). *Molecular Biology of the Cell.* New York, Garland.

Albrecht-Buehler, G. (1977). *Phagokinetic Tracks of 3T3 Cells: Parallels Between the Orientation of Track Segments and of Cellular Structures Which Contain Actin or Tubulin.* Cell, 12, 333–339.

Albrecht-Buehler, G. (1985). *Is the Cytoplasm Intelligent Too?* Cell and Muscle Motility, 6, 1–21.

Albrecht-Buehler, G. (1981). *Autonomous Movement of Cytoplasmic Fragments.* Proc. Natl. Acad. Sci. USA, 77, 6639–6643.

Albrecht-Buehler, G. (1981). *Does the Geometric Design of Centrioles Imply Their Function?* Cell Motility, 1, 237–245.

Albrecht-Buehler, G. (1976). *Filopodia of Spreading 3T3 Cells. Do They Have a Substrate Exploring Function?* J. Cell Biol., 69, 275–286.

Albrecht-Buehler, G. (1979). *The Angular Distribution of Directional Changes of Guided 3T3 Cells.* J. Cell Biol., 80, 53–60.

Albrecht-Buehler, G. (1977). *Daughter 3T3 Cells, Are They Mirror Images of Each Other?* J. Cell Biol., 72, 595–603.

Albrecht-Buehler, G. and A. Bushnell (1982). *Reversible Compression of Cytoplasm.* Exp. Cell Res., 140, 173–189.

Albrecht-Buehler, G. and A. Bushnell (1979). *The Orientation of Centrioles in Migrating 3T3 Cells.* Exp. Cell Res., 120, 111–118.

Alkon, D. L. (1983). *Sensory Function of Cilia.* J. Submicrosc. Cytol., 15(1), 173–176.

Allen, R. D. (1987). *The Microtubule as an Intracellular Engine.* Sci. Am., 255, 42–49.

Allen, R. D., L. R. Zacharski, S. T. Widirstky, R. Rosenstein, L. M. Zaitlin and D. R. Burgess (1979). *Transformation and Motility of Human Platelets, Details of the Shape Change and Release Reaction Observed by Optical and Electron Microscopy.* J. Cell Biol., 83, 126–142.

Allen, R. D., D. G. Weiss, J. H. Hayden, D. T. Brown, H. Fujiwake and M. Simpson (1985). *Gliding Movement of and Bidirectional Transport Along Single Native Microtubules From Squid Axoplasm: Evidence For An Active Role of Microtubules in Cytoplasmic Transport.* Cell Biol., 100(5), 1736–1752.

Allen, R. D. (1985). *New Observations on Cell Architecture and Dynamics by Video Enhanced Contrast Optical Microscopy.* Ann. Rev. Biophys., Biophys. Chem., 14, 265–290.

Allison, A. C., J. F. Nunn (1968). *Effects of General Anaesthetics on Microtubules. A Possible Mechanism of Anaesthesia.* The Lancet, December

21, Volume 2, 1326–1329.

Allman, W. F. (1986). *Mindworks.* Science 86, May, 23–31.

Alvarez, J. (1979). *Microtubules in Non-Medullated Fibres: Effect of the Action Potential.* Neuroscience Letters, 15, 15–17.

Alvarez, J. and B. U. Ramirez (1979). *Axonal Microtubules: Their Regulation by the Electrical Activity of the Nerve.* Neuroscience Letters, 15, 19–22.

Anderson, J. A. (1981). *Parallelism and Distribution in the Mammalian Nervous System.* In: *Parallel Modes of Associative Memory.* J. A. Anderson and G. E. Hinton eds., Hillsdale, N.J. Lawrence Erlbaum, 9–45.

Artusio, J. F. (1955). *Ether Analgesia During Major Surgery.* JAMA, 157, 33–36.

Ashmore, J. F. and I. J. Russell (1983). *Sensory and Effector Functions of Vertebrate Hair Cells.* J. Submicrosc. Cytol., 15, 163–166.

Asimov, I. (1981). *Change!* Boston, Houghton Mifflin, 147–149.

Atema, J. (1973). *Microtubule Theory of Sensory Transduction.* J. Theor. Biol., 38, 181–190.

Athenstaedt H. (1974). *Pyroelectric and Piezoelectric Properties of Vertebrates.* Ann. NY Acad. Sci., 238, 68–93.

Atkins, P. W. (1982). *Physical Chemistry. Second Edition.* W. H. Freeman Company, San Francisco.

Aufderheide, K. J., J. Frankel and N. E. Williams (1977). *Formation and Positioning of Surface-Related Structures in Protozoa.* Microbiol. Rev., 44, 252–302.

Avolio J., A. N. Glazzard, M. E. J. Holwill and P. Satir (1986). *Structures Attached to Doublet Microtubules of Cilia: Computer Modeling of Thin Section and Negative Stain Stereo Images.* Proc. Natl. Acad. Sci. USA, 83, 4804–4808.

Baines A. J. and V. Bennett (1986). *Synapsin I is a Microtubule Bundling Protein.* Nature, 319, 145–147.

Balasubramanium, D. and D. B. Wetlaufer (1966). *Reversible Alteration of the Structure of Globular Proteins by Anesthetic Agents.* Proc. Natl. Acad. Sci., 55, 762–765.

Bamburg, J. R., D. Bray and K. Chapman (1986). *Assembly of Microtubules at the Tip of Growing Axons.* Nature 321, 788–790.

Bardele, C. F. (1976). *Particle Movement in Heliazoan Axopods Associated with Lateral Displacement of Highly Ordered Membrane Domains.* Z. Natur. Forsch., 31C, 190–194

Barnes, D. M. (1986). *Brain Architecture: Beyond Genes.* Science, 233, 155–156.

Barnett, M. P. (1987). *Molecular Systems to Process Analog and Digital Data Associatively.* Proceedings of 3rd Molecular Electronic Device Conference, Forrest Carter editor, Naval Research Laboratory, Washington, D.C., in press.

Barra H. S., C. A. Arce, J. A. Rodriguez and R. Caputto (1974). *Some Common Properties of the Protein that Incorporates Tyrosine as a Single Unit into Microtubule Proteins.* Biochem. Biophys. Res. Comm., 60, 1384–1390.

Barrow, G. M. (1961). *Physical Chemistry.* New York, McGraw-Hill.

Bayley, P. M. and E. J. Manser (1985). *Assembly of Microtubules From Nucleotide-Depleted Tubulin.* Nature, 318, 683–685.

Becker, J. S., J. M. Oliver and R. D. Berlin (1975). *Flourescence Techniques for Following Interactions of Microtubule Subunits and Membranes.* Nature, 254, 152–154.

Becker, R. O. (1974). *The Basic Biological Data Transmission and Control System Influenced by Electrical Forces.* Ann. NY Acad. Sci., 238, 236–241.

Benioff, P. (1980) *The Computer as a Physical System: A Quantum Mechanical Hamiltonian Model of Computers as Represented by Turing Machines.* J. Stat. Phys., 22, 563–591.

Benioff, P. (1982) *Quantum Mechnical Models of Turing Machines That Dissipate No Energy.* Phys. Rev. Lett., 48, 1581–1585.

Bennett, C. H. (1982). *The Thermodynamics of Computation—a Review.* Internat. J. Theoret. Physics, 21, 905–940.

Bernard, C. (1875). *Lecons Sur Les Anesthesiques et Sur L'asphyxie.* Paris, J. B. Bailliere.

Bessen, M., R. B. Fay and G. B. Witman (1980). *Calcium Control of Waveform in Isolated Flagellar Axonemes of Chlamydomonas.* J. Cell Biol., 86, 446–455.

Beurle, R. L. (1956). *Properties of a Mass of Cells Capable of Regenerating Pulses.* Phil. Trans. Royal Soc., Series B, 240, 55–94.

Bhargava P. M. (1985). *Is the "Soluble" Phase of Cells Structured?* Biosystems, 18, 135–139.

Binder L. I., A. Frankfurter and L. I. Rebhun (1985). *The Distribution of Tau in the Mammalian Nervous System.* J. Cell Biol., 101(4), 1371–1378.

Bitnar R. L. (1983) *Awareness During Anesthesia.* In: *Complications in Anesthesiology.* F. K. Orkin and L. H. Cooperman eds., Philadephia, J. B.

Lippincott.

Blum H. (1967). *A New Model of Global Brain Function.* Perspect. Biol. Med., 10, 381–408.

Bohm D. (1973) *Quantum Theory as an Indication of a New Order in Physics. Part B. Implicate and Explicate Order in Physical Law.* Foundations of Physics, 3, 139–168.

Bohr N. (1958). *Atomic Physics and Human Knowledge.* New York, Wiley.

Boitieux, J. L., G. Desmet and D. Thomas (1979). *An "Antibody Electrode"—Preliminary Report on a New Approach in Enzyme Immunoassay.* Clin. Chem. 25, 318–321.

Bolterauer, H., R. Henkel and M. Opper (1986). *Multiquanta Excitations in the Alpha-Helix.* MIDIT Workshop on Structure, Coherence and Chaos in Dynamical Systems, Technical University of Denmark.

Bornens M. and I. E. Karsenti (1984). *The Centrosome.* In: *Membrane Structure and Function.* Vol. 6, E. E. Bittar ed., New York, Wiley and Sons, 99–171.

Bornens, M. (1979). *The Centriole as a Gyroscopic Oscillator: Implications for Cell Organization and Some Other Consequences.* Biological Cellulaire, 35(11), 115–132.

Bornens, M. (1983). *The Radial Spoke System and Bending Production: A Plausible Mechanism.* J. Submicrosc. Cytol., 15(1), 281–294.

Boveri, T. (1929). *The Origin of Malignant Tumors.* London, J. B. Bailliere.

Brammall, A., D. J. Beard and G. H. Hulands (1974). *Inhalation Anaesthetics and Their Interaction In Vitro with Glutamate Dehydrogenase and Other Enzymes.* Br. J. Anaesth., 46, 643–652.

Breuer A. C., C. M. Christian, M. Henkart and P. G. Nelson (1975). *Computer Analyses of Organelle Translocation and Primary Neuronal Cultures in Continuous Cell Lines.* J. Cell Biol., 65, 652–576.

Breuer A. C., R. E. Dayhoff, R. S. Ledley, A. W. Dudley and K. E. Marks (1983). *Computer Analysis of Organelle Traffic in Human Axons.* In: Proceedings of the Seventh Annual Symposium on Computer Applications in Medical Care, Silver Spring, IEEE Computer Society Press, 759–762.

Brinkley, B. R. (1982). *The Cytoskeleton: A Perspective.* In: *Methods in Cell Biology, Volume 24: The Cytoskeleton, Part A–Cytoskeletal Proteins Isolation and Characterization.* L. Wilson ed., New York, Academic Press, 1–8.

Brinkley, B. R. and E. Stubblefield (1970). *Ultrastructure and Interactions of the Kinetochore and Centriole in Mitosis and Meiosis.* Adv. Cell Biol., 1, 119–185.

Brodersen, R. W. and R. M. White (1977). *New Technologies for Signal Processing.* Science, 195, 1216–1221.

Brucke, E. (1861). *Die Elementarorganismen Sitzungsber Kaisherl.* Akad., Wiss, Vienna, Wien Math-Nautrwiss, KL, 44(2), 381.

Bruner, J. (1986). *Actual Minds, Possible Worlds.* Cambridge, Harvard University Press.

Bryan, J. (1972). *Definition of Three Classes of Binding Sites in Isolated Microtubule Crystals.* Biochem., 11, 2611–2616.

Bunge, M. B. (1986). *The Axonal Skeleton: Its Role in Generating and Maintaining Cell Form.* TINS, October, 447–482.

Burch, M. D. and H. J. Marchant (1983). *Motility and Microtubule Stability of Antarctic Algae at Sub Zero Temperatures.* Protoplasma, 115, 240–242.

Burgoyne, R. D. and R. Cumming (1983). *Characterization of Microtubule Associated Proteins at the Synapse - Absence of MAP 2.* Eur. J. Cell Biol., 30, 154–158.

Burgoyne, R. D. (1986). *Microtubule Proteins in Neuronal Differentiation.* Comp. Biochem. Physiol., 83B(1), 1–8.

Burns, R. G. (1978). *Spatial Organization of the Microtubule Associated Proteins of Reassembled Brain Microtubules.* J. Ultrastruct. Res., 65, 73–82.

Burnside, B. (1974). *The Form and Arrangement of Microtubules: An Historical, Primarily Morphological Review.* Ann. NY Acad. Sci., 253, 14–26.

Butler, S. T. (1965). *Light and Life in the Universe.* In: *Selected Lectures in Physics, Biology, and the Origin of Life.* S. T. Butler and H. Messel eds., Oxford, Pergamon Press.

Cairns-Smith, A. G. (1982). *Genetic Takeover and the Mineral Origins of Life.* Cambridge, University Press.

Cairns-Smith, A.G. (1985). *The First Organisms.* Sci. Am., 252, 90–100.

Cairns-Smith, A. G. and H. Hartman (1986). *Clay Minerals and the Origin of Life.* Cambridge, Cambrige University Press.

Calhoun, D. B., J. M. Vanderkooi and S. W. Englander (1983). *Penetration of Small Molecules Into Proteins Studied by Quenching of Phosphorescence and Fluorescence.* Biochem., 22, 1533–1539.

Capra, F. (1975). *The Tao of Physics.* Berkeley, Shambhala.

Careri, G., U. Buontempo, F. Galluzzi, A. C. Scott, E. Gratton and E. Shyamsunder (1984). *Spectroscopic Evidence for Davydov-Like Solitons in Acetanilide.* Physical Review B, 30(8), 4689–4702.

Carlier, M. F. and D. Pantaloni (1981). *Kinetic Analysis of Guanosine 5 Triphosphate Hydrolysis Associated with Tubulin Polymerization.* Biochem., 20, 1918-1924.

Carlier, M. F. and D. Pantaloni (1985). *Role of Nucleoside-Triphosphate Hydrolysis in the Dynamics of Microtubules and Microfilaments Assembly.* In: *Microtubules and Microtubule Inhibitors.* M. DeBrabander and J. DeMey eds., Amsterdam, Elsevier, 61–69.

Carlisle, E. M. (1972). *Silicon—An Essential Element for the Chick.* Science, 178, 619–621.

Carlson, J. G. and R. E. Stephens (1974). *Mitotic Acceleration Induced in Grasshopper Neuroblasts by Small Doses of Monochromatic Ultraviolet Radiation.* Photochem. Photobiol., 20, 1–7.

Carter, F. L. (1984). *The Molecular Device Computer: Point of Departure for Large Scale Cellular Automata.* Physica 10D, 175–194.

Celaschi, S. and S. Mascarenhas (1977). *Thermal Stimulated Pressure and Current Studies of Bound Water in Lysozyme.* Biophys. J., 29)2), 273–277.

Cerf, R. (1985). *Absolute Measurement of Enhanced Fluctuations in Assemblies of Biomolecules by Ultrasonic Techniques.* Biophys. J., 47, 751–756.

Cerf R., B. Michels, J. A. Schulz, J. Witz, P. Pfieffer and L. Hirth (1979). *Ultrasonic Absorption Evidence of Structural Fluctuations in Viral Capsids.* Proc. Natl. Acad. Sci. USA, 76, 1780–1782.

Chalfie, M. J. N. Thomson and J. E. Sulston (1983). *Induction of Neuronal Branching in Caenorhabditis elegans.* Science 221, 61–63.

Changeux, J. P., A. Devillers-Thiery and P. Chemouilli (1984). *Acetylcholine Receptor: An Allosteric Protein.* Science, 225, 1335–1345.

Changeux, J. P. (1985). *Neuronal Man: The Biology of Mind.* New York, Pantheon Books.

Chatterjee, R., J. A. Tuszynski and O. G. Fritz (1983). *Theory of Long Range Coherence in Biological Systems. 1. The Anomalous Behavior of Human Erythrocytes.* J. Theor. Biol., 104, 169–185.

Chen, Y. and T. L. Hill (1985). *Monte Carlo Study of the GTP Cap in a Five-Start Helix Model of a Microtubule.* Proc. Natl. Acad. Sci. USA, 82, 1131–1135.

Cherkin, A. and P. Harroun (1971). *Anesthesia and Memory Processes.* Anesthesiology, 34, 469–474.

Choi, M. Y. and B. A. Huberman (1983). *Dynamic Behavior of Nonlinear Networks.* Physical Review A, 28(2), 1204–1206.

Chung, S. H., S. A. Raymond and J. Y. Lettvin (1970). *Multiple Meaning in Single Visual Units.* Brain Behavior Evolution, 3, 72-101.

Clarkson, T. W. (1986). *The Cytoskeleton: A Target for Toxic Agents.* Edited by T. W. Clarkson, P. R. Sager, T. L. M. Syversen, New York, Plenum Press.

Clegg, J. S. (1981). *Intracellular Water, Metabolism, and Cellular Architecture.* Collect. Phenom., 3, 289–312.

Clegg, J. S. (1983). *Intracellular Water, Metabolism and Cell Architecture.* In: *Coherent Excitations in Biological Systems.* H. Fröhlich and F. Kremer eds., Berlin, Springer-Verlag, 162–175.

Clegg, J. S. (1979a). *Metabolism and the Intracellular Environment–Vicinal Water Network Model.* In: *Cell-Associated Water.* W. Drost-Hansen and J. S. Clegg, eds., New York, Academic Press, 363–413.

Condeelis, J. S. (1981). *Reciprocal Interactions Between the Actin Lattice and Cell Membrane.* Neurosci. Res. Prog. Bull., 19(1), 83–99.

Connolly, J. A. and V. I. Kalnins (1980). *The Distribution of Tau and HMW Microtubule-Associated Proteins in Different Cell Types.* Exp. Cell Res., 127, 341–350.

Conrad, M. and F.T. Hong (1985). *Molecular Computer Design and Biological Information Processing: An Electrochemical and Membrane Reconstitution Approach to the Synthesis of a Cellular Automaton.* International Symposium of Future Electron Devices—Bioelectronic and Molecular Electronic Devices, November 20–21, Tokyo, 88–94.

Conrad, M. (1986). *The Lure of Molecular Computing.* IEEE Spectrum, October, 55–60.

Conrad M., W. Guttinger and M. Dal Cin (1973). *Lecture Notes in Biomathematics 4/ Physics and Mathematics of the Nervous System.* Proceedings of a Summer School organized by the International Centre for Theoretical Physics, Trieste, and the Institute for Information Sciences, University of Tubingen, held at Trieste, August 21–31.

Conrad, M. and E. A. Liberman (1982). *Molecular Computing as a Link Between Biological and Physical Theory.* J. Theor. Biol., 98, 239–252.

Conrad, M. (1985). *On Design Principles for a Molecular Computer.* Communications of the ACM, 28(5), 464–480.

Cook, R. H., B. E. Ward and J. H. Austin (1979). *Studies in Aging of the Brain: IV. Familial Alzeheimer Disease: Relation to Transmissible Dementia, Aneuploidy, and Microtubular Defects.* Neurology, 29, 1402–1412.

Cooke, R. and I. D. Kuntz (1974). *The Properties of Water in Biological Systems*. Ann. Rev. Biophys. Bioeng., 3, 95–126.

Cooper, M.S. and N. M. Amer (1983). *Long Range Dielectric Aspects of the Eukaryotic Cell Cycle*. Physiol. Chem., 11(5), 435–443.

Cope, F. W. (1976). *A Primer of Water Structuring and Cation Association in Cells. I. Introduction: The Big Picture*. Physiol. Chem. Phys., 8, 479–483.

Cope, F. W. (1976). *A Primer of Water Structuring and Cation Association in Cells. II. Historical Notes, Present Status, and Future Directions*. Physiol. Chem. Phys., 8, 569–574.

Cope, F. W. (1974). *Solid State Physical Mechanisms of Biological Energy Transduction*. Ann. NY. Acad. Sci., 227, 74–97.

Corcoran, E. (1987). *Japan to Launch 'Sixth Generation Project'*. The Institute, News Supplement to IEEE Spectrum, 1–8.

Courtin, R. F., R. G. Bickford and A. Faulconer (1950). *The Classification and Significance of Electroencephalographic Patterns Produced by Nitrous Oxide-Ether Anesthesia During Surgical Operations*. Proc. Staff Meet. Mayo Clin., 25, 197–206.

Crick, F. (1984). *Memory and Molecular Turnover*. Nature, 312, 101.

Crick, F. (1982). *Do Dendritic Spines Twitch?* TINS 44-46.

Cronin, J. R. (1984). *Thinking About the Unthinkable: Life Without Carbon*. Origins of Life, 15, 71–72.

Cronly-Dillon, J., D. Carden and C. Birks (1974). *The Possible Involvement of Brain Microtubules in Memory Fixation*. J. Exp. Biol., 61, 443–454.

Dafu, D. and X. Jinghua (1985). *The Symmetry Breaking in Living Systems—The Selection Rules, Sensitivity and Chirality Matching*. Scientia Sinica, Series B, 28(3), 250–263.

Dafu, D. and X. Jinghua (1985). *The Symmetry Breaking in Living Systems II. Molecular Chirality and Morphological Chirality*. Commun. in Theor. Phys., Vol. 4, No. 1, 129–144.

Darnell, J. E. and W. F. Doolittle (1986). *Speculations on the Early Course of Evolution*. Proc. Natl. Acad. Sci., USA, 83, 1271–1275.

Davydov, A. S. (1982). *Biology and Quantum Mechanics*. Oxford, Pergamon Press.

Davydov, A. S. (1982). *The Migration of Energy and Electrons in Biological Systems*. In: *Quantum Mechanics in Biology*. A. S. Davydov ed., Oxford, Pergamon Press, 185–212.

Davydov, A. S. (1977). *Solitons and Energy Transfer Along Protein Molecules*. J. Theor. Biol. 66, 379–387.

Davydov, A. S. (1973). *The Theory of Contraction of Proteins Under Their Excitation.* J. Theor. Biol., 38, 559–569.

Davydov, A. S. (1985). *Solitons in Molecular Systems.* Dordrecht, Boston, D. Reidel.

DeBrabander, M. (1982). *A Model for the Microtubule Organizing Activity of the Centrosomes and Kinetochores in Mammalian Cells.* Cell Biol. Intern. Rep., 6(10), 901–915.

DeBrabander, M., F. Aerts, J. DeMey, G. Geuens, M. Moermans, R. Nuydens R and R. Willebrords (1985). *Microtubule Dynamics in the Mitotic Cycle, A Model.* In: *Aneuploidy.* Vicki L. Dellarco, Peter E. Voytek and Alexander Hollender eds., New York, Plenum Press.

DeBrabander, M. and J. DeMey (1985). *Microtubules and Microtubule Inhibitors.* Amsterdam, Elsevier.

DeBrabander, M., J. DeMey, G. Geuens, R. Nuydens, F. Aerts, R. Willebrords R. and M. Moermans (1985). *Microtubules and Cell Organization in Motility.* In: *Hormones and Cell Regulation, Vol. 9, Inserm European Symposium.* J. E. DuMont, B. Hamprecht and J. Nunez, eds., Amsterdam, Elsevier, 85–104.

DeBrabander, M., G. Geuens, G. Nuydens, R. Willebrords and J. DeMey (1981). *Microtubule Assembly in Living Cells After Release from Nocodazole Block, the Effects of Metabolic Inhibitors Taxol and pH.* Cell Biol. Intern. Rep., 5, 913–920.

DeBrabander, M., R. Nuydens, G. Geuens, M. Moermans and J. DeMey (1986). *The Use of Submicroscopical Particles Combined with Video Contrast Enhancement as a Simple Molecular Probe for the Living Cell.* In: *Cell Motility in the Cytoskeleton.* 6, 105–113.

DeBrabander, M., J. DeMey, R. Van deVeire, F. Aerts and G. Geuens (1975). *Microtubules in Mammalian Cell Shape and Surface Modulation: An Alternative Hypothesis.* Cell Biol., Intern. Rep., 1(5), 453–461.

DeBrabander, M., G. Geuens, J. DeMey and M. Joniav (1986). *The Organized Assembly and Function of the Microtubule System Throughout the Cell Cycle.* In: *Cell Movement and Neoplasia.* M. DeBrabander ed., Oxford, Pergamon Press.

DeCallatay, A. M. (1986). *Natural and Artificial Intelligence: Processor Systems Compared to the Human Brain.* Amsterdam, Elsevier.

Deery, M. J. and B. R. Brinkley (1983). *Cytoplasmic Microtubule Assembly Disassembly from Endogenous Tubulin in a Cell Model.* J. Cell Biol., 96, 1631–1641.

Del Giudice, E. (1986). *Collective Properties of Biological Systems.* In:

*Modern Bioelectrochemistry.* F. Gutmann and H. Keyzer eds., New York, Plenum Press.

Del Giudice, E., S. Doglia and M. Milani, (1982). *Self Focusing of Fröhlich Waves and Cytoskeleton Dynamics.* Phys. Letts., 90A, 104–106.

Del Giudice, E., S. Doglia and M. Milani (1983). *Actin Polymerization in Cell Cytoplasm.* In: *The Application of Laser Light Scattering to the Study of Biological Motion.* J. C. Earnshaw and M.W. Stear eds., New York, Plenum Press, 493–497.

Del Giudice, E., S. Doglia, M. Milani and G. Vitiello (1983). *Spontaneous Symmetry Breakdown and Boson Condensation in Biology.* Phys. Rev. Letts., Vol. 95(a), 508–510.

Del Giudice, E., S. Doglia and M. Milani (1981). *Nonlinear Properties of Coherent Electric Vibrations in Living Cells.* Phys. Letts. 85A, 402–404.

Del Giudice, E., S. Doglia and M. Milani (1984). *Order and Structure in Living Systems.* In: *Nonlinear Electrodynamics in Biological Systems.* W. R. Adey and A. F. Lawrence eds., New York, Plenum, 477–488.

Del Giudice, E., S. Doglia, M. Milani and G. Vitiello (1986). *Collective Properties of Biological Systems—Solitons and Coherent Electric Waves in a Quantum Field Theoretical Approach.* In: *Modern Bioelectrochemistry.* F. Gutmann and H. Keyzer eds., New York, London, Plenum Press, 263–287.

Del Giudice, E., S. Doglia and M. Milani (1982). *A Collective Dynamics in Metabolically Active Cells.* Physica Scripta, 26, 232–238.

Del Giudice, E., S. Doglia and M. Milani (1983). *Self Focusing and Ponderomotive Forces of Coherent Electric Waves: A Mechanism for Cytoskeleton Formation and Dynamics.* In: *Coherent Excitations in Biological Systems.* H. Fröhlich and F. Kremer eds., Berlin, Springer-Verlag, 123–127.

Demongeot, J. E. Goles and M. Tchuente (1985). *Dynamical Systems and Cellular Automata.* London, Academic Press.

DeVault, D., J. H. Parkes and B. Chance (1967). *Electron Tunneling in Cytochromes.* Nature, 215, 642.

Dewdney, A. K. (1987). *Computer Recreations.* Sci. Am., 256, February, 16–24.

Dewdney, A. K. (1985). *Computer Recreations.* Sci. Am., 252, May, 18–30.

Dewson, J. H. and A. Cowey (1969). *Discrimination of Auditory Sequences by Monkeys.* Nature 222, 695–697.

DeYoung, H. G. (1983). *Biosensors, The Mating of Biology and Electronics.* High Tech., Nov., 41–49.

Dolgoff, E. J. (1985). *A Holographic Brain Model.* In: *Holography in*

*Medicine: Proceedings of the International Symposium on Holography in Biomedical Sciences.* P. Greguss, Guildford (England), IPC Science and Technology Press, 143–148.

Donati F. and Kunov H. (1976). *A Model for Studying Velocity Variations in Unmyelinated Axons.* Trans. IEEE, BME-23, 23–28.

Doolittle, W. F. (1980). *Revolutionary Concepts in Evolutionary Cell Biology.* In: *Trends in Biochemical Science.* Amsterdam, Elsevier North Holland, 146–149.

Dreyfus, H. L. (1972). *What Computers Can't Do.* New York, Harper and Row.

Drexler, K. E. (1981). *Molecular Engineering: An Approach to the Development of General Capabilities for Molecular Manipulation.* Proc. Natl. Acad. Sci. USA, 78(9), 5275–5278.

Drexler, K. E. (1986). *Engines of Creation.* Garden City, Anchor Press.

Drost-Hansen, W. (1973). *Phase Transitions in Biological Systems: Manifestations of Cooperative Processes in Vicinal Water.* Ann. NY. Acad. Sci., 204, 100–112.

Dustin, Pierre (1984). *Microtubules (2nd Revised Edition).* Berlin, Springer–Verlag, 442.

Eccles, J. C. (1977). *The Understanding of the Brain.* New York, McGraw-Hill.

Eckert, R., Y. Naitoh and H. Machemer (1976). *Calcium in the Bioelectric and Motor Functions of Paramecium.* Symposium of the Society of Experimental Biology, 30, 233–255.

Eckholm, E. (1986). *Search for the Origins of Life is Widening.* International Herald Tribune, July 7.

Edelman, G. E. and G. N. Reeke (1982). *Selective Networks Capable of Representative Transformations, Limited Generalization, and Associative Memory.* Proc. Natl. Acad. Sci., USA, 79, 2091–2095.

Edelman, G. M. (1976). *Surface Modulation and Cell Recognition on Cell Growth.* Science, 192, 218–226.

Edwards, G. S., C. C. Davis, J. D. Saffer, M. L. Swicord (1984). *Microwave Absorption of Selected DNA Molecules.* Phys. Rev. Lett., 53(13), 1284–1287.

Eftink, M. R. and C. A. Ghiron (1975). *Dynamics of a Protein Matrix Revealed by Fluorescence Quenching.* Proc. Natl. Acad. Sci. USA, 72, 3290–3294.

Eger, E. I., C. Lundgren, S. L. Miller and W. C. Stevens (1969). *Anesthetic Potencies of Sulfur Hexafluoride, Carbon Tetrafluoride, Chloroform and Ethrane in Dogs: Correlation with the Hydrate and Lipid Theories of Anesthetic Action.* Anesthesiology, 30, 129-135.

Eigen, M. (1971). *Selforganization of Matter and the Evolution of Biological Macromolecules.* Die Naturwissenschaften, 33a, 465–523.

Eigen, M. and P. Schuster (1979). *The Hypercycle: A Principle of Natural Self-Organization.* Berlin, Springer.

Eilbeck, J. C. and A. C.Scott (1979). *Davydov Solitons.* 16 mm silent film—available from Swift Film Productions, 1 Wool Road, Wimbledon, England.

Eilbeck, J. C., P. S. Lomdahl and A. C. Scott (1984). *Soliton Structure in Crystalline Acetanilide.* Physical Review B, 30(4), 4703–4712.

Ellisman, M. H. (1981). *Beyond Neurofilaments and Microtubules.* In: *Cytoskeleton and Nervous System Architecture.* Neurosciences Res. Prog. Bull., 19(1), 43–58.

Elul, R. (1972). *The Genesis of the EEG.* Intern. Rev. Neur. Biol., 15, 227–272.

Engelborghs, Y. and J. Eccleston (1982). *Fluorescence Stopped Flow Study of the Binding of S6-GTP to Tubulin.* FEBS Letters, 141, 78–81.

Engelborghs, Y. and A. van Houtte (1981). *Temperature Jump Relaxation Study of Microtubule Elongation in the Presence of GTP /GDP Mixtures.* Biophys. Chem., 14, 195–202.

Engelborghs, Y., K. A. H. Heremans, L. C. M. DeMaeyer and J. Hoebeke (1976). *The Effect of Temperature and Pressure in Polymerization Equilibrium of Neuronal Microtubules.* Nature, 259, 686–689.

Engelborghs, Y. and G. Ide (1981). *Fluorescence Quenching and Induced Association of a Tubulin Colchicine Complex by Iodide.* J. Biol. Chem., 256, 11684–11687.

Erickson, R.O. (1973). *Tubular Packing of Spheres in Biological Fine Structure.* Science, 181, 705–716.

Ettinger, R. C. W. (1972). *Man Into Superman.* New York, St. Martin's.

Eyring, H., J. W. Woodbury and J. S. D'Arrigo (1973). *A Molecular Mechanism of General Anesthesia.* Anesthesiology, 38, 415–424.

Faivre, C., C. H. Legrand and A. Rabie (1983). *Effects of Thyroid Deficiency and Corrective Effects of Thyroxin on Microtubules and Mitochondria in Cerebellar Purkinje Cell Dendrites of Developing Rats.* Dev. Brain Res., 8, 21–30.

Farber, J. L. (1981). *The Role of Calcium in Cell Death.* Life Sciences, 29, 1289-1295.

Featherstone, R. M., C. A. Muelbacher, F. L. DeBon and J. A. Forsaith (1961). *Interactions of Inert Anesthetic Gases with Proteins.* Anesthesiology, 22, 977–981.

Feinberg, G. (1985). *Solid Clues: Quantum Physics, Molecular Biology, and the Future of Science.* New York, Touchstone Books, Simon and Schuster.

Feldman, J. A. (1985). *Connections-Massive Parallelism in Natural and Artificial Intelligence.* Byte 16, 4, 277–284.

Fermi, E., J. R. Pasta, S. M. Pasta and S. M. Ulam (1965). *Collected Works of Fermi.*, Vol. 2, Chicago, University of Chicago Press, 978–988.

Ferry, G. (1986). *Egalitarian Brain.* New Scientist, 9 January, 41–43.

Feynman, R. P. (1961). *There's Plenty of Room at the Bottom.* (Reprinted in Schneiker, 1986 and as Chapter 16 in Gilbert, 1961).

Feynman, R. P. (1986). *Quantum Mechanical Computers.* Foundations of Physics, 16(6), 507–531.

Finkelstein, A. and O. S. Andersen (1981). *The Gramicidin A Channel: A Review of Its Permeability Characteristics with Special Reference to the Single-File Aspect of Transport.* J. Membr. Biol., 59, 155–171.

Fischer, A. (1899). *Fixierung, Farbung und Bau des Protoplasmas.* G. Fischer ed., Germany, Jena.

Fischer, A. and W. B. Hardy (1899). J. Physiology, 24, 158.

Folda, T., L. Gros and H. Ringsdorf (1982). J. Macromol. Chem. Rapid. Commun., 3, 167.

Fox, S. W. (1972). *Molecular Evolution and the Origin of Life.* San Francisco, W. H. Freeman.

Fountain, B. S. (1984). *Solid State Memory Problems Support Mode Softening for Larger Assemblies of Biological Dipole Oscillation.* Phys. Lett., 102A, 77–79.

Franks, N. P. and W. R. Lieb (1982). *Molecular Mechanisms of General Anesthesia.* Nature, 300, 487–493.

Freeman, W. J. (1972). *Waves, Puzzles, and the Theory of Neural Masses.* Prog. Theoret. Biol., 2, 87–165.

Freeman, W. J. (1975). *Mass Action in the Nervous System.* New York, Academic Press.

Freeman, W. J. (1983). *The Physiological Basis of Mental Images.* Biol. Psychiatry 18(10), 1107–1125.

Fröhlich, H. (1977). *Dielectric Theory and Ion Channels in Nerve Membranes.* Biosystems, 8, 193–194.

Fröhlich, H. (1975). *The Extraordinary Dielectric Properties of Biological Materials and the Action of Enzymes.* Proc. Natl. Acad. Sci., USA, 72(11), 4211–4215.

Fröhlich, H. (1970). *Long Range Coherence and the Actions of Enzymes.* Nature, 228, 1093.

Fröhlich, H. (1982). *Conditions for Coherent Excitations in Biological Systems.*, Phys. Lett., 93A, 105–106.

Fröhlich, H. (1983). *Coherence in Biology.* In: *Coherent Excitations in Biological Systems.* H. Frohlich and F. Kremer eds., Berlin, Springer-Verlag, 1–5.

Fröhlich, H. (1984) *General Theory of Coherent Excitations on Biological Systems.* In: *Nonlinear Electrodynamics in Biological Systems.* W. R. Adey and A. F. Lawrence eds., New York, Plenum Press.

Fröhlich, H. (1986). *Coherent Excitations in Active Biological Systems.* In: *Modern Bioelectrochemistry.* F. Gutmann and H. Keyzer eds., New York, Plenum Press, 241–261.

Fuller, R. B. (1975). *Synergetics.* New York, Macmillan.

Fulton, C. (1971). *Centrioles.* In: *Origin and Continuity of Cell Organelles.* J. Reinert and H. Urspring eds., Springer-Verlag, Heidelberg, 170.

Fung, B. M. and T. W. McGaughy (1979). *Study of Spin–Lattice and Spin–Spin Relaxation Times of* $^1H$, $^2H$, *and* $^{17}O$ *in Muscle Water.* Biophys. J., 28, 293–303.

Furakawa, T. and E. J. Furshpan (1963). *Two Inhibitory Mechanisms in the Mauthner Neurons of Goldfish.* J. Neurophys., 26, 140–176

Gabor, D. (1948). *A New Microscopic Principle.* Nature, 161, 777–778.

Gal, V. and D. Trajkovic (1984). *Effect of Gamma Irradiation on Microtubule Assembly In Vitro.* Int. J. Radiat. Biol., 46(4), 435–442.

Galla, S. J., A. G. Rocco and L. D. Vandam (1958). *Evaluation of the Traditional Signs and Stages of Anesthesia: An Electroencephalographic and Clinical Study.* Anesthesiology, 19, 328–338.

Galvani, L. (1953). *Commentary on the Effects of Electricity on Muscular Motion.* Translated by Margaret Glover Foley, Norwalk, Conn., Burndy Library.

Gardner, M. (1970). *Mathematical Games: The Fantastic Combinations of John Conway's New Solitaire Game "Life."* Sci. Am., 223, 120–123.

Garlin, K. D. (1979). *Cell-Associated Water.* W. Drost-Hansen and J. S. Clegg, eds., New York, Academic Press, 293–362.

Gavish, B. (1978). *The Role of Geometry and Elastic Strains in Dynamic States of Proteins.* Biophys. Struct. Mechanism, 4, 37–52.

Gazzaniga, M. S. (1985). *The Social Brain—Discovering the Networks of the Mind.* New York, Basic Books.

Gershon, N. D., K. R. Porter and B. L. Trus (1985). *The Cytoplasmic Matrix: Its Volume and Surface Area and the Diffusion of Molecules Through It.* Proc. Natl. Acad. Sci. USA, 82, 5030–5034.

Geuens, G., G. G. Gundersen, R. Nuydens, F. Cornelissen, J. C. Bulinski and M. DeBrabander (1986). *Ultrastructural CoLocalization of Tyrosinated and Nontyrosinated Alpha Tubulin in Interphase and Mitotic Cells.* J. Cell Biol., 103(5), 1883–1893.

Gibbons, B. H., I. R. Gibbons and B. Baccetti (1983). *Structure and Motility of the 9+0 Flagellum of Eel Spermatazoa.* J. Submicr. Cytol. 15, 15–20.

Gibbons, I. R. (1968). *The Biochemistry of Motility.* Ann. Rev. Biochem., 37, 521–546.

Gilbert, H. D. (1961). *Miniaturization.* New York, Reinhold.

Goldman, R. D., R. E. Pollack and N. H. Hopkins (1973). *Preservation of Normal Behavior by Enucleated Cells in Culture.* Proc. Natl. Acad. Sci. USA, 70, 750–754.

Goldman, R., T. D. Pollard, J. Rosenbaum (1976). *Cell Motility.* Cold Spring Harbor, New York, Cold Spring Harbor Laboratory.

Golgi, C. (1918). *Dedicated to Professor Camillo Golgi.* J. Comp. Neurol., 30, 8.

Good, I. J. (1965). *Speculations Concerning the First Ultra Intelligent Machine.* Advance Computers, 6, 31–88.

Gomatam, J. (1982). *Pattern Synthesis from Singular Solitons in the Debye Limit: Helical Waves and Twisted Toroidal Scroll Structures.* J. Phys., A15, 1463–1476.

Gotlieb, A. I., L. M. May, L. Subrahamanyn and V. I. Kalnins (1981). *Distribution of Microtubule Organizing Centers—Migrating Sheets of Endothelial Cells.* J. Cell Biol., 91, 589–594.

Gower, M. (1985). *Dynamic Holograms from Crystals.* Nature, 316, 12–14.

Gransdorff, P. and I. Prigogine (1971). *Thermodynamic Theory of Structure Stability and Fluctuations.* New York.

Grant, E. H. (1966). *Dielectric Dispersion in Bovine Serum Albumin.* J. Mol. Biol., 19, 133–139.

Grantham, K. and S. R. Hameroff (1985). *Monitoring Anesthetic Depth.* In: *Monitoring in Anesthesia and Critical Care.* C. D. Blitt editor, Churchill Livingstone, 427–440.

Grassberger, P. (1984). *Chaos and Diffusion in Deterministic Cellular Automata.* Physica 10D, 52–58.

Green, D. (1970). *The Conformational Basis of Energy Transduction in Biological Systems.* Proc. Natl. Acad. Sci. USA, 67, 544–549.

Green, D. E. (1974). *A Framework of Principles for the Unification of Bioenergetics.* Ann. NY Acad. Sci., 227, 6–45.

Green, H. S. and T. Triffet (1985). *Extracellular Fields Within the Cortex.* J. Theor. Biol., 115(1), 43–64.

Griffith, J. S. (1971). *Mathematical Neurobiology.* New York, Academic Press.

Grossberg, S. (1978). *Communication, Memory, and Development.* In: *Progress in Theoretical Biology.* Vol. 5., R. Rosen and F. M. Snell eds., New York, Academic Press, 183–231.

Grossman, Y., E. Spira and I. Parnas (1973) *Differential Flow of Information Into Branches of a Single Axon.* Brain Research, 64, 379–386.

Grundler, W. and F. Keilmann (1983). *Sharp Resonances in Yeast Growth Proved Nonthermal Sensitivity to Microwaves.* Phys. Rev. Letts., 51, 1214–1216.

Gubkin, A. and W. Sovokin (1960). Trans. Bull. Acad. Sci., SSSR, 24, 246.

Gundersen, G. G., M. H. Kalnoski and J. C. Bulinski (1984). *Distinct Populations of Microtubules: Tyrosinated and Nontyrosinated Alpha Tubulin are Distributed Differently In Vivo.* Cell, 38, 779–789.

Gutfreund, H. (1972). *Enzymes: Physical Principles.* London, Wiley–Interscience.

Gutmann, F. (1986). *Some Aspects of Charge Transfer in Biological systems.* In: *Modern Bioelectrochemistry.* F. Gutmann and H. Keyzer eds., New York, Plenum Press, 177–197.

Haddon, R. C. and A. A. Lamola (1985). *The Molecular Electronic Device and the Biochip Computer: Present Status.* Proc. Natl. Acad. Sci., USA, 82, 1874–1878.

Haken, H. (1978). *Synergetics, Introduction Second Edition.* Berlin, Springer-Verlag.

Haken, H. (1976). *Quantum Field Theory of Solids. An Introduction.* Amsterdam, Elsevier North Holland.

Haldane, J. B. S. (1947). *What Is Life?* New York, Boni and Gaer.

Halsey, M. J. (1976). *Mechanisms of General Anesthesia.* In: *Anesthesia Uptake and Action.* E. I. Eger II ed., Baltimore, Williams and Wilkins.

Halsey, M. J. and E. B. Smith (1970). *Effects of Anaesthetics on Luminous Bacteria.* Nature, 227, 1363–1365.

Hamel, E., A. A. del Campo, J. Lustbader and C. C. Chin (1983). *Modulation of Tubulin-Nucleotide Interactions by Microtubule Associated Proteins. Polymerization with Ribose Modified Analogs of Guanosine 5'-triphosphate.* Biochem., 22, 1271–1279.

Hamel, E., A. del Campo and C. M. Lynn (1983). *Microtubule Assembly with a Guanosine 5' diphosphate Analog, 2',3'-dideoxyguanosine - 5' diphosphate.* Biochem., 22, 3664–3670.

Hameroff, S. R. (1974). *Chi: A Neural Hologram?* Am. J. Chi. Med., 2(2), 163–170.

Hameroff, S. R. and R. C. Watt (1982). *Information Processing in Microtubles.* J. Theor. Biol., 98, 549–561.

Hameroff, S. R. and R. C. Watt (1983). *Do Anesthetics Act by Altering Electron Mobility?* Anesth. Analg., 936–940.

Hameroff, S. R. and R. C. Watt (1983). *Microtubules: Biological Microprocessors?* In: *Molecular Electronic Devices.* F. L. Carter, ed. New York: Marcel Dekker, 341–356.

Hameroff, S. R., Steven A. Smith, and Richard C. Watt (1984). *Nonlinear Electrodynamics in Cytoskeletal Protein Lattices.* In: *Nonlinear Electrodynamics in Biological Systems.* W. R. Adey and A. F. Lawrence, New York, Plenum Press, 567–583.

Hameroff, S. R., S. A. Smith and R. C. Watt (1986). *Automaton Model of Dynamic Organization in Microtubules.* Ann. NY Acad. Sci., 466, 949–952.

Harmon, LD. and Lewis, E. R. (1966). *Neural Modeling.* Physiol. Rev., 46, 513–591.

Harper, H. A. (1969). *Review of Physiological Chemistry.* Los Altos, Lange Medical Publications.

Hart R., E. David, D. A. Norman and K. Lashley (1950). *In Search of the Engram.* In: *Symposia for the Society for Experimental Biology, No. 4, Physiological Mechanisms in Animal Behavior.* New York, Academic Press.

Hartman, H. (1975). *The Centriole and the Cell.* J. Theor. Biol., 51, 501–509.

Harvey, E. N. (1915). *The Effect of Certain Organic and Inorganic Substances upon Light Production by Luminous Bacteria.* Biol. Bull., 29, 308–311.

Hayashi, M. and F. Matsumura (1975). *Calcium Binding to Bovine Brain Tubulin.* FEBS Letters, 58(1), 222-226.

Haydon, D. A., J. Requena and B.W. Urban (1980). *Some Effects of*

*Aliphatic Hydrocarbons on the Electrical Capacity and Ionic Currents of the Squid Giant Axon Membrane.* J. Physiol., 309, 229–245.

Hebb, D. O. (1949). *The Organization of Behavior.* New York, Wiley.

Helmholtz, H. (1889). *Popular Scientific Lectures.* London, Longmans.

Henkel, R. and Bolterauer (1986). *Solitons in the Alpha-Helix.* MIDIT Workshop on Structure, Coherence and Chaos in Dynamical Systems, Technical University of Denmark.

Henry, B. M., B. W. Urban and D. A. Haydon (1978). *The Blockage of the Electrical Conductance in a Pore-Containing Membrane by the n-Alkanes.* Biochimica et Biophysica Acta, 513, 106–116.

Hesketh, J. (1985). *Glycoproteins Associated with Brain Microtubules.* Cell Biol. Int. Rep., 9(8), 675.

Hewitt, Carl (1985). *The Challenge of Open Systems.* Byte 16, 4, 223–242.

Hill, T. L. and M. F. Carlier (1983). *Steady State Theory of the Interference of GTP Hydrolysis in the Mechanism of Microtubule Assembly.* Proc. Natl. Acad. Sci. USA, 80, 7234–7238.

Hill, T. N. and Y. D. Chen (1984). *Phase Changes at the End of a Microtubule with a GTP Cap.* Proc. Natl. Acad. Sci. USA, 81, 5772–5776.

Hill, T. L. and M. W. Kirschner (1982). *Bioenergetics and Kinetics of Microtubule and Actin Filament Assembly-Disassembly.* Int. Rev. Cytol., 78, 1–125.

Hille, B. (1970). *Ionic Channels in Nerve Membranes.* Prog. Biophys. Molec. Biol., 21, 3–32.

Hinke, J. A. M., J. P. Caille and D. C. Gayton (1973). *Distribution and State of Monovalent Ions in Skeletal Muscle Based on Ion Electrode, Isotope, and Diffusion Analysis.* Ann. NY Acad. Sci., 204, 274–296.

Hinton, G. and J. A. Anderson, eds., (1981). *Parallel Models of Associative Memory.* Hillsdale, NJ: Lawrence Erlbaum.

Hinton, G. (1985). *Learning in Parallel Networks.* Byte 16, 4, 265–273.

Hinton, G. F., T. J. Sejnowski and D. H. Ackley (1984). *Boltzmann Machines: Constraint Satisfaction Networks That Learn.* CMU Technical Report: CMU-CS-84-119.

Hirokawa, N. (1986). *Quick Freeze, Deep-Etch Visualization of the Axonal Cytoskeleton.* TINS, February, 67–71.

Hol W. G. J., L. M. Halie and C. Sander (1981). *Dipoles of the Alpha Helix and Beta Sheet and Their Role in Protein Folding.* Nature, 294, 532–536.

Hollenbeck, P. J. (1985) *A Third Front for Cell Motility.* Nature, 317(5), 17–18.

Hopfield, J. J. (1982). *Neural Networks and Physical Systems with Emergent Collective Computational Abilities.* Proc. Natl. Acad. Sci. 79, 2554–2558.

Horio, T. and H. Hotani (1986). *Visualization of the Dynamic Instability of Individual Microtubules by Dark Field Microscopy.* Nature, 321, 605–607.

Horowitz, S. B. and P. L. Paine (1979). *Reference Phase Analysis of Free and Bound Intracellular Solutes. II. Isothermal and Isotropic Studies of Cytoplasmic Sodium, Potassium and Water.* Biophys. J., 25, 45–62.

Huang, B., Z. Ramanis, S. K. Dutcher and D. G. L. Luck (1982). *Uniflagellar Mutants of Chlamydomonas: Evidence for the Role of Basal Bodies in Transmission of Positional Information.* Cell, 29, 745–753.

Huber, R. (1979). *Conformational Flexibility in Protein Molecules.* Nature, 280, 538–539.

Huberman, B.A. and T. Hogg (1985). *Dynamic Associations in Nonlinear Computing Arrays.* Physica Scripta, 32, 271–273.

Hudspeth, W. J. and G. B. Jones (1975). *Stability of Neural Interference Patterns.* In: *Holography in Medicine: Proceedings of the International Symposium on Holography in Biomedical Sciences.* P. Greguss ed., Guildford (England), IPC Science and Technology Press.

Hunding, A. (1982). *Spontaneous Biological Pattern Formation in the 3-Dimensional Sphere: Prepatterns in Mitosis and Cytokinesis.* In: *Evolution of Order and Chaos.* H. Haken ed., Berlin, Springer-Verlag.

Huxley, H. E. (1964). *Evidence for Continuity of the Central Elements of The Triads and Extracellular Space in Frog Sartorius Muscle.* Nature, 202, 1067–1071.

Hyden, H. (1973). *Changes in Brain Protein During Learning.* In: *Macromolecules and Behavior.* G. B. Ansell and P. B. Bradley eds., Baltimore, University Park Press, 3–26.

Hyden, H. (1977). *The Differentiation of Brain Cell Protein, Learning, and Memory.* Biosystems, 8, 213–218.

Hyman, J. M., D. W. McLaughlin and A. C. Scott (1981). *On Davydov's Alpha Helix Solitons.* Physica D, 1 and 2, 23–44.

Hymans, J. (1981). *Hooked on Microtubules.* Nature, 291, 107–108.

Indi, S. G. Wakley, and Stebbings (1985). *Does Freeze-Substitution Reveal the Hydrophobic Nature of Interprotofilament Bonding in Microtubules.* Cell Biol. Int. Rep., 9(9), 859–865.

Itoh, K. and T. Shimanouchi (1970). *Vibrational Frequencies and Modes of Alpha Helix.* Biopolymers, 9, 383–399.

Jacobson, M. (1974). *Through the Jungle of the Brain: Neuronal Specificity and Topology Revisited.* Ann. NY Acad. Sci., 228, 63–67.

Jaffe, L. F. and R. Nuccitelli (1977). *Electrical Controls of Development.* Ann. Rev. Biophys. Bioeng., 6, 445–476.

Jantsch, E. (1980). *The Self-Organizing Universe.* New York: Pergamon Press.

Jarosch, R. (1986). *A Model for the Molecular Basis of Filament Contractility and Sliding as Demonstrated by Helix Models.* Cell Motil., 6(2), 229–236.

Jarosch, R. (1986). *The Mechanical Behaviour of Doublet Microtubules Simulated by Helical Models.* Cell Motil., 6(2), 209–216.

Jarosch, R. and I. Fossner (1982). *A Rotation Model for Microtubule and Filament Sliding.* Eur. J. Cell, 26(2), 295–302.

Javan, A. (1985). *Interview.* Lasers and Applications, October, 52.

Javan, A. (1986). *Personal Communication to Conrad Schneiker.*

Jaynes, J. (1976). *The Origin of Consciousness and the Breakdown of the Bicameral Mind.* London, Alan Layne Penguin Books.

Jennings, H. S. (1906). *Behavior of the Lower Organisms* New York, Macmillan.

Ji, S. (1974). *A General Theory of ATP Synthesis and Utilization.* Ann. NY Acad. Sci., 227, 211–226.

Jinghua, X. and D. Dafu (1981). *Molecular Chirality and Biological Time Arrow.* Scientia Sinica, 24(7), 997–1009.

Jinghua, X. and D. Dafu (1979). *Replication of Biological Macromolecules and Formation of Dissipative Structures.* Scientia Sinica, 22(10), 1206–1219.

Job D., M. Pabion and R. L. Margolis (1985). *Generation of Microtubules Stability Subclasses by Microtubule Associated Proteins, Implications for the Microtubule Dynamic Instability Model.* J. Cell Biol., 101, 1680–1689.

John, E. R. (1967). *Mechanisms of Memory.* New York, Academic Press.

John, E. R. (1980). *Multipotentiality: A Statistical Theory of Brain Function.* In: *The Psychobiology of Consciousness.* R. J. Davidson and J. M. Davidson, Eds. New York, Plenum Press.

John, E. R., Y. Tang, A. B. Brill, R. Young and K. Ono (1986). *Double-Labeled Metabolic Maps of Memory.* Science, 233, 1167–1175.

Joshi, H. C., D. Chu, R. E. Buxbaum and S. R. Heidemann (1985). *Tension and Compression in the Cytoskeleton of PC 12 Neurites.* J. Cell Biol., 101, 697–705.

Kaiser, F. In: *Coherent Excitations in Biological Systems.* H. Fröhlich and F. Kremer, 120–133.

Kandel, E. R. (1976). *Cellular Basis of Behavior.* San Francisco, W. H. Freeman.

Kang, S. and J. P. Green (1970). *Steric and Electronic Relationships Among Some Hallucinogenic Compounds.* Proc. Natl. Acad. Sci. USA, 67(1), 62–67.

Kang, S. and J. P. Green (1970). *Correlations Between Activity and Electronic State of Hallucinogenic Amphetamines.* Nature, 226, 645.

Katchalsky, A. (1971). *Biological Flow Structures and Their Relation to Chemicodiffusional Coupling.* Neurosci. Res. Prog. Bull., 9, 397–413.

Katchalsky, A., V. Rowland and R. Blumenthal (eds.) (1974). *Dynamic Patterns of Brain Cell Assemblies.* Neurosci. Res. Prog. Bull., 12, 3–187.

Kakiuchi, S., T. Sobue and K. Morimoto (1983). *Cytoskeletal Control by Calmodulin Binding Protein Linked Flip-Flop Mechanism.* In: *Calcium Binding Proteins.* B. deBernard Ed., Amsterdam, Elsevier, 449–460.

Karplus, M. and J. A. McCammon (1983). *Protein Ion Channels, Gates Receptors.* In: *Dynamics of Proteins: Elements and Function.* Ann. Rev. Biochem., J. King Ed., Menlo Park, Benjamin/Cummings Publishing Co., 53, 263–300.

Karplus, M. and J. A. McCammon (1979). *Protein Structural Fluctuations During a Period of 100 ps.* Nature, 277, 578.

Kato, S., M. Aizawa and S. Suzuki (1978). *Effects of Chemical Modifications of Membranes on Transmembrane Potential.* J. Membr. Sci., 3, 29–38.

Kaufman, R. D. (1977). *Biophysical Mechanisms of Anesthetic Action: Historical Perspective and Review of Current Concepts.* Anesthesiology, 46, 49–62.

Khuri, S. F., W. Marston, M. Josa, W. Crowe, R. Macera, M. Zierler, R. Lange and W. Miller (1984). *First Report of Intramyocardial pH in Man: I. Methodology and Initial Results.* Med. Instr., 18(3), 167–171.

King, J. (1985). *Protein and Nucleic Acid Structure and Dynamics.* Menlo Park, Benjamin/Cummings Publishing.

Kirschner, M., Mitchison T. (1986). *Beyond Self Assembly: From Microtubules to Morphogenesis.* Cell, 45, 329–342.

Kohonen, T. (1984) *Self-Organization and Associative Memory.* Berlin, Springer–Verlag.

Kolata, G. (1984). *Gene Therapy Method Shows Promise.* Science, 223, 1376–1379.

Koplin, D. D. and E. I. Eger (1979). *Theories of Narcosis*. N. Engl. J. Med., 301, 1222–1224.

Koppenol, W. H. (1980). *Effect of a Molecular Dipole on the Ionic Strength Dependence of a Bimolecular Rate Constant. Identification of the Site of Reaction*. Biophys. J., 29, 493–507.

Koruga, D. (1974). *Microtubule Screw Symmetry: Packing of Spheres as a Latent Bioinformation Code*. Ann. NY Acad. Sci., 466, 953–955.

Koshland D. E., H. Warrick, B. Taylor and J. Spudich (1976). *The Control of Flagellar Rotation in Bacterial Behavior*. In: *Cell Motility*. R. D. Goldman, T. Pollard, J. Rosenbaum Eds., Cold Spring Harbor New York, Cold Spring Harbor Laboratory.

Kossiakoff, A. A. (1982). *Protein Dynamics Investigated by the Neutron Diffraction-Hydrogen Exchange Technique*. Nature, 296, 713–721.

Krauss, S. and C. T. O'Konski (1959). *Electric Properties of Macromolecules. III. Kerr Constants and Rotational Diffusion of Bovine Serum Albumin in Aqueous Solutions*. J. Phys. Chem., 81, 5082–5088.

Kuffler, S. W., J. G. Nicholls and A. R. Martin (1984). *From Neuron to Brain: A Cellular Approach to the Function of the Nervous System*. Sunderland, Massachusetts, Sinauer Pub.

Kumar, N. and M. Flavin (1982). *Modulation of Some Parameters of Assembly of Microtubules In Vitro by Tyrosinylation of Tubulin*. Euro. J. Biochem., 128, 215–222.

Kumar, N. and M. Flavin (1981). *Preferential Action of a Brain Detyrosylating Carboxypeptidase on Polymerized Tubulin*. J. Biol. Chem., 256, 7678–7686.

Kung, C., S. Y. Chang, W. Y. Sato, J. Van Houten and H. Hansma (1975). *Genetic Dissection of Behavior in Paramecium*. Science, 188, 898–904.

Kuntz, L. D. and W. Kauzmann (1974). In: *Advances in Protein Chemistry*. New York, Academic Press, 239–345.

Lacy, M. E. (1984). *Phonon-Electron Coupling as a Possible Transducing Mechanism in Bioelectronic Processes Involving Neuromelanin*. J. Theor. Biol., 111, 201–204.

Lakowicz, J. R. and G. Weber (1973). *Quenching of Protein Fluorescence by Oxygen. Detection of Structural Fluctuations in Proteins on the Nanosecond Time Scale*. Biochem., 12, 4171–4179.

Lakowicz, J. R., H. Cherek and D. R. Bevan (1980). *Demonstration of Nanosecond Dipolar Relaxation in Biopolymers by Inversion of Apparent Fluorescence Phase Shift and Demodulation Lifetimes*. J. Biol. Chem., 255,

4403–4406.

Lambeir, A. and Y. Engelborghs (1980). *A Quantitative Analysis of Tubulin Colchicine Binding to Microtubules.* Eur. J. Biochem., 109, 619–624.

Landauer, R. (1982). *Uncertainty Principle and Minimal Energy Dissipation in the Computer.* Int. J. Theoret. Phys., 21, 283–297.

Langton, C. G. (1986). *Studying Artificial Life With Cellular Automata.* Physica 22D, 120–149.

Larsson, H., M. Wallin and A. Edström (1976). *Induction of a Sheet Polymer of Tubulin by* $Zn^{2+}$. Exp. Cell Res., 100, 104–110.

Lasek, R. J. (1981). *The Dynamic Ordering of Neuronal Cytoskeletons.* Neurosci. Res. Prog. Bull., 19(1), 7–31.

Lashley, K. (1950). *In Search of the Engram.* In: *Symposium of the Society for Experimental Biology.* No.4, Cambridge, Cambridge University Press, 454–482.

Lashley, K. S. (1929). *Brain Mechanisms and Intelligence: A Quantitative Study of Injuries to the Brain.* Chicago, University of Chicago Press.

Lauger, P. (1977). *Ion Transport and Rotation of Bacterial Flagella.* Nature, 268, 360–362.

Lawrence, A. F. and W. R. Adey (1982). *Nonlinear Wave Mechanisms in Interactions Between Excitable Tissue and Electromagnetic Fields.* Neurol. Res., 4, 115–153.

Layne, S. P. (1984). *The Modification of Davydov Solitons by the Extrinsic H-N-C=O Group.* Los Alamos National Laboratory unclassified release LA-UR-83-2253 and In press: *Nonlinear Electrodynamics in Biological Systems.* Plenum Press.

Layne, S. P., I. J. Bigio, A. C. Scott and P. S. Lomdahl (1985). *Transient Fluorescence in Synchronously Dividing* Escherichia coli. Proc. Natl. Acad. Sci., USA, 82, 7599–7603.

Lazarides, E. and J. P. Revel (1979). *The Molecular Basis of Cell Movement.* Sci. Am., 240, May, 100–113.

Lazarides, E. (1980). *Intermediate Filaments as Mechanical Integrators of Cellular Space.* Nature, 283, 249–256.

Lee, J. C., D. J. Field, H. J. George and J. Head (1986). *Biochemical and Chemical Properties of Tubulin Subspecies.* Ann. NY Acad. Sci., 466, 111–128.

Lerner, E. J. (1984). *Why Can't a Computer Be More Like a Brain?.* High Tech., August, 34–41.

Lettvin, J. Y. and R. C. Gestalind (1965). *Speculations on Smell.* Cold Spring Harbor Symposia on Quantitative Biology, 30, 217–225.

Lever, M. J., K. W. Miller, W. D. M. Paton and G. B. Smith (1971). *Pressure Reversal of Anesthesia.* Nature 231, 368–371.

Levinthal, Cyrus (1984). *The Formation of 3-Dimensional Biological Structures; Computer Uses and Future Needs.* Ann. NY Acad Sci., 426, 171–180.

Levitt, M., C. Sander and P. S. Stern (1985). *Protein Normal Mode Dynamics: Trypsin Inhibitor, Crambin, Ribonuclease and Lysosyme.* J. Molec. Biol., 181, 423–447.

Liberman, Ye A., S. V. Minina, N. Ye Shklovskii-Kordi and M. Conrad (1982). *Change in the Mechanical Parameters as a Possible Mode of Information Processing by the Neurone.* Biophysics 27(5), 906–915.

Liebovitch, L. S., and J. Fischbarg (1986). *Membrane Pores: a Computer Simulation of Interacting Pores Analyzed by g1(T) and g2(T) Correlation Functions.* J. Theor. Biol., 119, 287–297.

Lindemann and Rikmensboel (1972). *Sperm Flagella: Autonomous Oscillations of the Contractile System.* Science, 175, 337–338.

Lindsay, S. M. (1987). *Personal Communication.*

Lindsay, S. M. and B. Barris (1987). *A Scanning Tunneling Microscope for Biology.* Biophys. J. (Abstracts), 51(2), 444a.

Little, M. (1982). *Microtubules and Microrganisms.* P. Cappuccinelli, R. N. Morris, M. Little, K. Krauhse, H. Ponstingl, R. Luduena and N. Rice, eds., New York, Marcel Dekker Pub., 1–14.

Llinas, R. and C. Nicholson (1975). *Calcium Role in Depolarization Secretion Coupling: An Aequorin Study in Squid Giant Synapse.* Proc. Natl. Acad. Sci., USA, 72(1), 187–190.

Loeb, L. B. (1965). *Electrical Coronas–Their Basic Physical Mechanisms.* Berkeley, University of California Press.

Loizzi, R. F. (1983). *Ovariectomy Induced Tubulin Polymerization in Pregnant Rat.* Proc. Soc. Exper. Biol. Med., 173, 252–263.

Lomdahl, P. S., L. MacNeil, A. C. Scott, M. E. Stoneham and S. J. Webb. (1982). *An Assignment to Internal Soliton Vibrations of Laser-Raman Lines from Living Cells.* Phys. Letts., 92A, 4, 207–210.

Longuet-Higgins, H. C. (1968). *The Non-Local Storage of Temporal Information.* Proc. Roy. Soc B, 171, 327–334.

Lowe, C. R. (1979). *The Affinity Electrode.* FEBS Letters, 106(2), 405–408.

Lowenstein, O., M. P. Osborne and J. Warshall (1964). *Structure and Innvervation of the Sensory Epithelia of the Labyrinth in the Thornback ray (Raja clavata).* Proc. Royal Soc. of London, B 160, 1–12.

Luduena, R. F., E. M. Shooter and L. Wilson (1976). *Structure of Tubulin Dimer.* J. Biol. Chem., 252, 7006–7114.

Mackey, S., G. Schuessler, D. J. Goldberg and J. H. Schwartz (1981). *Dependence of Fast Axonal Transport on the Local Concentration of Organelles.* Biophys. J., 36, 455–459.

Mackintosh, A. R. (1987). *The First Electronic Computer.* Physics Today, March, 25–32.

Macnab, R.M. and D.E. Koshland (1972). *The Gradient Sensing Mechanism in Bacterial Chemotaxis.* Proc. Natl. Acad. Sci. USA, 69, 2509.

Maddox, J. (1987). *Liquid-Nitrogen-Temperature Superconductors Arrive.* Nature, 325, 756.

Malawista, S. E., De Boisfleury-Chevance A., (1982). *The Cytokineplast: Purified, Stable, and Functional Motile Machinery from Human Blood Polymorphonuclear Leukocytes.* J. Cell Biol., 95, 960–973.

Mandelbrot, B. (1982). *The Fractal Geometry of Nature.* San Francisco, W. H. Freeman.

Mandelkow, E. and J. Bordas (1983). *Structure of Tubulin Rings Studied by X-Ray Scattering Using Synchotron Radiation.* J. Mol. Biol., 167, 179–196.

Mandler, G. (1975). *Consciousness: Respectable, Useful, and Probably Necessary.* In: *Information Processing and Cognition—The Loyola Symposium.* Hillsdale, New Jersey, Lawrence Erlbaum Pub.

Mandoli, D. F. and W. Briggs (1982). *Optical Properties of Etiolated Plant Tissues.* Proc. Natl. Acad. Sci. USA, 79, 2902–2906.

Mareck, A., Fellous, J. Francon and J. Nunez (1980). *Changes in Composition and Activity of Microtubule Associated Proteins During Brain Development.* Nature, 284, 353–355.

Margolis, R. L. and L. Wilson (1978). *Opposite End Assembly and Disassembly of Microtubules at Steady State In Vitro.* Cell, 13, 1–8.

Margolis, R. L. and L. Wilson (1981). *Microtubule Treadmilling—Possible Molecular Machinery.* Nature, 293, 705–711.

Margulis, L. and D. Sagan (1986). *Strange Fruit on the Tree of Life.* The Sciences, May/June, 38–45.

Margulis, L. and D. Sagan (1986). *Origins of Sex: Three Billion Years of Genetic Recombination.* New Haven and London, Yale University Press.

Margulis, J. and D. Sagan (1986). *Microcosmos: Four Billion Years of Microbial Evolution.* New York, Summit Books.

Margulis, L. (1975). *Origin of Eukaryotic Cells.* New Haven, Yale University Press.

Margulis, L., L. To and D. Chase (1978). *Microtubules in Prokaryotes.* Science, 200, 1118–1124.

Markin, V. S. and Y. A. Chizmadzhev (1972). *Properties of a Multi-Component Excitable Medium.* J. Theor. Biol., 36, 61–80.

Markin, V. S. (1973). *Interaction in Bundles.* Biophys., 18, 324–332.

Mascarenhas, S. (1974). *The Electret Effect in Bone and Biopolymers and the Bound Water Problem.* Ann. NY Acad. Sci., 238, 36–52.

Mascarenhas, S. (1975). *Electrets in Biophysics.* J. Electrostat., 1, 141–146.

Masurovsky, E. B., E. R. Peterson, S. M. Crain and S. B. Horwitz (1983). *Morphological Alterations in Dorsal Root Ganglion Neurons and Supporting Cells of Organotypic Mouse Spinal Cord Ganglion Cultures Exposed to Taxol.* Neuroscience, 10(2), 491–509.

Matsumoto, G. and H. Sakai (1979). *Microtubules Inside the Plasma Membrane of Squid Giant Axons and Their Possible Physiological Function.* J. Membr. Biol., 50, 1–14.

Matus, A., M. Ackermann, G. Pehling, H. R. Byers and K. Fujiwara (1982). *High Actin Concentrations in Brain Dendritic Spines and Postsynaptic densities.* Proc. Natl. Acad. Sci., 79: 7590–7594.

Matus, A., R. Bernhardt, R. Bodmer and D. Alaimos (1986). *Microtubule Associated Protein 2 and Tubulin Are Differently Distributed in the Dendrites of Developing Neurons.* Neuroscience, 17(2), 371–389.

Mayer, R. J. (1985). *Molecular Turnover and Memory.* Nature, 313, 636.

McAuliffe, K. (1981). *Computer-Brain Implant.* Omni, Dec., 54–58.

McBride (1978). *Ion-Selective Fluid Effect Transistors with Polymeric Membranes.* Analytica Chimica Acta, 101, 239–245.

McCammon, J. A. (1981). *Gated Binding of Ligands to Proteins.* Nature, 293, 316–317.

McCammon, J. A., B. R. Gelin, M. Karplus (1977). *Dynamics of Folded Proteins.* Nature, 267, 585–590.

McClare, C. W. F. (1974). *Resonance in Bioenergetics.* Ann. NY Acad. Sci., 227, 74–115.

McCulloch, W. S. and W. H. Pitts, (1943). *A Logical Calculus of the Ideas Immanent in Nervous Activity.* Bull. Math. Biophys., 5, 115–133.

McIntosh, J. R. (1984). *Microtubule Catastrophe*. Nature, 312, 196–197.

McNair, H. M. and E. J. Bonelli (1969). *Basic Gas Chromatography, 5th Edition*. Walnut Creek, Varian Aerograph, 105.

Medvinsky, A. B., Pertsov, A. M., Polishuk, G. A., Fast, V. G., (1983). In: *Electric Field of the Heart*. O. Baum, M. Roschevsky and L. Titomir, Nauka, Moscow, 38–51.

Meindl, J. D. (1980). *Biomedical Implantable Microelectrodes*. Science, 210, 263–267.

Meyer, H. H. (1899). *Welche Eigenschaft der Anesthetica Bedingt Ihre Narkitishe Wirkung*. Arch. Exp. Pathol. Pharmacol., 42, 109–118.

Michels, B., Y. Dormoy, R. Cerf and J.A. Schulz (1985). *Ultrasonic Absorption in Tobacco Mosaic Virus and its Protein Aggregates*. J. Mol. Biol., 181, 103–110.

Milch, J. R. (1987). *Computer Based on Molecular Implementation of Cellular Automata*. Proc. Third Molecular Electronics Device Workshop, Naval Research Laboratory, Forrest Carter editor, Washington, D. C., in press.

Mileusnic, R., S. P. Rose and P. Tillson (1980). *Passive Avoidance Learning Results in Region Specific Changes in Concentration of, and Incorporation Into, Colchicine Binding Proteins in the Chick Forebrain*. J. Neur. Chem., 34(4), 1007–1015.

Millar, D. P., R. J. Robbins and A. H. Zewail (1980). *Direct Observation of the Torsional Dynamics of DNA and RNA by Picosecond Spectroscopy*. Proc. Natl. Acad. Sci., 77, 5593–5597.

Miller, S. (1961). *A Theory of Gaseous Anesthetics*. Proc. Natl. Acad. Sci., USA, 47(9), 1515.

Minsky, M. (1986). *The Society of Mind.*. New York, Simon and Schuster.

Minsky, M. and S. Papert (1972). *Perceptrons: An Introduction to Computational Geometry*. Cambridge, MIT Press.

Moran, D. T. and Varela F. G. (1971). *Microtubules and Sensory Transduction*. Proc. Natl. Acad. Sci. USA, 68, 757–760.

Moravec, H. P. (1987). *Mind Children*. San Francisco, University Press.

Murphy, D. B. and G. G. Borisy (1975). *Association of High Molecular-Weight Proteins with Microtubules and Their Role in Microtubule Assembly In Vitro*. Proc. Natl. Acad. Sci., USA, 72(7), 2696–2700.

Narayanamurti, V. (1981). *Phonon Optics and Propagation in Semiconductors*. Science, 213, 717–723.

Negendank, W. and G. Karreman (1979). *Rate of Potassium-Sodium Exchange by Human Lymphocytes: Prediction of the Cooperative Adsorption Model.* J. Cell Physiol., 98, 107–112.

Nelson, C. (1986). *Personal Communication.*

Ng, S. F. and J. Frankel (1977). *180 Degree Rotation of Cilia Role and Its Morphogenetic Implications in Tetrahymena Pyriformus.* Proc. Natl. Acad. Sci. USA, 74, 1115.

Nicolas, G. and I. Prigogine (1977). *Self Organization in Nonequilibrium Systems.* New York, Wiley.

Nietzsche, F. W. (1957). *The Urge to Mass Destruction.* New York, Grune and Stratton.

Nirenberg, M. W. and J. H. Matthaei (1961). *The Dependence of Cell-Free Protein Synthesis in E coli Upon Naturally Occurring or Synthetic Polyribonucleotides.* Proc. Natl. Acad. Sci. USA, 47, 1588–1602.

Nirenberg, M. W. and P. Leder (1964). *RNA Codewords and Protein Synthesis.* Science, 145, 1399–1407.

Ochs, S. (1982). *Axoplasmic Transport and Its Relation to Other Nerve Functions.* New York, Wiley Interscience.

O'Keefe, J. and A. Speakman (1986). *Material Presented for the University of Arizona "Cognitive Science Symposium."* Organized by L. Nadel, February.

O'Konski, C. T., K. Yoshioka and W. H. Orttung (1959). *Electric Properties of Macromolecules. IV. Determination of Electric and Optical Parameters from Saturation of Electric Birefringence in Solutions.* J. Phys. Chem., October, 1558–1565.

Oparin, A. I. (1938). *The Origin of Life.* Translated by Sergius Margulis, New York, Macmillan.

Orbach, R. (1986). *Dynamics of Fractal Networks.* Science, 231, 814–819.

Orgel, L. E. (1968). *Evolution of the Genetic Apparatus.* J. Molec. Biol., 38, 381–393.

Overton, F. (1901). *Studien Uber Die Narkose.* Jena, Germany, Fisher.

Pagels H.R., H. L. Dreyfus, J. McCarthy, M. L. Minsky, S. Papert and J. Searle *Panel Discussion: Has Artificial Intelligence Research Illuminated Human Thinking?* Ann. NY Acad. Sci., 426, 138–160.

Pagels, H. R. (1984). *Computer Culture: The Scientific, Intellectual, and Social Impact of the Computer.* Ann. NY Acad. Sci., 246.

Pain, R. H. (1983). *Dynamic Proteins.* Nature, 305, 581–582.

Panteloni, D., M. E. Calier and E. D. Korn (1985). *The Interactions Between ATP Actin and ADP Actin, A Tentative Model for Actin Polymerization.* J. Biol. Chem., 260, 6572–6578.

Parrett, T. (1982). *Membranes, Succeeding by Separating* Technology, March/April, 16–29.

Paula-Barbosa, M. M. and M. A. Tavares (1985). *Long Term Alcohol Consumption Induces Microtubular Changes in the Adult Rat Cerebellar Cortex.* Brain Res., 339, 195–199.

Pauling, L. (1961). *A Molecular Theory of General Anesthesia.* Science, 134, 15.

Pauling, L. (1964). *The Hydrate Microcrystal Theory of General Anesthesia.* Anesth. Analg., 43, 1.

Pavlov, I. P. (1963). *Lectures On Conditioned Reflexes.* London, Lawrence and Wishart (Reprint of 1928 Edition).

Payne, M. R., L. I. Binder and O. Steward (1983). *Immunocytochemical Localization of Actin and Microtubule Associate Protein MAP 2 in Dendritic Spines.* Proc. Nat. Acad. Sci. USA, 80, 1738–1742.

Pfister K., J. H. R. Kagi and P. Christen (1978). *Syncatalytic Conformational Changes in Aspartate Amino Acid Transfer is Determined by Hydrogen Deuterium Exchange.* Proc. Natl. Acad. Sci., USA, 75, 145–148.

Pickett-Heaps, J. D., D. H. Tippit, S. A. Cohn and T. P. Spurck (1986). *Microtubule Dynamics in the Spindle. Theoretical Aspects of Assembly/Disassembly Reactions In Vivo.* J. Theor. Biol., 118, 153–169.

Pickup, J. C. (1985). *Biosensors: A Clinical Perspective.* The Lancet, October, 817–820.

Poffenbarger, M. and G. M. Fuller (1977). *Effects of Psychotropic Drugs on Neurotubule Assembly.* J. Neurochem., 28, 1167–1174.

Platt, J.R. (1966). *The Step to Man.* Wiley, New York, 141–155.

Pohl, D. (1987). *Remarks at "Workshop on Biomedical Applications of STM."* University of Arizona College of Medicine, Departments of Anesthesiology, Anatomy, Advanced Biotechnology Laboratory, and Optical Sciences Center, January 21.

Pohl, H. (1983). *Natural Oscillating Fields of Cells.* In: *Coherent Excitations in Biological Systems.* H. Frohlich and F. Kremer, eds., Berlin, Springer-Verlag, 199–210.

Popp, F. A., W. Nagel, K. H. Lee, W. Cholz, O. Weingartner and R. Wolf (1984). Cell Biophys., 6, 33–52.

Popper, K. R. and J. C. Eccles (1977). *The Self and Its Brain.* Berlin, Springer-Verlag.

Porter, K. R. and L. G. Tilney (1965). *Microtubules and Intracellular Motility*. Science, 150, 382.

Porter, K. R. and J. B. Tucker (1981). *The Ground Substance of the Living Cell.* Sci. Amer., 56–67.

Pribram, K. H. (1966). *Some Dimensions of Remembering: Steps Towards a Neuropsychological Model of Memory.* In: *Macromolecules and Behavior*. Edited by J. Gaito, New York, Appleton Century Crofts.

Pribram, K. H. (1971). *Languages of the Brain: Experimental Paradoxes and Principles.* In: *Neurophysiology*. New York, Brandon House.

Pribram, K. H. (1986). *The Cognitive Revolution and Mind/Brain Issues.* American Psychologist, May, 507–520.

Prigogine, I. and C. George (1983). *The Second Law as a Selection Principle: The Microscope Theory of Dissipative Processes in Quantum Systems.* Proc. Natl. Acad. Sci., USA, 80, 4590–4594.

Puck, T. T. (1977). *Cyclic AMP, the Microtubule-Microfilament System, and Cancer.* Proc. Natl. Acad. Sci. USA, 74, 4491–4495.

Pullman, B. and A. Pullman (1963). *Quantum Biochemistry.* New York, Interscience.

Purich, D. L. and D. Kristofferson (1984). *Microtubule Assembly: A Review of Progress, Principles, and Perspectives.* In: *Advances in Protein Chemistry.* Vol. 36, C. B. Anfinsen, J. T. Edsall and F. M. Richards eds., Orlando, Academic Press, 133–212.

Purpura, D. P., N. Bodick, L. K. Suzuki, I. Rapin and S. Wurzelmann (1982). *Microtubule Disarray in Cortical Dendrites and Neurobehavioral Failure.* Dev. Brain Res., 5, 287–297.

Purpura, D. P. (1974). *Dendritic Spine Dysgenesis and Mental Retardation* Science 186, 1126–1128.

Raes, M., G. Geuens, M. DeBrabander and J. Remacle (1983). *Microtubules and Microfilaments in Ageing Hamster Embryo Fibroblasts In Vitro.* Exp. Gerontol., 18, 241–254.

Raes, M., M. DeBrabander and J. Remacle (1984). *Polyploid Cells in Ageing Hamster Fibroblasts In Vitro: Possible Implications of the Centrosome.* Eur. J. Cell Biol., 35, 70–80.

Rall, W. and J. Rinzel (1973). *Branch Input Resistance and Steady Attenuation for Input to One Branch of a Dendritic Neuron Model.* Biophys. J., 13, 648–688.

Ramon y Cajal, S. (1955). *Studies On the Cerebral Cortex.* London, Llyod-Luke (Translated by Lisbeth M. Kroft).

Rampil, I. J., J. A. Holzer, D. O. Quest, S. H. Rosenbaum and J. W.

Correll (1983). *Prognostic Value of Computerized EEG Analysis During Carotid Endarterectomy.* Anesth. Analg., 62, 186–192.

Rebhum, L. I. (1972). *Polarized Intracellular Particle Transport, Saltatory Movements and Cytoplasmic Streaming.* Int. Rev. Cytol., 32, 93–137.

Record, R. (1986). Personal Communication.

Reeke, G. R. and G. M. Edelman (1984). *Selective Networks and Recognition Automata.* In: *Computer Culture: The Scientific, Intellectual, and Social Impact of the Computer.* Heinz R. Pagels ed., Ann. NY Acad. Sci., 426, 181–201.

Richards, C. D. (1974). *The Action of General Anaesthetics on Synaptic Transmission Within the Central Nervous System.* In: *Molecular Mechanisms in General Anesthesia.* M. J. Halsey, R. A. Millar and J. A. Sutton eds., Edinburgh, Churchill Livingstone.

Robach, Y., B. Michels, R. Cerf, J. Braunwald and F. Tripier-Darcy (1983). *Ultrasonic Absorption Evidence for Structural Fluctuations in Frog Virus III and its Subparticles.* Proc. Natl. Acad. Sci. USA, 80, 3981–3985.

Robinson, A. L. (1986). *The Return of the Vacuum Tube.* Science, 231, 218.

Robinson, A. L. (1987). *Superconductor Claim Raised to 94k.* Science 235, 1137–1138.

Robinson, J. and Y. Engelborghs (1982). *Tubulin Polymerization and Dimethylsulfoxide.* J. Biol. Chem., 257, 5367–5371.

Robinson, K. R. (1985). *The Responses of Cells to Electrical Fields: A Review.* J. Cell Biol., 101 2023–2027.

Roos, U. P. (1984). *From Protomitosis to Mitosis. An Alternative Hypothesis on the Origin and Evolution of the Mitotic Spindle.* Origins of Life, 13, 183–193.

Rosenblatt, F. (1962). *Principles of Neurodynamics.* New York, Spartan Books.

Roth L. E., D. J. Pihlaja and Y. Shigenaka (1970). *Microtubules in the Heliazoan Axopodium. I. The Gradion Hypothesis of Allosterism in Structural Proteins.* J. of Ultrastr. Res., 30, 7–37.

Roth L.E. and D. J. Pihlaja (1977). *Gradionation: Hypothesis for Positioning and Patterning.* J. Protozoology, 24, 1, 2–9.

Rothwell, S. W., W. A. Grasser and D. B. Murphy (1986). *End to End Annealing of Microtubules In Vitro.* J. Cell Biol., 102, 619–627.

Rowlands, S. (1983). *Coherent Excitations in Blood.* In: *Coherent Excitations in Biological Systems.* H. Fröhlich and F. Kremer eds., Berlin, Springer-Verlag.

Rowlands, S., L. Sewchand and L. Skibo (1983). *Conversion of Albumin into a Transmitter of the Ultra Long Range Interaction of Human Erythrocytes.* Cell Biophys., 5, 197–203.

Rowlands, S., L. S. Sewchand and E. G. Enns (1982). *Further Evidence for a Fröhlich Interaction of Erythrocytes.* Phys. Lett., 87a, 256–260.

Sagan, L. (1967). *On the Origin of Mitosing Cells.* J. Theor. Biol. 14, 225–274.

Sander, L. M. (1986). *Fractal Growth Processes.* Nature, 322, 780–793.

Satir, P. and B. Satir (1964). *A Model for Nine Fold Symmetry in Alpha Keratin and Cilia.* J. Theor. Biol., 7, 123–128.

Satir, V., C. Schooley and D. Satir (1973). *Membrane Fusion in a Model System. Mucocyst Secretion in Tetrahymena.* J. Cell Biol., 56, 153–176.

Satir, P. (1984). *Cytoplasmic Matrix: Old and New Questions.* J. Cell Biol., 99(1), 235–238.

Schliwa, M., K. R. Pryzwansky and U. Euteneur (1982). *Centrosome Splitting in Neutrophils: An Unusual Phenomenon Related to Cell Activation and Motility.* Cell, 31, 705–717.

Schmitt, F. O., P. Dev and B. H. Smith (1976). *Electrotonic Processing of Information by Brain Cells.* Science, 193, 114–120.

Schneiker, C. W. (1986). *NanoTech: Science, Engineering and Related Topics.* This book is a technical report of the Department of Anesthesiology, College of Medicine, University of Arizona, Tucson, Arizona, 85724.

Schneiker, C. W. (1986). *NanoTechnology with STMs, Feynman Machines, and von Neumann Machines.* Presented at 1986 SPAIN STM Conference, July 14–18.

Schneiker, C. W. and S. R. Hameroff (1986). *NanoTechnology Workstations Based on Scanning Tunneling/Optical Microscopy: Applications to Molecular Scale Devices.* To be published in: *Proceedings of the Third International Symposium on Molecular Electronic Devices, Oct. 6–8, 1986, Arlington, VA.* F. L. Carter and H. Wohltjen, eds. (1987), Elsevier North-Holland.

Schultheiss, R. and E. Mandelkow (1983). *Three Dimensional Reconstruction of Tubulin Sheets and Reinvestigation of Microtubule Surface Lattice.* J. Molec. Biol., 170, 471–496.

Schwan, H. P. (1977). *Field Interaction with Biological Matter.* Ann. NY Acad. Sci., 303, 198–213.

Schwan, H. P. and K. R. Foster (1977). *Microwave Dielectric Properties of Tissue, Some Comments on the Rotational Mobility of Tissue Water.* Biophys. J., 17, 193–197.

Schwan, H. P. and L. D. Sher (1969). *Alternating Current Field Induces Forces and Their Biological Implications.* J. Electro. Chem. Soc., 116, 170–174.

Schwartz, A. W. and L. E. Orgel (1985). *Template Directed Synthesis of Novel Nucleic Acid Structures.* Science, 228, 585–587.

Schwarz, G. (1967). *A Basic Approach to a General Theory for Cooperative Intramolecular Conformation Changes of Linear Biopolymers.* Biopolymers, 5, 321–324.

Schwarz, G. (1977). *Chemical Transitions of Biopolymers Induced by an Electric Field and Their Effects in Dielectrics and Birefringence.* Ann. NY Acad. Sci., 303, 190–197.

Scott, A. C. (1985). *Anharmonic Analysis of Resonant Microwave Absorption in DNA.* Physica Scripta 32, 617–623.

Scott, A. C. (1982). *Dynamics of Davydov Solitons.* Phys. Rev. A, 26, 578–595.

Scott, A. C. (1973). *Information Processing in Dendritic Trees.* Math Bio. Sci., 18, 153–160.

Scott, A. C. (1977). *Neurophysics.* New York, Wiley Interscience Publications.

Scott, A. C. (1984) *Solitons and Bioenergetics.* In: *Nonlinear Electrodynamics in Biological Systems.* W. R. Adey and A. F. Lawrence eds., New York, Plenum Press.

Scott, A. C. (1985). *Davydov Solitons in Polypeptides.* Phil. Trans. R. Soc. Lond., A, 315, 423–436.

Scott, A. C., E. Gratton, E. Shyamsunder and G. Careri (1985). *IR Overtone Spectrum of the Vibrational Soliton in Crystalline Acetanilide.* Phys. Rev. B, 32(8), 5551–5553.

Scott, Andrew (1985). *Pirates of the Cell: The Story of Viruses from Molecule to Microbe.* Oxford, Basil Blackwell.

Seeman, P. (1975). *The Membrane Expansion Theory of Anesthesia.* In: *Molecular Mechanisms of Anesthesia: Progress in Anesthesiology.* Volume 1, B. R. Fink ed., New York, Raven Press, 143–252.

Seifriz, W. (1950). *The Effects of Various Anesthetic Agents on Protoplasm.* Anesthesiology, 11, 24.

Serpersu, E. H. and T. Y. Tsong (1984). *Activation of Electrogenic $Rb^+$ Transport of ($Na_1$ K) ATPase by an Electric Field.* J. Biol. Chem., 259, 7155–7162.

Serpersu, E. H. and T. Y. Tsong (1983). *Stimulation of a Ouabain-Sensitive $Rb^+$ Uptake in Human Erthrocytes With an External Field.* J.

Membr. Biol., 74, 191–201.

Shannon, C. E. (1948). *A Mathematical Theory of Communication.* Bell System Technology Journal, 27, 379.

Shaw, G. and D. Bray (1977). *Movement and Extension of Isolated Growth Cones.* Exp. Cell Research, 104, 55–62.

Sherrington, C. S. (1933). *The Brain and Its Mechanism.* London, Cambridge Univ. Press.

Sherrington, C.S. (1947). *The Integrative Action of the Nervous System.* New Haven, Yale Univ. Press.

Shimizu, H. and H. Haken (1983). *Co-Operative Dynamics in Organelles.* J. Theor. Biol., 104, 261–273.

Shohet, J. L. and S. A. Reible (1974). *Models for Energy Transduction and Transfer in Biological Systems.* Ann. NY. Acad. Sci., 227, 641–650.

Shoulders, K. R. (1962). *On Microelectronic Components, Interconnections, and System Fabrication.* In: *Aspects of the Theory of Artificial Intelligence.* C. A. Muses, ed., New York, Plenum Press, 217–235.

Shoulders, K. R. (1965). *Toward Complex Systems.* In: *Microelectronics and Large Systems.* Spartan Books, 97–128.

Shoulders, K. R. (1986). Personal Communication to Conrad Schneiker.

Simons, Geoff (1983). *Are Computers Alive?* Santa Cruz, Harvester Press.

Slopek, S., I. Durlakow, B. Weberdab, A. Kucharew, M. Dabrowsk and R. Bisikiew (1983). *Results of Bacteriophage Treatment of Suppurative Bacterial Infections.* Arch. Immun., 31(3), 267–291.

Smith, S. A., R. C. Watt, and S. R. Hameroff (1984). *Cellular Automata in Cytoskeletal Lattices.* Physica 10D, 168–174.

Smoluchowski, M. (1917). *Investigation into a Mathematical Theory of the Kinetics of Coagulation of Colloidal Solutions.* Zeit Phys Chim, 92, 129–168.

Sobue, T. Tanaka, N. Ashino and S. Kakiuchi (1985). *$Ca^{2+}$ and Calmodulin Regulate Microtubule-Associated Protein-Actin Filament Interaction in a Flip-Flop Switch.* Biochim. Biophys. Acta., 845, 366–372.

Soifer, D. (1986). *Factors Regulating the Presence of Microtubules in Cells.* In: *Dynamic Aspects of Microtubule Biology.* D. Soifer ed., Ann. NY. Acad. Sci., 466, 1–7.

Solsky, R. L. and G. A. Rechnitz (1979). *Antibody-Selective Membrane Electrodes.* Science, 204, 1308–1309.

Somjen, G. C. (1983). *Neurophysiology, The Essentials.* Baltimore, Williams and Wilkins.

Somogyi, B., J. Matko, S. Papp, J. Hevessy, G. R. Welch and S. Damjanovich (1984). *Forster-Type Energy Transfer as a Probe for Changes in Local Fluctuations of the Protein Matrix.* Biochem., 23, 3403–3411.

Sonneborn, T. M. (1970). *Gene Action and Development.* Proc. R. Soc. Lon. B, 176, 347.

Sperry, R. W. (1980). *Mind-Brain Interaction—Mentalism, Yes—Dualism, No.* Neuroscience, 5(2), 195–206.

Sperry, R. W. (1969). *Neurology and the Mind-Brain Problem.* American Scientist, 40, 291–312.

Sperry, R. W. (1969). *A Modified Concept of Consciousness.* Psychological Review, 76, 532–636.

Springer, M. S., M. F. Goy and J. Adler (1979). *Protein Methylation in Behavioral Control Mechanisms and Signal Transduction.* Nature, 280, 279.

Stearns, M. E. (1980). *Lattice Dependent Regulation of Axonal Transport by Calcium Ions.* In: *Microtubules and Microtubule Inhibitors.* M. DeBrabender and J. DeMey eds., Amsterdam, Elsevier North Holland, 17–30.

Stebbings, H. and C. Hunt (1982). *The Nature of the Clear Zone Around Microtubules.* Cell Tissue Res., 227, 609–617.

Stephens, R. E. and K. T. Edds (1976). *Microtubules: Structure, Chemistry, and Function.* Physiological Reviews, 56(4), 709–777.

Stevens, John K. (1985). *Reverse Engineering the Brain.* [Artificial Intelligence] Byte 16, 4, 287-299.

Syversen, T. L. M., P. R. Sager, T. W. Clarkson, J. B. Cavanagh, A. Elgsaeter, H. C. Guldberg, S. D. Lee, M. A. Lichtman, N. K. Mottet and J. B. Olmsted (1986). *The Cytoskeleton as a Target for Toxic Agents.* In: *The Cytoskeleton.* T. W. Clarkson, P. R. Sager and T. L. M. Syversen eds., New York, Plenum, 25–34.

Szent-Gyorgyi, A. (1960). *Introduction to a Submolecular Biology.* New York, Academic Press.

Tauc L. and G. M. Hughes (1963). *Modes of Initiation and Propagation of Spikes in the Branching Axons of Molluscan Central Neurons.* J. Gen. Physiol., 46, 533–549.

Taylor, F. and J. Condeelis (1981). *Microfilament Membrane Interactions in Cell Shape and Surface Architecture.* Intern. Cell Biol., H. G. Schweiger ed., Berlin, Springer-Verlag, 306–320.

Taylor, F. J. R. (1976). *Dissipativity and Organization in Living Matter: A Quantum Field Theory Approach.* Taxon, 23, 229, 197.

Telesnin, V.R. (1969). *Stability of the Steady State in a Homogeneous Excitable Ring.* Biophys., 14, 140–149.

Terakawa, S. and T. Nakayama (1985). *Are Axoplasmic Microtubules Necessary for Membrane Excitation?* J. Membr. Biol, 85, 65–77.

Thyberg, J. (1984). *The Microtubular Cytoskeleton and the Initiation of DNA Synthesis.* Exp. Cell. Res., 155, 1–8.

Thom, R. (1972). *Structural Stability and Morphogenesis. An Outline of a General Theory of Models.* (English Translation by D. Fowler), Reading Mass.

Toda, M. (1979). *Solitons and Heat Conduction.* Phys Scripta, 20(3-4), 424–430.

Toffoli, T. (1984). *Cellular Automata as an Alternative to (Rather Than an Approximation of) Differential Equations in Modeling Physics.* Physica 10D, 117–127.

Tomita, K. (1982). *Structure and Macroscopic Chaos in Biology.* J. Theor. Biol., 99, 111–118.

Troshin, A. B. (1966). *Problems of Cell Permeability.* New York, Pergamon Press.

Trudell, J. R. (1977). *A Unitary Theory of Anesthesia Based on Lateral Phase Separations in Nerve Membranes.* Anesthesiology, 46, 5–10.

Trump, B. F., I. K. Berezsky, K. U. Laiho, A. R. Osornio, W. J. Mergner and M. W. Smith (1980). *The Role of Calcium in Cell Injury.* Purity Review, (Scanning Electron Microscopy), Chicago, AMF O'Hare.

Tuszynski, J. A., R. Paul, R. Chatterjee and S. R. Sreenivasa (1984). *Relationship Between Fröhlich and Davydov Models of Biological Order.* Phys. Rev. A, 30(5), 2666–2675.

Ueda, I. (1965). *Effects of Diethylether and Halothane on Firefly Luciferin Luminescence.* Anesthesiology, 26, 603–606.

Ueda, I. and H. Kamaya (1973). *Kinetic and Thermodynamic Aspects of the Mechanisms of General Anesthesia in a Model System of Firefly Luminescence In Vitro.* Anesthesiology, 38, 425–436.

Urey, H. C. (1939). *The Separation of Isotopes and Their Use in Chemistry and Biology.* Science in Progress, 1, 35–77.

Vale, R. D., T. S. Reese and M. P. Sheetz (1985). *Identification of a Novel Force-Generating Protein, Kinesin, Involved in Microtubule-Based Motility.* Cell, 42, 39–50.

Vallano, M. L., J. R. Goldenring, R. S. Lasher and R. Delorenzo (1986). *Association of Calcium/Calmodulin Dependent Kinase With Cytoskeletal*

*Preparations: Phosphorylation of Tubulin, Neurofilament, and Microtubule-Associated Proteins*. Ann. NY Acad. Sci., 466, 357–374.

Vallee, R. B. (1982). *A Taxol-Dependent Procedure for the Isolation of Microtubules and Microtubule-Associated Proteins (MAPs)*. J. Cell Biol., 92, 435–442.

Van Brunt, E. E., M. D. Sheperd, J. R. Wall, W. F. Ganong and M. T. Clegg (1964). *Penetration of Light into the Brain of Mammals*. in: *Photo-Neuro-Endocrine Effects in Circadian Systems with Particular Reference to the Eye*. Ann. NY Acad. Sci., Vol. 117, 217–227.

Van Brunt, J. (1985). *Biochips: The Ultimate Computer*. Bio/Technology, 3, 209–215.

Van Heerden, P. J. (1968). *The Foundation of Empirical Knowledge With a Theory of Artificial Intelligence*. Wassenaar Netherlands, Wistik.

Vasiliev, J. M., I. M. Gelfand (1976). *Effects of Colcemid on Morphogenetic Processes and Locomotion of Fibroblasts*. In: Cell Motility, edited by Goldman R., T. Pollard and J. Rosenbaum, Cold Spring Harbor, 279–304.

Vasiliev, J. M. and H. T. Tien, (1984). *Structure and Properties of Cell Membranes*. Volume 2, G. Benga, Editors, Boca Raton, CRC Press.

Vassilev, P., M. Kanazirska, and H. T. Tien (1985). *Intermembrane Linkage Mediated by Tubulin*. Biochem. Biophys. Res. Comm., 126(1), 559–565.

Vassilev, P. M., R. T. Dronzine, M. P. Vassileva and G. A. Georgiev (1982). *Parrallel Arrays of Microtubules Formed in Electric and Magnetic Fields*. Bioscience Reports 2, 1025–1029.

Volkenstein, M. V. (1972). *The Conformon*. J. Theor. Biol., 34, 193–195.

Von Hippel, A. R. (1962). *Molecular Designing of Materials*. Science, 138(12), 91–108.

Von Neumann, J. (1958). *The Computer and the Brain*. New Haven, Yale University Press.

Von Neumann, J. (1966). *Theory of Self-Reproducing Automata*. A. W. Burks Ed., Urbana, University of Illinois Press.

Waltz, D. L. (1982). *Artificial Intelligence*. Sci. Am., 247(4), 118.

Watt, R. C. and S. R. Hameroff (1987). *Phase Space Analysis of Human EEG During General Anesthesia*. Ann. NY Acad. Sci., in press.

Waxman, S. G. and R. J. Melker (1971). *Closely Spaced Nodes of Ranvier in the Mammalian Brain*. Brain Research, 32, 445–448.

Waxman, S. G, G. D. Pappas and M. V. L. Bennett (1972). *Morphological Correlates of Functional Differentiation of Nodes of Ranvier Along Single*

*Fibers in the Electric Organ of the Knifefish Sternarchus.* J. Cell Biol., 53, 210–224.

Waxman, S. G.(1972). *Regional Differentiation of the Axon: A Review with Special Reference to the Concept of the Multiplex Neuron.* Brain Research, 47, 269–288.

Webb, S. J. (1980). *Laser Raman Spectroscopy of Living Cells.* Phys. Rep., 60, 201–224.

Weber, G. (1974). *Addition of Chemical and Osmotic Energies by Ligand-Protein Interactions.* Ann. NY Acad. Sci., 227, 486–496.

Weber, R. (1982). *The Enfolding-Unfolding Universe: A Conversation with David Bohm.* in: *The Holographic Paradigm.* Ken Wilber ed., Boulder, Shambhala, New Science Library, 44–104.

Wei, L. Y. (1974). *Dipole Mechanisms of Electrical, Optical and Thermal Energy Transductions in Nerve Membrane.* Ann. NY Acad. Sci., 227, 285–293.

Weisenberg, R. C. (1980). *Role of Co-operative Interactions, Microtubule-associated Proteins and Guanosine Triphosphate in Microtubule Assembly: A Model.* J. Molec. Biol., 139, 660–677.

Weisenberg, R. C., G. B. Borisy and E. W. Taylor (1968). *The Colchicine Binding Protein of Mammalian Brain and Its Relation to Microtubules.* Biochem., 7(12), 4466–4479.

Weisenberg, R. C. and S. N. Timasheff (1970). *Aggregation of Microtubule Subunit Protein. Effects of Divalent Cations, Colchicine and Vinblastine.* Biochem., 9(21), 4110–4116.

Weisenberg, R. C., and W. J. Weery (1976). *The Role of Nucleotide Hydrolysis in Microtubule Assembly.* Nature, 263, 792–793.

Weiss, D. G. and R. D. Allen (1986). *The Organization of Force Generation in Microtubule Based Motility.* M. DeBrabander and J. DeMey eds., *Microtubules and Microtubule Inhibitors.* Amsterdam, Elsevier North Holland, 233–240.

Weiss, D. G., F. Keller, J. Gulden and W. Mailee (1986). *Towards a New Classification of Intracellular Particle Movements Based on Quantitative Analysis.* Cell Motility and the Cytoskeleton, 6, 128–135.

Weiss, P. (1961). *Guiding Principles in Cell Locomotion and Cell Aggregation.* Exp. Cell Res. Supp., 8, 260.

Westerhoff, H. V., T. Y. Tsong, P. B. Chock, Yd. Chen and R. D. Astumian (1986). *How Enzymes Can Capture and Transmit Free Energy From an Oscillating Electric Field.* Proc. Natl. Acad. Sci. USA, Volume 83, 4734–4738.

Westlake, P. R. (1970). *Possibilities of Neural Holographic Processes Within the Brain.* Kybernetik, 7(4), 129.

Westrum, L. E., D. H. Jones, E. G. Gray and J. Barron (1980). *Microtubules, Dendritic Spines and Spine Apparatuses.* Cell Tissue Res., 208, 171-181.

Wheatley, J., D. S. Buchanan, G. W. Swift, A. Migliori and T. Hofler (1985). *Nonlinear Natural Engine: Model for Thermodynamic Processes in Mesoscale Systems.* Proc. Natl. Acad. Sci., USA, 82, 7805–7809.

Wheatley, D.N. (1985). *On the Possible Importance of an Intracellular Circulation.* Life Science, 36, 299–307.

Wheatley, D. N. (1982). *The Centriole: A Central Enigma of Cell Biology.* Amsterdam, Elsevier.

White, D. C. and C. R. Dundas (1970). *Effects of Anaesthetics on Emission of Light by Luminous Bacteria.* Nature, 226, 456–458.

White, D. C. and M. J. Halsey (1974). *Effect of Changes in Temperature and Pressure During Experimental Anaesthesia.* Br. J. Anaesth., 46, 196–201.

White, J. (1969). *Viral Induced Repair of Damaged Neurons with Preservation of Long Term Information Content.* Second Annual Cryonics Conference, April 11.

Whitham, G. B. (1974). *Linear and Nonlinear Waves.* New York, Wiley Interscience.

Wilber, Ken (1982). *The Holographic Paradigm and Other Paradoxes.* Boulder, London, Shambhala, New Science Library.

Williams, H. Smith and M. B. Huggins (1983). *Effectiveness of Phages in Treating Experimental E coli Diarrhea in Calves, Piglets and Lambs.* J. Gen. Microbiol., 128, 2659.

Winfree, A. T. (1967). *Biological Rhythms and the Behavior of Coupled Oscillators.* J. Theor. Biol., 16, 15–42.

Winfree, A. T. and S. H. Strogatz (1984). *Organizing Centers for Three Dimensional Chemical Waves.* Nature, 311, 611–615.

Wisniewski, H., W. Karczewski and K. Wisniewska (1966). *Neurofibrillary Degeneration of Nerve Cells After Intracerebral Injection of Aluminum Cream.* Acta Neuropathol., 6, 211–219.

Wisniewski, H., M. L. Shelanski and R. D. Terry (1967). *Experimental Colchicine Encephalopathy. I. Induction of Neurofibrillary Degeneration.* Lab. Invest., 17, 577–587.

Wolfram, S. (1984) *Cellular Automata as Models of Complexity.* Nature, 311, 419–424.

Wolfram, S. (1984). *Universality and Complexity in Cellular Automata.* Physica 10D, 1–35.

Woodbridge, P. D. (1957). *Changing Concepts Concerning Depth of Anesthesia.* Anesthesiology, 18, 536–550.

Woodbury, J. W., J. S. D'Arrigo and H. Eyring (1975). *Molecular Mechanism of General Anesthesia.* Progress in Anesthesiology, Volume I, B. R. Fink ed., New York, Raven Press, 253–276.

Wright, R. (1985). *The On/Off Universe. The Information Age.* The Sciences, May/June, 7–9.

Wu, T. M. and S. Austin (1978). *Bose-Einstein Condensation in Biological Systems.* J. Theor. Biol., 71, 209–214.

Wulf, R. J. and R. M. Featherstone (1957). *A Correlation of Van der Waals constants with Anesthetic Potency.* Anesthesiology, 18, 97–105.

Yamada, K. M., B. S. Spooner and N. K. Wessells (1970). *Axon Growth: Roles of Microfilaments and Microtubules.* Proc. Natl. Acad. Sci., USA, 66, 1206–1212.

Yamamoto, G. (1985). *Molecular Gears With Two–Toothed and Three–Toothed Wheels.* J. Molec. Struct., 126, 413–420.

Yamamoto, N., Y. Nagasawa, M. Sawai, T. Sudo and H. Tsubomura (1978). *Potentiometric Investigations of Antigen-Antibody and Enzyme-Enzyme Inhibitor Reactions Using Chemically Modified Metal Electrodes.* J. Immunolog. Meth., 22, 309–317.

Yano M., Y. Yamamoto and H. Shimizu (1980). *An Actomyosin Motor.* Nature, 299, 557–559.

Yates, F. E. (1984). *Report on Conference on Chemically–Based Computer Designs.* Crump Institute for Medical Engineering Report CIME TR/84/1, University of California, Los Angeles, CA.

Young, J. Z. (1951). *Doubt and Certainty in Science. A Biologist's Reflections on the Brain.* New York, Oxford, University Press.

Yue-xian, L. and D. Dafu (1984). *Chaos and Other Temporal Self-Organization Patterns in Coupled Enzyme-Catalyzed Systems.* Commun. in Theor. Phys., Vol. 3, No. 5, 629–638.

Zabusky, N. J. and M. D. Kruskal (1965). *Interactions of Solitons in Collisionless Plasma and Recurrence of Initial States.* Phys. Rev. Lett., 15, 240.

Zamperoni, P. (1986). *Some Reversible Image Operators from the Point of View of Cellular Automata.* Biol. Cybern., 54, 253–261.

Zaug, A. J., and T. R. Cech (1986). *The Intervening Sequence RNA of Tetrahymena is an Enzyme.* Science, 231, 470–475.

Zieve, G. and F. Solomon (1984). *Direct Isolation of Neuronal Microtubule Skeletons.* Molec. and Cell. Biol., 4(2), 371–374.

# Appendix A

# STM References

Abraham, D. W., H. J. Mamin, E. Ganz and J. Clarke (1986). *Surface Modification with the Scanning Tunneling Microscope.* IBM J. Res. Develop., 30(5), Sept., 492–499.

Aguilar, M., P. J. Pascual and A. Santisteban (1986). *Scanning Tunneling Microscope Automation.* IBM J. Res. Develop., 30(5), Sept., 525–532.

Amer, N. M., A. Skumanich and D. Ripple (1986). *Photothermal Modulation of the Gap in Scanning Tunneling Microscopy.* Appl. Phys. Letts., 49(3), 21 July, 137–139.

Arvia, A. J. (1986). *Scanning Tunneling Microscopy and Electrochemistry: Surface Topography of (100)-Type Electrofaceted Platinum.* In Press.

Baratoff, A. (1984). *Theory of Tunneling Microscopy Methods and Approximations.* Physica, 127B, 134–150.

Baratoff, A., G. Binnig and H. Rohrer (1983). *Summary Abstract: Scanning Tunneling Microscopy of Semiconductor Surfaces.* J. Vac. Sci. Technol. B, 1(3), July-Sept., 703–704.

Baratoff, A., G. Binnig, H. Fuchs, F. Salvan and E. Stoll (1986). *Tunneling Microscopy and Spectroscopy of Semiconductor Surfaces and Interfaces.* Surface Science, 168, 734–743.

Baratoff, A., G. Binnig, H. Fuchs, F. Salvan, E. Stoll (1986). *Tunneling Microscopy and Spectroscopy of Semiconductor Surfaces and Interfaces.* Surface Sci., 168, 734–743.

Baro, A. M., G. Binnig, H. Rohrer, C. Gerber, E. Stoll, A. Baratoff and F. Salvan (1984). *Real-Space Observation of the 2×1 Structure of Chemisorbed Oxygen on Ni(110) by Scanning Tunneling Microscopy.* Phys. Rev. Lett., 52(15), 1304–1307.

Baro, A. M., R. Miranda, J. Alaman, N. Garcia, G. Binnig, H. Rohrer, C.

Gerber and J. L. Carrascosa (1985). *Determination of Surface Topography of Biological Specimens at High Resolution by Scanning Tunneling Microscopy.* Nature, 315, 16 May, 253–254.

Baro, A., R. Miranda and J. L. Carrascosa (1986). *Application to Biology and Technology of the Scanning Tunneling Microscope Operated in Air at Ambient Pressure.* IBM J. Res. Develop, 30(4), 4 July, 380–386.

Becker, R. S., J. A. Golovchenko and B. S. Swartzentruber (1985a). *Tunneling Images of Germanium Surface Reconstructions and Phase Boundaries.* Phys. Rev. Lett., 54(25), June, 2678–2680.

Becker, R. S., J. A. Golovchenko and B. S. Swartzentruber (1985). *Electron Interferometry at Crystal Surfaces.* Phys. Rev. Lett., 55(9), 26 August, 987–990.

Becker, R. S., J. A. Golovchenko and B. S. Swartzentruber (1985b). *Tunneling Images of the 5×5 Surface Reconstruction on Ge-Si(111).* Phys. Rev. B, 32(12), December, 8445–8457.

Becker, R. S., J. A. Golovchenko and B. S. Swartzentruber (1987). *Atomic-Scale Surface Modifications Using a Tunnelling Microscope.* Nature, 325, 29 January, 419–421.

Behm, R. J., W. Hosler, E. Ritter and G. Binning (1986). *Summary Abstract: The Hexagonal Reconstruction of Pt(100): A Scanning Tunneling Microscopy Study.* J. Vac. Sci. Tech., 4(3), May/June, 1330–1331.

Benistant, P. A. M., G. F. A. van de Walle, H. van Kemptn and P. Wyder (1986). *Comparison of Transverse-Electron-Focusing and Scanning-Tunneling-Microscopy Measurements on Ag(001) and (011) Surfaces.* Phys. Rev. B, 33(2), 15 Jan., 690–694.

Berghaus, Th., H. Neddermeyer and St. Tosch (1986). *A Scanning Tunneling Microscope for the Investigation of the Growth of Metal Films on Semiconductor Surfaces.* IBM J. Res. Develop., 30(5), Sept., 520–524.

Binnig, G and H. Rohrer (1985a). *Scanning Tunneling Microscopy.* Surface Science, 152/153, 17–26.

Binnig, G. K. and H. Rohrer (1984). *Scanning Tunneling Microscopy.* Physica, 127B, 37–45.

Binnig, G. K., C. Gerber, H. Rohrer and E. Weibel (1985a). *Nano-Aperture.* IBM Tech. Disclosure Bul., 27(8), Jan., 4893.

Binnig, G. and C. Gerber (1979). *Piezo Drive With Coarse and Fine Adjustment.* IBM Tech. Disclosure Bul., 22(7), Dec., 2897–2898.

Binnig, G. and D. P. E. Smith (1986). *Single-Tube Three-Dimensional Scanner for Scanning Tunneling Microscopy.* Rev. Sci. Instrum., 57(8), August, 1688–1689.

Binnig, G. and H. Rohrer (1982). *Scanning Tunneling Microscopy.* Helvetica Physica Acta, 55, 726–735.

Binnig, G. and H. Rohrer (1983a). *Scanning Tunneling Microscopy.* Surface Science, 126, 236–244.

Binnig, G. and H. Rohrer (1983b). *Scanning Tunneling Microscopy.* Helevetica Physica Acta, 56, 481–482.

Binnig, G. and H. Rohrer (1983c). *Surface Imaging by Scanning Tunneling Microscopy.* Ultramicroscopy, 11, 157–160.

Binnig, G. and H. Rohrer (1983d). *Scanning Tunneling Microscopy, An Atomic Probe.* SEM Inc., 1079–1082.

Binnig, G. and H. Rohrer (1985b). *The Scanning Tunneling Microscope.* Sci. Am., 253(2), August, 50–56.

Binnig, G. and H. Rohrer (1986). *Scanning Tunneling Microscopy.* IBM J. Res. Develop., 30(4), 4 July, 355–369.

Binnig, G., C. Gerber, H. Rohrer and E. Weibel (1985). *Sputter Tip.* IBM Tech. Discl. Bull., 27(8), January, 4890.

Binnig, G., H. Rohrer, C. Gerber and E. Weibel (1982a). *Tunneling Through A Controllable Vacuum Gap.* Appl. Phys. Lett., 40(2), 178–179.

Binnig, G., H. Rohrer, C. Gerber and A. Baro (1985b). *Revisiting the 7×7 Reconstruction of Si(111).* Surface Science, 157, L353–L378.

Binnig, G., H. Fuch, C. Gerber, H. Rohrer, E. Stoll and E. Tosatti (1986). *Energy-Dependent State-Density Corrugaion of a Graphite Surface as Seen by Scanning Tunneling Microscopy.* In Press.

Binnig, G., H. Rohrer, C. Gerber and E. Weibel (1982b). *Surface Studies by Scanning Tunneling Microscopy.* Phys. Rev. Lett., 49(1), 57–61.

Binnig, G., H. Rohrer, C. Gerber and E. Weibel (1982c). *Vacuum Tunneling.* Physica, 109 and 110B, 2075–2077.

Binnig, G., H. Rohrer, C. Gerber and E. Weibel (1983a). *(111) Facets as the Origin of Reconstructed Au(110) Surfaces.* Surface Science, 131, L379–L384.

Binnig, G., H. Rohrer, C. Gerber and E. Weibel (1983b). *7×7 Reconstruction on Si(111) Resolved in Real Space.* Phys. Rev. Lett., 50(2), 120–123.

Binnig, G., K. H. Frank, H. Fuchs, N. Garcia, B. Reihl, H. Rohrer, F. Salvan and A. R. Williams (1985). *Tunneling Spectroscopy and Inverse Photoemission: Image and Field States.* Phys. Rev. Lett., 55(9), August, 991–994.

Binnig, G., N. Garcia and H. Rohrer (1984). *Conductivity Sensitivity of Inelastic Scanning Tunneling Microscopy.* Physical Review B, 32(2), July,

1336–1338.

Binnig, G., N. Garcia, H. Rohrer, J. M. Soler and F. Flores (1984). *Electron-Metal-Surface Interaction Potential With Vacuum Tunneling: Observation of the Image Force.* Physical Review B, 30(8), 4816–4818.

Bono, J. and R. H. Good, Jr. (1985). *Theoretical Discussions of the Scanning Tunneling Microscope.* Surf. Sci., 151, 543–552.

Bryant, A., D. P. E. Smith, G. Binnig, W. A. Harrison and C. F. Quate (1986). *Anomalous Distance Dependence in Scanning Tunneling Microscopy.* Appl. Phys. Lett., 49(15), 13 Oct., 936–938.

Bryant, A., D. P. E. Smith and C. F. Quate (1986). *Imaging in Real Time with the Tunneling Microscope.* Appl. Phys. Lett., 48(13), March, 832–834.

Buttiker, M. and R. Landauer (1986). *Traversal Time for Tunneling.* IBM J. Res. Develop., 30(5), Sept., 451–454.

Chaing, S. and R. J. Wilson (1986). *Scanning Tunneling Microscopy.* In: R. Feder, J. W. McGowan and D. M. Shinozaki, *Examining the Submicron World.* Plenum Press, New York, 1–19.

Chiang, S. and R. J. Wilson (1986b). *Construction of a UHV Scanning Tunneling Microscope.* IBM J. Res. Develop., 30(5), Sept., 515–519.

Coleman, R. V., B. Drake, P. K. Hansma and G. Slough (1985). *Charge Density Waves Observed with a Tunneling Microscope.* Phys. Rev. Lett., 55(4), July, 394–397.

Coombs, J. H. and J. B. Pethica (1986). *Properties of Vacuum Tunneling Currents: Anomalous Barrier Heights.* IBM J. Res. Develop., 30(5), Sept., 455–459.

Corb, B. W., M. Ringger and H.-J. Guntherodt (1987). *Microstructural Investigation of Nd-Fe-B with the Scanning Tunneling Microscope.* Appl. Phys. Lett., 50(6), 9 Feb., 353–355.

Demuth, J. E., R. J. Hamers, R. M. Tromp and M. E. Welland (1986). *A Scanning Tunneling Microscope for Surface Science Studies.* IBM J. Res. Develop., 30(4), July, 396–402.

Demuth, J. E., R. J. Hamers, R. M. Tromp and M. E. Welland (1986). *A Simplified Scanning Tunneling Microscope for Surface Science Studies.* J. Vac. Sci. Technol. A, 4(3), May/June, 1320–1323.

Dietrich, H. P., M. Lanz and D. F. Moore (1984). *Ion Beam Machining of Very Sharp Points.* IBM Tech. Disclosure Bul., 27(5), Oct., 3039–3040.

Dragoset, R. A., R. D. Young, H. P. Layer, S. R. Mielczarek, E. C. Teague and R. J. Celotta (1986). *Scanning Tunneling Microscopy Applied to Optical Surfaces.* Optics Letters, 11(9), Sept., 560–562.

Drake, B., R. Sonnenfeld, J. Schneir, P. K. Hansma, G. Solugh R. V. Coleman (1986). *A Tunneling Microscope for Operation in Air or Fluids.* Rev. Sci. Instr., 57(3), March, 441–445.

Durig, U., J. K. Gimzewski and D. W. Pohl (1986). *Experimental Observation of Forces Acting During Scanning Tunneling Microscopy.* Amer. Phys. Soc., 57(19), 10 Nov., 2403–2406.

Elrod, S. A., A. Bryant, A. L. de Lozanne, S. Park, D. Smith, C. F. Quate (1986). *Tunneling Microscopy From 300 to 4.2 K.* IBM J. Res. Develop., 30(4), July, 387–395.

Elrod, S. A., A. L. de Lozanne and C. F. Quate (1984). *Low-Temperature Vacuum Tunneling Microscopy.* Appl. Phys. Lett., 45(11), 1 Dec., 1240–1242.

Farrell, H. H., M. Levinson (1985). *Scanning Tunneling Microscope as a Structure-Modifying Tool.* Physical Review B, 31(6), March, 3593–3598.

Feenstra, R. M. and G. S. Oehrlein (1985). *Surface Morphology of Oxidized and Ion-Etched Silicon by Scanning Tunneling Microscopy.* Applied Physics Letters, 47(2), 15 July, 97–99.

Feenstra, R. M. and A. P. Fein (1986). *Scanning Tunneling Microscopy of Cleaved Semiconductor Surfaces.* IBM J. Res. Develop., 30(5), Sept., 466–471.

Feenstra, R. M., W. A. Thompson and A. P. Fein (1986). *Real-Space Observation of $\pi$-Bonded Chains and Surface Disorder on Si(111) 2$\times$1.* Phys. Rev. Lett., 56(6), 10 February, 608–611.

Feuchtwang, T. E., P. H. Cutler and N. M. Miskovsky (1983). *A Theory of Vacuum Tunneling Microscopy.* Physics Letters, 99A(4), November, 167–171.

Fink, H. W. (1986). *Mono-Atomic Tips for Scanning Tunneling Microscopy.* IBM J. Res. Develop., 30(5), Sept., 460–465.

Flores, F. and N. Garcia (1984). *Voltage Drop in the Experiments of Scanning Tunneling Microscopy for Si.* Phys. Rev. B, 30(4), 15 August, 2289–2291.

Garcia, R., J. J. Saenz, J. M. Solar and N. Garcia (1986). *Distance-Voltage Characteristics in Scanning Tunneling Microscopy.* J. Phys., 19, L131–L134.

Garcia, N. (1986). *Theory of Scanning Tunneling Microscopy and Spectroscopy: Resolution, Image and Field States, and Thin Oxide Layers.* IBM J. Res. Develop., 30(5), Sept., 533–542.

Garcia, N. and F. Flores (1984). *Theoretical Studies for Scanning Tunneling Microscopy.* Physica 127B, 137–142.

Garcia, N., A. M. Baro, R. Miranda, H. Rohrer, C. Gerber, R. Garcia Cantu and J. L. Pena (1985). *Surface Roughness Standards, Obtained with the Scanning Tunneling Microscope Operated at Atmospheric Air Pressure.* Metrologia, 21, 135–138.

Garcia, N., B. Reihl, K. H. Frank and A. R. Williams (1985). *Image States: Binding Energies, Effective Masses, and Surface Corrugation.* Phys. Rev. Lett., 54(6), 11 Feb., 591–594.

Garcia, N., C. Ocal and F. Flores (1983). *Model Theory for Scanning Tunneling Microscopy: Application to Au(110)(1x2).* Phys. Rev. Lett., 50(25), 20 June, 2002–2005.

Garcia, R. (1986). *Tunneling Current Through Localized Surface States Via Inelastic Processes.* In Press.

Garcia, R., J. J. Saenz and N. Garcia (1986). *Conductivity and Structure of Thin Oxide Layers Grown on a Metal Substrate: Scanning-Tunneling Microscopy in NiO on Ni(100).* Physical Review B, 33(6), March, 4439–4442.

Gerber, C., G. Binnig, H. Fuchs, O. Marti and H. Rohrer (1985). *Scanning Tunneling Microscope Combined With a Scanning Electron Microscope.* Rev. Sci. Instrum., 57(2), 221–223.

Gimzewski, J. K. and A. Humbert (1986). *Scanning Tunneling Microscopy of Surface Microstructure on Rough Surfaces.* IBM J. Res. Develop., 30(5), Sept., 472–477.

Gimzewski, J. K., A. Humbert, D. W. Pohl and S. Veprek (1986). *Scanning Tunneling Microscopy of Nanocrystalline Silicon Surfaces.* Surface Sci., 168, 795–800.

Gimzewski, J. K., A. Humbert, J. G. Bednorz and B. Reihl (1985). *Silver Films Condensed at 300 and 90 K: Scanning Tunneling Microscopy of Their Surface Topography.* Phys. Rev. Lett., 55(9), 26 August, 951–954.

Golovchenko, J. A. (1986). *The Tunneling Microscope: A New Look at the Atomic World.* Science, 232, 4 April, 48–53.

Gomer, R. (1986). *Extensions of the Field Emission Fluctuation Method for the Determination of Surface Diffusion Coefficients.* In Press: Applied Physics.

Gomer, R. (1986). *Possible Mechanisms of Atom Transfer in Scanning Tunneling Microscopy.* IBM J. Res. Develop., 30(4), July, 428–430.

Gomez, J., L. Vazquez, A. M. Baro, N. Garcia, C. L. Perdriel, W. E. Triaca and A. J. Arvia (1986). *Scanning Tunneling Microscopy and Electrochemistry: Surface Topography of (100)-Type Electrofaceted Platinum.* In press.

Hamers, R. J., R. M. Tromp and J. E. Demuth (1986). *Scanning Tunneling Microscopy of Si(001).* Phys. Rev. B, 34(8), Oct., 5343–5357.

Hansma, P. K. and J. Tersoff (1987). *Scanning Tunneling Microscopy.* J. Appl. Phys., 61(2), 15 Jan., R1–R23.

Hosler, W., R. J. Behm, E. Ritter (1986). *Defects on the Pt(100) Surface and Their Influence on Surface Reactions—A Scanning Tunneling Microscopy Study.* IBM J. Res. Develop., 30(4), July, 403-410.

Hsue-Yang Liu, Fu-Ren F. Fan, Charles W. Lin and A. J. Bard (1986). *Scanning Electrochemical and Tunneling Ultramicroelectrode Microscope for High-Resolution Examination of Electrode Surfaces in Solution.* J. Am. Chem. Soc., 108, 3838–3839.

Humbert, A., J. K. Gimzewski and B. Reihl (1985). *Postannealing of Coldly Condensed Ag Films: Influence of Pyridine Preadsorption.* Phys. Rev. B, 32(6), 15 Sept., 4252–4253.

IBM (1985a). *Fast Scan Piezo Drive.* IBM Technical Disclosure Bulletin, 27(10b), March, 5976–5977.

IBM (1985b). *Magnetostrictive Positioner.* IBM Technical Disclosure Bulletin, 27(11), April, 6373.

IBM (1985c). *Replacement Mechanism for the Tips of Scanning Electron, Tunneling and Optical Microscopes.* IBM Technical Disclosure Bulletin, 28(1), June, 435–436.

IBM (1985d). *Piezo-electrostatic Rotary Drive.* IBM Technical Disclosure Bulletin, 28(2), July, 504–505.

Kaiser, W. J., R. C. Jaklevic (1986). *Spectroscopy of Electronic States of Metals with a Scanning Tunneling Microscope.* IBM J. Res. Develop., 30(4), July, 411–416.

Kuk, Y. and P. J. Silverman (1986). *Role of Tip Structure in Scanning Tunneling Microscopy.* Appl. Phys. Lett., 48(23), June, 1597–1599.

Lang, N. D. (1985) *Vacuum Tunneling Current from an Absorbed Atom.* Phys. Rev. Lett., 55(2), 8 July, 230–233.

Lang, N. D. (1986a). *The Theory of Single-Atom Imaging in the Scanning Tunneling Microscope.* Phys. Rev. Lett., 56(11), 17 March, 1164–1167.

Lang, N. D. (1986b). *Electronic Structure and Tunneling Current for Chemisorbed Atoms.* IBM J. Res. Develop., 30(4), July, 374–379.

Lang, N. D. (1986c). *Spectroscopy of Single Atoms in the Scanning Tunneling Microscope.* Phys. Rev. Lett., 34(8), 15 Oct., 5947–5950.

Lang, N. D. (1987). *Apparent Size of an Atom in the Scanning Tunneling Microscope as a Function of Bias.* Phys. Rev. Lett., 58(1), 5 Jan., 45–48.

Liu, H.-Y., F.-R. Fan, C. W. Lin and A. J. Bard (1986). *Scanning Electrochemical and Tunneling Ultramicroelectrode Microscope for High-Resolution Examination of Electrode Surfaces in Solution.* J. Am. Chem. Soc., 108(13), 3838–3839.

Louis, E., F. Flores and P. M. Echenique (1986). *Current Saturation Through Image Surface States in Scanning Tunneling Microscopy.* Solid State Comm., 59(7), 453–455.

Mamn, H. J., D. W. Abraham, E. Ganz and J. Clarke (1985). *2-Dimensional, Remote Micropositioner for a Scanning Tunneling Microscope.* Rev. Sci. Instrum., 56(11), November, 2168–2170.

McCord, M. A., and R. F. W. Pease (1985). *High Resolution, Low-Voltage Probes From a Field Emission Source Close to the Target Plane.* J. Vac. Sci. Technol. B, 3(1), Jan./Feb., 198–201.

McCord, M.A. and R. F. W. Pease (1986). *Lithography with the Scanning Tunneling Microscope.* J. Vac. Sci. Technol. B, 4(1), Jan/Feb., 86–88.

Miranda, R., N. Garcia, A. Baro, R. Garcia, J. L. Pena and H. Rohrer (1985) *Technological Applications of Scanning Tunneling Microscopy at Atmospheric Pressure.* Appl. Phys. Lett., 47(4), 15 August, 367–369.

Moreland, J., S. Alexander, M. Cox, R. Sonnenfeld and P. K. Hansma (1983). *Squeezable Electron Tunneling Junctions.* Appl. Phys. Lett., 43(4), 15 August, 387–388.

Moreland, J. and J. W. Ekin (1985). *Electron Tunneling into Superconducting Filaments using Mechanically Adjustable Barriers.* Appl. Phys. Lett., 47(2), July, 175–177.

Moreland, J. and P. K. Hansma (1984). *Electromagnetic Squeezer for Compressing Squeezable Electron Tunneling.* Rev. Sci. Instrum., 55(3), March, 399–403.

Morita, S. K. Itaya and N. Mikoshiba (1986). *Images of Barrier Layer of Anodic Aluminum Oxide in Air Obtained with Scanning Tunneling Microscope.* Japanese J. of Applied Physics, 25(9), Sept., L743–L745.

Morita, S., T. Okada, Y. Ishigame and N. Mikoshiba (1986). *Scanning Tunneling Microscopy of Metal Surfaces in Air.* In Press; presented at Scanning Tunneling Microscopy '86, Santiago de Compostela, (Spain), July 14–18, 1–8.

Morita, S., T. Okada, Y. Ishigame, C. Sato and N. Mikoshiba (1986). *Voltage-Dependence of Scanning Tunneling Microscopy on Titanium Surface in Air.* Japanese J. of Applied Physics, 25(6), L516–L518.

NASA (1986). *Electrochemical Process Makes Fine Needles.* NASA Tech Briefs, May/June, 135.

Newmark, P. and L. Garwin (1986). *Electron Microscopy Acclaimed.* Nature, 323, 23 Oct., 663.

Park, S. L. and C. F. Quate (1986). *Tunneling Microscopy of Graphite in Air.* Appl. Phys. Lett., 48(2), 13 January, 112–114.

Pashley, M. D., J. B. Pethica and J. Coombs (1985). *Scanning Tunneling Microscope Studies.* Surface Science, 152/153, 27–32.

Persson, B. N. J., and J. E. Demuth (1986). *Inelastic Electron Tunnelling from a Metal Tip.* Solid State Communications, 57(9), 769–772.

Pethica, J. B. and M. D. Pashley (1983). *Scanning Tunneling Microscopy.* Nature, 305, October, 666.

Pohl, D. W. (1986). *Some Design Critieria in Scanning Tunneling Microscopy.* IBM J. Res. Develop., 30(4), July, 417–427.

Pohl, D. W., J. K. Gimzewski, A. Humbert and S. Veprek (1985). *Surface Topography Studies of Nanocrystalline Si by STM.* Proc. SPIE, 565, *Micron and Submicron Integrated Circuit Metrology,* 98–101.

Quate, C. F. (1986). *Vacuum Tunneling: A New Technique for Microscopy.* Physics Today, Aug., 26–33.

Ringger, M., B. W. Corb, H.-R. Hidber, R. Schlogl, R. Wiesendanger, A. Stemmer, L. Rosenthaler, A. J. Brunner, P. C. Oelhafen, H.-J. Guntherodt (1986). *STM Activity at the University of Basel.* IBM J. Res. Develop., 30(5), Sept., 500–508.

Ringger, M., H. R. Hidber, R. Schlögl, P. Oelhafen and H. J. Güntherodt (1985). *Nanometer Lithography With the Scanning Tunneling Microscope.* Appl. Phys. Lett., 46(9), 1 May, 832-834.

Ringger, M., H. R. Hidber, R. Schlögl, P. Oelhafen, H. J. Güntherodt, K. Wandelt and G. Ertl (1985). *The Surface Topography of a Pd(100) Single Crystal and Glassy $Pd_{81}Si_{19}$ Studied by Scanning Tunneling Microscopy.* In: M. A. Van Hove and S. Y. Tong, *The Structure of Surfaces,* Springer-Verlag, 48–53.

Ritter, E., R. J. Behm, G. Potschke and J. Wintterlin (1987). *Direct Observation of a Nucleation and Growth Process on an Atomic Scale.* In Press.

Robinson, A. L. (1983a). *IBM Images Surfaces by Electron Tunneling.* Science, 220, 1 April, 43–44.

Robinson, A. L. (1985a). *A Spatially Resolved Surface Spectroscopy.* Science, 229, 1074–1076.

Scheinfein, M. and M. Isaacson (1986). *Electronic and Chemical Analysis of Fluoride Interface Structures at Subnanometer Spatial Resolution.* J. Vac. Sci. Technol. B, 4(1), Jan/Feb., 326–332.

Schneiker, C. (1986). *NanoTech: Science, Engineering and Related Topics.* Department of Anesthesiology Advanced Biotechnology Laboratory, University of Arizona, Tucson, Arizona 85724 USA.

Schneiker, C. and S. R. Hameroff (1987). *NanoTechnology Workstation Based on Scanning Tunneling/Optical Microscopy: Applications to Molecular Scale Devices.* To be published in Proceedings of the Third International Workshop on Molecular Electronic Devices (Elsevier North-Holland).

Schneir, J., R. Sonnenfeld and P. K. Hansma (1986). *Tunneling Microscopy Study of the Graphite Surface in Air and Water.* Phys. Rev. B, 34(8), 15 Oct., 4979–4984.

Schroer, P. H, and J. Becker (1986). *Computer Automation for Scanning Tunneling Microscopy.* IBM J. Res. Develop., 30(5), Sept., 543–552.

Selloni, A., P. Carnevali, E. Tosatti and C. D. Chen (1985). *Voltage-Dependent Scanning-Tunneling Microscopy of a Crystal Surface: Graphite.* Physical Review B, 31(4), 15 February, 2602–2605.

Smith, D. E. and G. Binnig (1986). *Ultrasmall Scanning Tunneling Microscope for Use in a Liquid-Helium Storage Dewar.* Rev. Sci. Instrum., 57(10), Oct., 2630–2631.

Smith, D. P. E. and G. Binnig (1986). *Ultrasmall Scanning Tunneling Microscope for Use in a Liquid-Helium Storage Dewar.* Rev. Sci. Instrum., 57(10), Oct., 2360–2361.

Smith, D. P. E., G. Binnig and C. F. Quate (1986). *Atomic Point-Contact Imaging.* Appl. Phys. Lett., 49(18), 3 Nov., 1166–1168.

Smith, D. P. E., A. Bryant, C. F. Quate, J. P. Rabe, C. Gerber and J. D. Swalen (1986). *Images of a Lipid Bilayer at Molecular Resolution by Scanning Tunning Microscopy.* Submitted to Proc. Nat. Acad. Sci. USA.

Smith, D., P. E. Scott and A. Elrod (1985) *Magnetically Driven Micropositioners.* Rev. Sci. Instru., 56(10), October, 1970–1971.

Smith, D. P. E. and C. F. Quate (1986). Material Presented at the STM 86 Meeting in Santiago de Campostela, Spain, August, 1986.

Soler, J. M., A. M. Baro and N. Garcia (1986). *Interatomic Forces in Scanning Tunneling Microscopy: Giant Corrugations of the Graphite Surface.* Phys. Rev. Lett., 57(4), 28 July, 444–447.

Sonnenfeld, R., J. Moreland and P. K. Hansma (1985). *Contactless Tunneling to Semiconductors.* J. Appl. Phys., 58(1), July, 392–396.

Sonnenfeld, R., and P. K. Hansma (1986). *Atomic-Resolution Microscopy in Water.* Science, 232, 11 April, 211–213.

Sonnenfeld, R., and B. C. Schardt (1986). *Tunneling Microscopy in an Electrochemical Cell: Images of Ag Plating.* Appl. Phys. Lett., 49(18), 3

Nov., 1172–1174.

Stoll, E. (1984). *Resolution of the Scanning Tunnel Microscope.* Surf. Sci. Lett., 143, L411–L416.

Stoll, E. (1986). *Information- and Image-Processing of Scanning Tunneling Microscope Data.* In: W. F. Fagan, ed., *Optics in Engineering Measurement, Proc. SPIE Vol. 599,* 442–450.

Stoll, E., A. Baratoff, A. Selloni and P. Carnevali (1984). *Current Distribution in the Scanning Vacuum Tunnel Microscope: A Free-Electron Model.* J. Phys. C, 17, 3073–3086.

Stroscio, J. A., R. M. Feenstra and A. P. Fein (1986). *Electronic Structure of the Si(111)2×1 Surface by Scanning-Tunneling Microscopy.* Phys. Rev. Lett., 57(20), 17 Nov., 2579–2582.

Tersoff, J. (1985). *Theory of the Scanning Tunneling Microscope.* In: M. A. Van Hove and S. Y. Tong, eds., *The Structure of Surfaces,* Springer-Verlag, 54.

Tersoff, J. and D. R. Hamann (1983). *Theory and Application for the Scanning Tunneling Microscope.* Phys. Rev. Lett., 50(25), 20 June, 1998–2001.

Tersoff, J. and D. R. Hamann (1985). *Theory of the Scanning Tunneling Microscope.* Phys. Rev. B, 31(2), 15 January, 805–813.

Thompson, D. E. (1984). *Atom Detection Improves, On The Surface.* Science News, 126, 18 Aug., 101.

Thompson, D. E. (1985). *Electron Tunneling for Ultrafine Detail.* Science News, 127(14), 215.

Tokumoto, H., et al. (1986). *Observation of Atomic Image of 2H-$NbSe_2$ Surface by Scanning Tunneling Microscope.* Japanese J. Appl. Phys., 25(8), Aug., L621–L623.

Travaglini, G., H. Rohrer, M. Amrein and H. Gross (1986). *Scanning Tunneling Microscopy on Biological Matter.* (Material presented at STM 86 Meeting in Santiago de Campostela, Spain, August, 1986).

Tromp, R. M., R. J. Hamers, and J. E. Demuth (1985). *Si(001) Dimer Structure Observed with Scanning Tunneling Microscopy.* Phys. Rev. Lett., 55(12), 16 Sept., 1303–1306.

Tromp, R. M., R. J. Hamers and J. E. Demuth (1986). *Quantum States and Atomic Structure of Silicon Surfaces.* Science, 234, 17 Oct., 304–309.

Van de Walle, G. F. A., H. van Kempen and P. Wyder (1986). *Tip Structure Determination by Scanning Tunneling Microscopy.* Surf. Science, 167, L219–L224.

Van de Walle, G. F. A., H. van Kempen and P. Wyder (1987). *Scanning*

*Tunneling Microscopy on Photoconductive Semi-insulating GaAs.* Appl. Phys. Lett., 50(1), 5 Jan., 22–24.

Van Kempen, H. (1986). *Construction and Applications of the Scanning Tunneling Microscope.* In press. Presented at 5th Phefferkorn Conf. on Physical Aspects of Materials Characteristics.

Van Kempen, H. and G. F. A. Van de Walle (1986). *Applications of a High-Stability Scanning Tunneling Microscope.* IBM J. Res. Develop., 30(5), Sept., 509–514.

Van Loenen, E. J., J. E. Demuth, R. M. Tromp and R. J. Hamers (1987). *Local electron States and Surface Geometry of Si(111)-($\sqrt{3} \times \sqrt{3}$) Ag.* Phys. Rev. Lett., 58(4), 26 Jan., 373–376.

Vieira, S. (1986). *The Behavior and Calibration of Some Piezoelectric Ceramics Used in the STM.* IBM J. Res. Develop., 30(5), Sept., 553–556.

West, P., J. Kramar, D. V. Baxter, R. J. Cave and J. D. Baldeschwieler (1986). *Chemical Applications of Scanning Tunneling Microscopy.* IBM J. Res. Develop., 30(5), Sept., 484–491.

Wilson, R. J. and S. Chiang (1987). *Structure of the Ag/Si(111) Surface by Scanning Tunneling Microscopy.* Phys. Rev. Lett., 58(4), 26 Jan., 369–372.

Young, R. D., J. Ward and F. Scire (1971). *Observation of Metal-Vacuum-Metal Tunneling, Field Emission, and the Transition Region.* Phys. Rev. Lett., 27, 922–nnn.

Young, R., J. Ward and F. Scire (1972). *The Topografiner: An Instrument for Measuring Surface Microtopography.* Rev. Sci. Instr., 43(7), July, 999–1011.

## A.1 Near-Field Scanning Optical Microscopes

Betzig, E., A. Lewis, A. Harootunian, N. Isaacson and E. Kratschmer (1986). *Near-Field Scanning Optical Microscopy (NSOM), Development and Biological Applications.* Biophysics Journal, 49, January, 269–279.

Binnig, G. K., C. Gerber, H. Rohrer and E. Weibel (1985a). *Nano-Aperture.* IBM Tech. Disclosure Bul., 27(8), Jan., 4893.

De Brabander, R. Nuydens, G. Geuens, M. Moeremans, J. De Mey and C. Hopkins (1988). *Nanovid Ultramicroscopy: A New Non-Destructive Apporach Providing New Insights in Subcellular Motility.* In: M. De Bravander and J. De Mey, eds., *Microtubules and Microtubule Inhibitors 1985.* Elsevier Science Publishers, North Holland, 187–196.

Durig, U., D. Pohl and F. Rohner (1986). *Near-Field Optical Scanning Microscopy with Tunnel-Distance Regulation.* IBM J. Res. Develop., 30(5), Sept., 478–483.

Harootunian, A., E. Betzig, M. Isaacson and A. Lewis (1986). *Super-Resolution Fluorescence Near-Field Scanning Optical Microscopy.* Appl. Phys. Lett., 49(11), 15 Sept., 674–676.

IBM (1985e). *Optical Microscope Beyond the Defraction Limit.* IBM Technical Disclosure Bulletin, 28(6), November, 2691–2693.

Lewis, A., M. Isaacson, A. Harootunian and A. Muray (1984). *Development of a 500 Å Spatial Resolution Light Microscope.* Ultramicroscopy, 13, 227–232.

McCutchen, C. W. (1967). *Superresolution in Microscopy and the Abbe Resolution Limit.* J. Optical Society of America, 57(10), October, 1190–1192.

Pohl, D. W., W. Denk and M. Lanz (1984). *Optical Stethoscopy: Image Recording with Resolution.* Appl. Phys. Lett., 44(7), 1 April, 651–653.

Pohl, D. W., W. Denk and U. Duerig (1985). *Optical Stethoscopy: Imaging with λ/20 Resolution.* Proc. of the SPIE, 565, Micron and Submicron Integrated Circuit Metrology, August, 56–61.

## A.2 Other STM-Related Instruments

Binnig, G., C. F. Quate and C. Gerber (1986). *Atomic Force Microscope.* Phys. Rev. Lett., 56(9), 3 March, 930–933.

Muralt, P. (1986). *GaAs pn Junction Studied by Scanning Tunneling Potentiometry.* Appl. Phys. Lett., 49(21), 24 Nov., 1441–1443.

Muralt, P. and D. W. Pohl (1986). *Scanning Tunneling Potentiometry.* Appl. Phys. Lett., 48(8), 24 February, 514–516.

Muralt, P., D. W. Pohl and W. Denk (1986). *Wide-Range, Low-Operating-Voltage, Bimorph STM: Application as Potentiometer.* IBM J. Res. Develop., 30(5), Sept., 443–450.

Smith, D. P. E., G. Binning and C. F. Quate (1986). *Atomic Point-Contact Imaging.* Appl. Phys. Lett., 49(18), 3 November, 1166–1168.

Williams, C. C. and H. K. Wickramasinghe (1986). *Scanning Thermal Profiler.* Appl. Phys. Lett., 49(23), 8 Dec., 1587–1589.

# Appendix B

# Replicating Systems References

Albus, J. S. (1981). *Brains, Behavior, and Robotics.* Peterborough, NY, Byte Press.

P. W. Anderson (1983). *Suggested Model for Prebiotic Evolution: The Use of Chaos.* Proc. Natl. Acad. Sci. USA, 80, June, 3386–3390.

Arbib, M. A. (1967a). *Some Comments on Self-Reproducing Automata.* Systems and Computer Science. J. F. Hart and Satoru Takasu, eds., Univ. of Toronto Press, 42–59.

Arbib, M. A. (1967b). *Automata Theory and Development: Part 1.* J. Theor. Biol., 14(2), 131–156.

Arbib, M. A. (1969). *Machines Which Compute and Construct.* In chapter 10 of: *Theories of Abstract Automata.* Prentice Hall, Englewood Cliffs, N. J.

Berry, A. (1975). *The Vision of John von Neumann.* In chapter 11 of: *The Next Ten Thousand Years, A Vision of Man's Future in the Universe.*

Burks, A. W. (1975). *Logic, Biology and Automata—Some Historical Reflections.* Int. J. Man-Machine Studies, 7(3), 297–312.

Burks, A. W., ed. (1970). *Essays on Cellular Automata.* Chicago, Univ. of Illinois Press.

Cairns-Smith, A. G. (1971). *The Life Puzzle.* Toronto: Univ. of Toronto Press.

Cairns-Smith, A. G. (1981). *Beginnings of Organic Evolution.* In The Study of Time IV, J. T. Fraser, N. Lawrence, D. Park, eds., 15–33. New York, Springer-Verlag.

Cairns-Smith, A. G. (1982). *Genetic Takeover.* Cambridge, Cambridge University Press.

Cairns-Smith, A. G. (1985). *The First Organisms.* Scientific American, June.

Cairns-Smith, A. G. and H. Hartman, eds. (1986). *Clay Minerals and the Origin of Life.* Cambridge, Cambridge Univ. Press.

Calder, N. (1978). *Spaceships of the Mind.* Penguin Books, 18–19.

Cliff, R. A. (1985). *An Hierarchical System Architecture for Automated Design, Fabrication, and Repair.* In: *Space Manufacturing 4,* J. Grey and L. A. Hamdan, eds., 121–126.

Codd, E. F. (1968). *Cellular Automata.* New York: Academic Press.

Dyson, F. J. (1970). *The Twenty-First Century.* Vanuxem Lecture, delivered at Princeton University, February 26, 1970. This was mostly published in: The Key Reporter, 42(3), 1977.

Dyson, F. (1981). *Disturbing the Universe.* New York: Harper Colophon.

Eigen, M. (1971). *Selforganization of Matter and the Evolution of Biological Macromolecules.* Naturwissenschaften, 58, 465–523.

Eigen, M. and P. Schuster (1979). *The Hypercycle, A Principle of Natural Self-Organization.* New York, Springer-Verlag.

Fox, S. W. (1973). *The Rapid Evolution of Complex Systems from Simple Beginnings.* Proc. 3rd Int. Conf. from Physics to Theoretical Biology, Versailles. Basel: Karger, 133–144.

Fox, S. W. (1980). *The Assembly and Properties of Protobiological Structures: The Beginnings of Cellular Peptide Synthesis.* BioSystems, 12(3), 155–166.

Fox, S. W. (1984). *Self-Sequencing of Amino Acids and Origins of Polyfunctional Protocells.* In: *Proceedings of the 7th International Conference on the Origins of Life,* July, 485–488.

Freitas, Jr., R. A. (1981). *A Self-Replicating, Growing Lunar Factory.* In: *Space Manufacturing 4, Proceedings of the Fifth Princeton/AIAA Conference,* J. Grey and L. A. Hamdan, eds., 109–119.

Freitas, Jr., R. A. and W. P. Gilbreath (1982). *Chapter 5: Replicating Systems Concepts: Self-Replicating Lunar Factory and Demonstration.* In: *Advanced Automation for Space Missions,* NASA Conference Publication 2255, 189–335.

Jacob, F. (1977). *Evolution and Tinkering.* Science, 196(4295), 1161–1169.

Jacobson, H. (1958). *On Models of Reproduction.* American Scientist, 46, 255–284.

Katz, M. J. and Y. Chow (1985). *Templeting and Self-Assembly*. J. Theor. Biol., 113, 1–13.

Katzir, A. and F. Vranty (1984) *Use of PVC for Replecating Submicron Features for Microelectronic Devices*. Rev. Sci. Inst., 55(4), April, 633–634.

Kemeny, J. G. (1955). *Man Viewed as a Machine*. Scientific American, 192(4), 58–67.

Laing, R. (1975). *Some Alternative Reproductive Strategies in Artifical Molecular Machines*. J. Theor. Biol., 54, 63–84.

Laing, R. (1976). *Automaton Inspection*. J. Comp. Sys. Sciences, 13, 172–183.

Laing, R. (1977). *Automaton Models of Reproduction by Self-Inspection*. J. Theor. Biol., 66, 437–456.

Laing, R. (1979). *Machines as Organisms: An Exploration of the Relevance of Recent Results*. Biosystems, 11, 201–215.

Langton, C. G. (1984). *Self-Reproduction in Cellular Automata*. Physica, 10D, 135–144.

Lehmann, U. and H. Kuhn (1984). *Model Approach of the Breakthrough of a Translation Machine and the Origin of the Genetic Code*. In: *Proceedings of the 7th International Conference on the Origins of Life*. July, 497–504.

Lofgren, L. (1961). *Kinematic and Tessellation Models of Self-Repair*. in: *Biological Prototypes and Synthetic Systems, Vol. 1, Proceedings of the Second Annual Bionics Symposium*. August 30–September 1, 342–369.

Lofgren, L. (1962). *Self-Repair as a Computability Concept in the Theory of Automata*. In: *Proceedings of the Symposium on Mathematical Theory of Automata*. New York, April, 205–232.

Long, J. E. and T. J. Healy (1980). *Advanced Automation for Space Missions—Technical Summary: A Report of the 1980 NASA/American Society for Engineering Education Summer Study on the Feasibility of Using Machine Intelligence in Space Applications, Univ. Santa Clara, Sept.*

Moore, E. F. (1956). *Artifical Living Plants*. Scientific American, 195(4), 118–126.

Moore, E. F. (1962). *Machine Models of Self-Reproduction*. 17–33.

Moore, E. F. (1964). *Mathematics in the Biological Sciences*. Scientific American, 211(3), 148–164.

Morowitz, H. J. (1959). *A Model of Reproduction*. American Scientist, 47, 261–263.

Morrison, P. (1964). *A Thermodynamic Characterization of Self-Reproduction*. Reviews of Modern Physics, April, 517–524.

Myhill, J. (1964). *The Abstract Theory of Self-reproduction.* In: *Views on General Systems Theory,* M. D. Mesarovic, ed., 106–118. New York, John Wiley and Sons.

Nickerson, K. W. (1984). *A Hypothesis on the Role of Pressure in the Origin of Life.* J. Theor. Biol., 110, 487–499.

Penrose, L. S. (1959). *Automatic Mechanical Self-Reproduction.* New Biology, Penguin Books, 92–117.

Penrose, L. S. and R. Penrose (1957). *A Self-Reproducing Analogue.* Letters to the Editors, Nature, June, 1183.

Penrose, L. S. (1959). *Self-Reproducing Machines.* Scientific American, 200(6), 105–114.

Price, C. C., ed. (1974). *Synthesis of Life.* Stroudsburg, Penn., Dowden, Hutchinson and Ross.

Schuster, P. (1979). *A Mathematical Model of the Hypercycle.* Proc. of the International Symposium on Synergetics, September 24-29, 170–178.

Schuster, P. (1981). *Self-Organization and Integration of Information Stored in Selfreplicative Units.* Self-Organizing Systems, Verlag, New York, 47–61.

Schuster, P. and K. Sigmund (1979). *Self-Organization of Biological Macromolecules and Evolutionary Stable Strategies.* Proc. of the International Symposium on Synergetics, Sept. 24-29, 156–169.

Schwartz, A. W., and L. E. Orgel (1985). *Template-Directed Synthesis of Novel, Nucleic Acid-Like Structures.* Science, 228, 3 May, 585–587.

Vitanyi, P. M. B. (1974). *Genetics of Reproducing Automata.* Proceedings of the 1974 Conference on Biologically Motivated Automata Theory, June, 166–171.

Von Neumann, J. (1966). *Theory of Self-Reproducing Automata.* (Completed and ed. by A. W. Burkes.) Urbana, Illinois: Univ. Illinois Press.

Weiss, A. (1981). *Replication and Evolution in Inorganic Systems.* Angew Chem. Int. Ed. Engl., 20, 850–860.

# Appendix C

# Collective Computing References

Carpenter, G. A. and S. Grossberg (1986). *A Massively Parallel Architecture for a Self-Organizing Neural Pattern Recognition Machine.* In press: *Computer Vision, Graphics and Image Processing.*

Dammasch, I. E., G. P. Wagner, and J. R. Wolff (1986). *Self-Stabilization of Neuronal Networks.* Biol. Cybern., 54, 211–222.

Denker, J. S., ed. (1986). *Neural Networks for Computing; AIP Conference Proceedings 151.* New York, American Institute of Physics.

d'Humieres, D. and B. A. Huberman (1984). *Dynamics of Self-Organization in Complex Adaptive Networks.* Journal of Statistical Physics, 34(3/4), 361–379.

Fisher, A. D., C. L. Giles and J. M. Lee (1984). *Associative Processor Architectures for Optical Computing.* J. Opt. Soc. Am. A, 1, 1337.

Gelperin, A., J. J. Hopfield and D. W. Tank (1986). *The Logic of Limax Learning.* 237–261.

Grossberg, S., ed. (1987). *The Adaptive Brain I; Cognition, Learning, Reinforcement, and Rhythm.* and *The Adaptive Brain II; Vision, Speech, Language, and Motor Control.* Amsterdam, North-Holland.

Hastings, H. M. (1985). *Stochastic Information Processing in Biological Systems II—Statistics, Dynamics, and Phase Transitions.* In: S. L. Mintz and A. Perlmutter, eds. *Information Processing in Biological Systems.* Plenum Press, 131–144.

Hastings, H. M. and S. Waner (1985a). *Low Dissipation Computing in Biological Systems.* BioSystems, 17, 241–244.

Hastings, H. M. and S. Waner (1986). *Biologically Motivated Machine Intelligence.* SIGART Newsletter, January, Number 95, 29–31.

Hogg, T. and Huberman B. A. (1984). *Understanding Biological Computation: Reliable Learning and Recognition.* Proc. Natl. Acad. Sci., 81, Nov., 6871–6875.

Hogg, T. and B. A. Huberman (1985). *Parallel Computing Stuuctures Capable of Flexible Associations and Recognition of Fuzzy Inputs.* J. Statistical Physics, 41(1/2), Oct., 115–123.

Hopfield, J. J. (1982). *Neural Networks and Physical Systems with Emergent Collective Computational Abilities.* Proc. Natl. Acad. Sci. USA, 79, 2554–2558.

Hopfield, J. J. (1982). *Brain, Computer, and Memory.* Engineering and Science, Sept., 2–7.

Hopfield, J. J. (1984a). *Collective Processing and Neural States.* In: *Modeling and Analysis in Biomedicine,* C. Nicolini, ed., 371–389.

Hopfield, J. J. (1984b). *Neurons with Graded Response have Collective Computational Properties like Those of Two-State Neurons.* Proc. Natl. Acad. Sci. USA, 81, 3088–3092.

Hopfield, J. J. amd D. W. Tank (1985). *"Neural" Computation of Decisions in Optimisation Problems.* Biological Cybernetics, 52, 141–152.

Huberman, B. A. and T. Hogg (1984). *Adaptation and Self-Repair in Parallel Computing Structures,* Phys. Rev. Letts., 52, 1048–1051.

Huberman, B. A. and T. Hogg (1985). *Dynamic Associations in Nonlinear Computing Arrays.* Physica Scripta 32, 271–273.

Huberman, B. A. and M. Kerszberg (1985). *Ultradiffusion: The Relaxation of Hierarchical Systems.* J. Phys. A, 18, L331–L336.

Jackel, L. D., R. E. Howard, H. P. Graf, B. Straughn, and J. S. Denker (1986). *Artificial Neural Networks for Computing.* J. Vac. Sci. Technol. B, 4(1), Jan/Feb., 61–63.

Keirstead, W. P. and Huberman, B. A. (1986). *Collective Detection of Motion in the Presence of Noise.* Phys. Rev. Lett., 56(10), 10 March, 1094–1097.

Kohonen, T. (1984) *Self-Organization and Associative Memory.* Springer-Verlag, New York.

Mandell, A. J., P. V. Russo and S. Knapp (1982). *Strange Stability in Hierarchically Coupled Neuropsychobiological Systems.* Evolution of Order and Chaos, Proc. of the International Symposium on Synergetics, April 26–May 1, H. Haken, ed., 270–286.

Milgram, M. and H. Atlan (1983). *Probabalistic Automata as a Model for Epigenesis of Cellular Networks.* J. Theor. Biol., 103, 523–547.

Ritter, H. and K. Schulten (1986). *On the Stationary State of Kohenen's Self-Organizing Sensory Mapping.* Biol. Cybern., 54, 99–106.

Silverman, D. J., G. L. Shaw and J. C. Pearson (1986). *Associative Recall Properties of the Trion Model of Cortical Organization.* Biol. Cybern., 53, 259–271.

Sivilotti, M., M. Emerling and C. A. Mead (1985). *A Novel Associative Memory Implemented Using Collective Computation.* In: *1985 Chapel Hill Conference on Very Large Scale Integration,* H. Fuchs, ed., Computer Science Press, 329–342.

Toulouse, G., S. Dehaene and J. Changeux (1986). *Spin Glass Model of Learning by Selection.* Proc. Natl. Acad. Sci. USA, 83, 1695–1698.

# Appendix D

# Quantum Computing References

Albert, D. Z. (1983). *On Quantum-Mechanical Automata.* Physics Letters, 98A(5,6), 24 Oct., 249–252.

Avery, J. and C. M. E. Pavlidou (1974). *A General Perspective of the Quantum Mechanical Foundations of Energy Transduction.* Ann. New York Acad. Sci., 227, 651–668.

Bate, R. T. (1982). *Quantum-Mechanical Limitations on Device Performance.* In: *VLSI Electronics Microstructure Science, Vol. 5.* N. G. Einspruch, ed., 359–386. New York, Academic Press.

Beinoff, P. (1980). *The Computer as a Physical System: A Microscopic Quantum Mechanical Hamiltonian Model of Computers as Represented by Turing Machines.* J. Stat. Phys., 22, 563–591.

Beinoff, P. (1982a). *Quantum Mechanical Models of Turing Machines That Dissipate No Energy.* Phys. Rev. Lett., 48, 1581–1585.

Beinoff, P. (1982b). *Quantum Mechanical Hamiltonian Models of Turing Machines.* J. Stat. Phys., 29, 515–546.

Beinoff, P. (1982c). *Quantum Mechanical Hamiltonian Models of Discrete Processes That Erase Their Own Histories: Application to Turing Machines.* Internat. J. Theoret. Phys., 21, 177–201.

Bennett, C. H. (1073). *Logical Reversibility of Computation.* IBM J. Res. Dev., 17, 525–532.

Bennett, C. H. (1982). *The Thermodynamics of Computation: A Review.* Int. J. Theor. Phys., 21, 905–940.

Bennett, C. H. (1984). *Thermodynamically Reversible Computation.*

Phys. Rev. Lett., September, 53(12), 1202–1206.

Bennett, C. H. (1986). *On the Nature and Origin of Complexity in Discrete, Homogeneous, Locally-Interacting Systems.* Foundations of Physics, 16(6), 585–592.

Bennett, C. H., G. Brassard, S. Breidbart and S. Wiesner (1983). *Quantum Cryptography, or Unforgeable Subway Tokens.* In: *Advances in Cryptography. Proc. of Crypto.* 1982, New York, Plenum.

Bergquist, J. C., R. G. Hulet, W. M. Itano and D. J. Wineland (1986). *Observation of Quantum Jumps in a Single Atom.* Phys. Rev. Lett., 57(14), 6 Oct., 1699–1702.

Carter, F. L., ed. (1982a). *Molecular Electronic Devices.* Marcel Dekker, New York.

Carter, F. L. (1982b). *Conformational Switching at the Molecular Level.* In: Carter (1982a).

Carter, F. L. (1983). *Molecular Level Fabrication Techniques and Molecular Electronic Devices.* J. Vac. Sci. Technol. B, 1(4), Oct.-Dec., 959–968.

Carter, F. L. (1984a). *Molecular Electronics: An Opportunity for a Biotechnical Synergisn.* In: Adey and Lawrence (1984).

Carter, F. L. (1984b). *The Molecular Device Computer: Point of Departure for Large Scale Cellular Automata.* Physica, 10D, 175–194.

Carter, F. L., (1986). *Chemistry and Microsturctures: Fabrication at the Molecular Size Level.* Superlattices and Microstructures, 2(2), 113–128.

Carter, F. L., (1986). *Molecular Computing and the Chemical Elements of Logic.* in SPIE's Advanced Institute Series: *Hybrid and Optical Computers.* Xerox International Center, Leesburg, VA.

Carter, F. L., ed. (1987). *Proceedings of the 2nd International Workshop on Molecular Electronic Devices (1983).* In preparation. Marcel Dekker, New York.

Carter, F. L., ed. (1987). *Proceedings of the 3rd International Workshop on Molecular Electronic Devices (1986).* In preparation. Elsevier North-Holland, Amsterdam.

Ceperley D. and B. Alder (1986). *Quantum Monte Carlo.* Science, 231, February, 555–560.

Cook, Richard J. and H. J. Kimble (1985). *Possibility of Direct Observation of Quantum Jumps.* Phys. Rev. Lett., 54(10), 11 March, 1023–1026.

Davydov, A. S. (1977). *Solitons and Energy Transfer Along Protein Molecules.* J. Theor. Biol., 66, 379–387.

Davydov, A. S. (1982). *Biology and Quantum Mechanics.* Oxford, Pergamon.

Davydov, A. S. (1985). *Solitons and Excitons in Quasi-One-Dimensional Systems.* In: Takeno (1985), 218–223.

Davydov, A. S. (1985). *Solitons in Molecular Systems.* D. Reidel Pub. Co., Boston.

DeArmond, M. K. and K. W. Hanck (1984). *Switching and Charge Storage in Transition Metal Complexes—Smart Molecules?* In: *IEEE COMPCON Conference Proceedings,* 115–118.

Deutsch, D. (1982). *Is There a Fundamental Bound on the Rate at Which Information Can Be Processed?* Phys. Rev. Lett., 48(4), 25 Jan., 286–288.

Deutsch, D. (1985). *Quantum Theory, the Church-Turing Principle and the Universal Quantum Computer.* Proc. R. Soc. Lond. A, 400, 97–117.

Deutsch, D. (1985). *Quantum Theory as a Universal Physical Theory.* Int. J. Theor. Phys., 24(1), 1–41.

DeVault, D. (1984). *Quantum-Mechanical Tunnelling in Biological Systems.* Cambridge University Press.

Erber, T. and S. Putterman (1985). *Randomness in Quantum Mechanics—Natures Ultimate Cryptogram?* Nature, 318(7), Nov., 41–43.

Erber, T., T. M. Rynne, W. F. Darsow and M. J. Frank (1983). *The Simulation of Random Processes on Digital Computers: Unavoidable Order.* J. Computational Physics, 49, 394–419.

Feynman, R. P. (1982). *Simulating Physics With Computers.* Int. J. Theor. Phys., 21(6/7), 467–488.

Feynman, R. P. (1985a). *Quantum Mechanical Computers.* Optics News, Feb., 11–20.

Feynman, R. P. (1985c). *Quantum Mechanical Computing.* A talk for a meeting of the Cal. Tech. Society of Physics Students on March 12.

Feynman, R. P. (1986a). *Quantum Mechanical Computers.* Foundations of Physics, 16(6), 507–531.

Fredkin, E. and T. Toffoli (1982). *Conservative Logic.* Internat. J. Theoret. Phys., 21, 219–253.

Freiser, M. J. and P. M. Marcus (1969). *A Survey of Some Physical Limitations on Computer Elements.* IEEE Tran. Magnetics, 5(2), 82–90.

Goto, E. and T. Soma (1986). *Quantum Flux Parametron Paves Way to Superhigh-Speed Computers.* JEE, May, 62–65.

Huang, G. M., T. J. Tarn and J. W. Clark (1983). *On the Controllability of Quantum-Mechanical Systems.* J. Math. Phys., 24, Nov., 2608–2618.

IBM (1985f). *Quantum Public Key Distribution System.* IBM Tech. Discl. Bulletin, 28(7), December, 3153–3163.

Keyes, R. W. (1969). *Physical Problems and Limits in Computer Logic.* IEEE Spectrum, May, 36–45.

Keyes, R. W. (1970). *Power Dissipation in Information Processing.* Science, 168, 796–802.

Keyes, R. W. (1972). *Physical Problems of Small Structures in Electronics.* Proc. IEEE, 60(9), 1055–1062.

Keyes, R. W. (1975). *Physical Limits in Digital Electronics.* Proc. IEEE, 63(5), 740–766.

Keyes, R. W. and R. Landauer (1970). *Minimal Energy Dissipation in Logic.* IBM J. Res. Dev., 14, 152–157.

Landauer, R. (1961). *Irreversibility and Heat Generation in the Computing Process.* IBM J. Res. and Dev., July, 183–191.

Landauer, R. (1982). *Uncertainty Principle and Minimal Energy Dissipation in the Computer.* Internat. J. Theoret. Phys., 21, 283–297.

Landauer, R. (1986). *Computation and Physics: Wheeler's Meaning Circuit?* Foundations of Physics, 16(6), 551–564.

Likharev, K. K. (1982). *Classical and Quantum Limitations on Energy Consumption in Computation.* Int. J. Theor. Phys., 21(3/4), 311–326.

Maddox, J. (1985a). *Monitoring Quantum Jumps.* Nature, 314, April, 493.

Maddox, J. (1985b). *Towards the Quantum Computer?.* Nature, 316, August, 573.

McClaire, W. F. (1971). *Chemical Machines, Maxwell's Demon and Living Organisms.* J. Theor. Biol., 30, 1–34.

Moracsik, M. J. (1986). *Detecting "Secret" Quantum Numbers.* Phys. Rev. Lett., 56(9), March, 908–911.

Ong, C. K., G. M. Huang, T. J. Tarn and J. W. Clark (1984). *Invertibility of Quantum-Mechanical Control Systems.* Math. Sys. Theory, 17, 335–350.

Peres, A. (1985). *Reversible Logic and Quantum Computers.* Phys. Rev. A, 32(6), 3266–3276.

Peres, A. (1986). *Existence of "Free Will" as a Problem of Physics.* Foundations of Physics, 16(6), 573–584.

Porod, W., R. O. Grondin, D. K. Ferry and G. Porod (1984). *Dissipation in Computation.* Phys. Rev. Lett., 52(3), 232–235.

Reiss, H. and C. Huang (1971). *Statistical Thermodynamic Formalism in the Solution of Information Theory Problems.* J. Stat. Phys., 3, 191–210.

Simon, J. (1978). *The Limit of Computing Speed.* Comm. ACM, 21(10), October, 876–877.

Stroscio, M. A. (1986a). *Quantum-Based Electronic Devices.* Superlattices and Microstructures, 2(1), 45–47.

Stroscio, M. A. (1986b). *Comment on "Microchips as Precision Quantum-Electrodynamic Probes."* Phys. Rev. Lett., 56(19), May, 2107.

Yates, F. E. (1984) *Report on Conference on Chemically-Based Computer Designs.* Krump Institute for Medical Engineering Report CIME TR/84/1, University of California, Los Angeles, California 90024.

# Index